Si Front-End Processing—
Physics and Technology of
Dopant-Defect Interactions III

MATERIALS RESEARCH SOCIETY
SYMPOSIUM PROCEEDINGS VOLUME 669

Si Front-End Processing— Physics and Technology of Dopant-Defect Interactions III

Symposium held April 17–19, 2001, San Francisco, California, U.S.A.

EDITORS:

Erin C. Jones
IBM T.J. Watson Research Center
Yorktown Heights, New York, U.S.A.

Kevin S. Jones
University of Florida
Gainesville, Florida, U.S.A.

Martin D. Giles
Intel Corporation
Portland, Oregon, U.S.A.

Peter Stolk
Philips Research Leuven
Leuven, Belgium

Jiro Matsuo
Kyoto University
Kyoto, Japan

Materials Research Society
Warrendale, Pennsylvania

CAMBRIDGE UNIVERSITY PRESS
Cambridge, New York, Melbourne, Madrid, Cape Town,
Singapore, São Paulo, Delhi, Mexico City

Cambridge University Press
32 Avenue of the Americas, New York NY 10013-2473, USA

Published in the United States of America by Cambridge University Press, New York

www.cambridge.org
Information on this title: www.cambridge.org/9781107412200

Materials Research Society
506 Keystone Drive, Warrendale, PA 15086
http://www.mrs.org

First published 2001
First paperback edition 2013

Single article reprints from this publication are available through
University Microfilms Inc., 300 North Zeeb Road, Ann Arbor, MI 48106

CODEN: MRSPDH

ISBN 978-1-107-41220-0 Paperback

CONTENTS

FUTURE DEVICE ISSUES

ADVANCES IN DOPANT PROFILING

*Invited Paper

DOPANT DIFFUSION ISSUES

POSTER SESSION

DOPANT DEFECT CLUSTERING

LASER ANNEALING

ADVANCES IN RTA

*Invited Paper

SIMULATION AND MODELING

Author Index

Subject Index

*Invited Paper

PREFACE

This volume contains the proceedings of Symposium J, "Si Front-End Processing—Physics and Technology of Dopant-Defect Interactions III," held April 17-19 at the 2001 MRS Spring Meeting in San Francisco, CA, the third MRS symposium on this topic. The foci of the symposium were the phenomena which control the three-dimensional dopant profile in deep submicron devices. As device sizes continue to shrink, increasing the dopant activation while simultaneously decreasing the junction depth requires increasing control and understanding of dopant movement. As dopant diffusion and activation are determined by interactions with defects, other atoms, and interfaces, control of dopant behavior requires specific knowledge of these processes. To take advantage of atomistic simulation methods which can be used to model not only the dopant behaviors but now the properties of whole devices, high-precision advanced characterization techniques (e.g., two-dimensional junction profiling) are essential. These problems are by nature cross-disciplinary, and therefore the goal of this symposium was to bring together materials scientists, silicon technologists, and TCAD researchers to share experimental results and physical models, demonstrate their importance to the technologies, and identify key issues for future research in this field.

The theme of the symposium was set on the first day by a series of invited papers, addressing the need for shallow junctions and dopant control for future silicon technologies. A considerable part of the symposium was aimed at unraveling the complex potpourri of atomistic processes that are activated during the annealing of ion-implanted dopant atoms. Various papers attested to the impressive level that has been reached over recent years in the understanding of how ion-generated vacancies and interstitials evolve through the formation of clusters and extended defects during annealing. These and other insightful experimental results have significantly improved the ability with which dopant, defect and impurity profiles can be predicted through numerical simulations. Transient enhanced diffusion and dopant clustering severely hinder the formation of shallow, low-resistive junctions, in particular for p-type dopants. Furthermore, sputtering effects have been found to restrict the efficiency with which dopants are introduced into silicon at low implantation energies (e.g., 0.2 keV B). At the symposium, various approaches to improving junction performance were presented and discussed, among them the co-implantation of impurities such as fluorine, the optimization of rapid thermal annealing conditions, and the use of pulsed-laser annealing. These methods clearly hold promise for meeting the future industry needs, but further tailoring of the involved dopant-defect interactions is needed to ensure successful implementation into silicon processing.

Erin C. Jones
Kevin S. Jones
Martin D. Giles
Peter Stolk
Jiro Matsuo

November 2001

ACKNOWLEDGMENTS

The organizers wish to thank the following sponsors for their generosity in supporting the symposium:

Agere Systems
Applied Materials, Inc.
Axcelis Technologies
Fujitsu, Ltd.
Intel Corporation
Varian Semiconductor Equipment Associates, Inc.

MATERIALS RESEARCH SOCIETY SYMPOSIUM PROCEEDINGS

MATERIALS RESEARCH SOCIETY SYMPOSIUM PROCEEDINGS

Volume 659— High-Temperature Superconductors—Crystal Chemistry, Processing and Properties,
U. Balachandran, H.C. Freyhardt, T. Izumi, D.C. Larbalestier, 2001, ISBN: 1-55899-569-2

Volume 660— Organic Electronic and Photonic Materials and Devices, S.C. Moss, 2001, ISBN: 1-55899-570-6

Volume 661— Filled and Nanocomposite Polymer Materials, A.I. Nakatani, R.P. Hjelm, M. Gerspacher,
R. Krishnamoorti, 2001, ISBN: 1-55899-571-4

Volume 662— Biomaterials for Drug Delivery and Tissue Engineering, S. Mallapragada, R. Korsmeyer,
E. Mathiowitz, B. Narasimhan, M. Tracy, 2001, ISBN: 1-55899-572-2

Volume 664— Amorphous and Heterogeneous Silicon-Based Films—2001, M. Stutzmann, J.B. Boyce,
J.D. Cohen, R.W. Collins, J. Hanna, 2001, ISBN: 1-55899-600-1

Volume 665— Electronic, Optical and Optoelectronic Polymers and Oligomers, G.E. Jabbour, B. Meijer,
N.S. Sariciftci, T.M. Swager, 2001, ISBN: 1-55899-601-X

Volume 666— Transport and Microstructural Phenomena in Oxide Electronics, D.S. Ginley, M.E. Hawley,
D.C. Paine, D.H. Blank, S.K. Streiffer, 2001, ISBN: 1-55899-602-8

Volume 667— Luminescence and Luminescent Materials, K.C. Mishra, J. McKittrick, B. DiBartolo,
A. Srivastava, P.C. Schmidt, 2001, ISBN: 1-55899-603-6

Volume 668— II-VI Compound Semiconductor Photovoltaic Materials, R. Noufi, R.W. Birkmire, D. Lincot,
H.W. Schock, 2001, ISBN: 1-55899-604-4

Volume 669— Si Front-End Processing—Physics and Technology of Dopant-Defect Interactions III, M.A. Foad,
J. Matsuo, P. Stolk, M.D. Giles, K.S. Jones, 2001, ISBN: 1-55899-605-2

Volume 670— Gate Stack and Silicide Issues in Silicon Processing II, S.A. Campbell, C.C. Hobbs, L. Clevenger,
P. Griffin, 2001, ISBN: 1-55899-606-0

Volume 671— Chemical-Mechanical Polishing 2001—Advances and Future Challenges, S.V. Babu, K.C. Cadien,
J.G. Ryan, H. Yano, 2001, ISBN: 1-55899-607-9

Volume 672— Mechanisms of Surface and Microstrucure Evolution in Deposited Films and Film Structures,
J. Sanchez, Jr., J.G. Amar, R. Murty, G. Gilmer, 2001, ISBN: 1-55899-608-7

Volume 673— Dislocations and Deformation Mechanisms in Thin Films and Small Structures, O. Kraft,
K. Schwarz, S.P. Baker, B. Freund, R. Hull, 2001, ISBN: 1-55899-609-5

Volume 674— Applications of Ferromagnetic and Optical Materials, Storage and Magnetoelectronics, W.C. Black,
H.J. Borg, K. Bussmann, L. Hesselink, S.A. Majetich, E.S. Murdock, B.J.H. Stadler, M. Vazquez,
M. Wuttig, J.Q. Xiao, 2001, ISBN: 1-55899-610-9

Volume 675— Nanotubes, Fullerenes, Nanostructured and Disordered Carbon, J. Robertson, T.A. Friedmann,
D.B. Geohegan, D.E. Luzzi, R.S. Ruoff, 2001, ISBN: 1-55899-611-7

Volume 676— Synthesis, Functional Properties and Applications of Nanostructures, H.W. Hahn, D.L. Feldheim,
C.P. Kubiak, R. Tannenbaum, R.W. Siegel, 2001, ISBN: 1-55899-612-5

Volume 677— Advances in Materials Theory and Modeling—Bridging Over Multiple-Length and Time Scales,
L. Colombo, V. Bulatov, F. Cleri, L. Lewis, N. Mousseau, 2001, ISBN: 1-55899-613-3

Volume 678— Applications of Synchrotron Radiation Techniques to Materials Science VI, P.G. Allen, S.M. Mini,
D.L. Perry, S.R. Stock, 2001, ISBN: 1-55899-614-1

Volume 679E—Molecular and Biomolecular Electronics, A. Christou, E.A. Chandross, W.M. Tolles, S. Tolbert.
2001, ISBN: 1-55899-615-X

Volume 680E—Wide-Bandgap Electronics, T.E. Kazior, P. Parikh, C. Nguyen, E.T. Yu, 2001,
ISBN: 1-55899-616-8

Volume 681E—Wafer Bonding and Thinning Techniques for Materials Integration, T.E. Haynes, U.M. Gösele,
M. Nastasi, T. Yonehara, 2001, ISBN: 1-55899-617-6

Volume 682E—Microelectronics and Microsystems Packaging, J.C. Boudreaux, R.H. Dauskardt, H.R. Last,
F.P. McCluskey, 2001, ISBN: 1-55899-618-4

Volume 683E—Material Instabilities and Patterning in Metals, H.M. Zbib, G.H. Campbell, M. Victoria,
D.A. Hughes, L.E. Levine, 2001, ISBN: 1-55899-619-2

Volume 684E—Impacting Society Through Materials Science and Engineering Education, L. Broadbelt,
K. Constant, S. Gleixner, 2001, ISBN: 1-55899-620-6

Volume 685E—Advanced Materials and Devices for Large-Area Electronics, J.S. Im, J.H. Werner, S. Uchikoga,
T.E. Felter, T.T. Voutsas, H.J. Kim, 2001, ISBN: 1-55899-621-4

Prior Materials Research Society Symposium Proceedings available by contacting Materials Research Society

Future Device Issues

Mat. Res. Soc. Symp. Proc. Vol. 669 © 2001 Materials Research Society

Doping process issues for Sub-0.1 μm generation MOSFETs

Toshihiro Sugii, Sergey Pidin, Youichi Momiyama, Ken- ichi Goto, Takuji Tanaka,
Tomonari Yamamoto, Toshirou Futatugi, and Masataka Kase[1]
Fujitsu Laboratories Ltd., 10-1 Morinosato-Wakamiya, Atsugi 243-0197, Japan
[1]Fujitsu Ltd., 1500, Mizono, Tado-cho, Kuwana-gun, Mie 511-0192, Japan

ABSTRACT

To meet the market demands for higher performance LSIs, traditional scaling has been aggressively pursued and has enjoyed great success over the 0.1-μm generations. To maintain continued growth of CMOS performance beyond the 0.1-μm generation, key issues originating from traditional scaling are addressed from the viewpoint of the doping processes (channel engineering, high-activation, and gate- electrode structure) in this paper.

To meet the acceleration in gate-length miniaturization, short-channel effects must be suppressed at a low threshold voltage by using aggressive channel engineering. A channel-impurity profile must be optimized two-dimensionally, not uniformly or one-dimensionally. Channel engineering using tilted-channel implantation (TCI) with Indium is demonstrated.

Traditional scaling results in large variations of threshold voltage due to the statistical-impurity variation in a channel region. We studied the effect of the above channel engineering on threshold voltage fluctuation caused by a statistical- dopant variation by measurement and simulation. It is reported that the two-dimensionally optimized channel profile enhances threshold-voltage fluctuation even if the implantation process variation is negligible.

As CMOS device scales, reduction of parasitic resistance becomes very important for a high performance operation. Resistance at extension egde and contact resistance at silicide-Si interface are dominant factors. The traditional approach is to use a higher RTA temperature and a shorter RTA time. The ultimate RTA is laser annealing. We will demonstrate the laser annealing process with an ultra-low contact resistance of 4×10^{-8} Ω-cm^2.

By integrating the above technologies, ultra-thin gate insulators, and reduction in gate to source/drain overlap length, we can establish front-end process for sub-0.1 μm generations.

INTRODUCTION

Over the last few years, the gate length of MOSFETs has been aggressively scaled because of market demands for processors and system LSIs that operate at higher speeds and consume less power. The history of the Semiconductor Industry Association (SIA) International Technology Roadmap for Semiconductors (ITRS) demonstrates how fast gate length was scaled in the past and expected pace of scaling in the future. Figure 1 shows the SIA roadmap for gate length from the 1994, 1997, 1999 versions of the ITRS [1] and LSI makers. The expected NAND gate delay corresponding to the gate length has been added to the graph. The practical scaling trend of LSI makers is further along that the SIA roadmap. We can see that with the acceleration in gate-length miniaturization, the speed of improvements is expected to keep pace with that of prior generations. This is what fuels the acceleration. LSIs with 0.13-μm technology will start to be mass-produced this year, and current development is focusing on technologies for 0.1-μm and sub-0.1-μm generations in which the gate length will be around 50 nm. The time required for developing each generation is becoming shorter and shorter. An important strategy under these circumstances is to choose the optimum technology based on the schedule, performance, and cost [2]. In this paper, we review the challenges to achieving continued growth in CMOS performance beyond the 0.1-μm generation from the point of doping related issues.

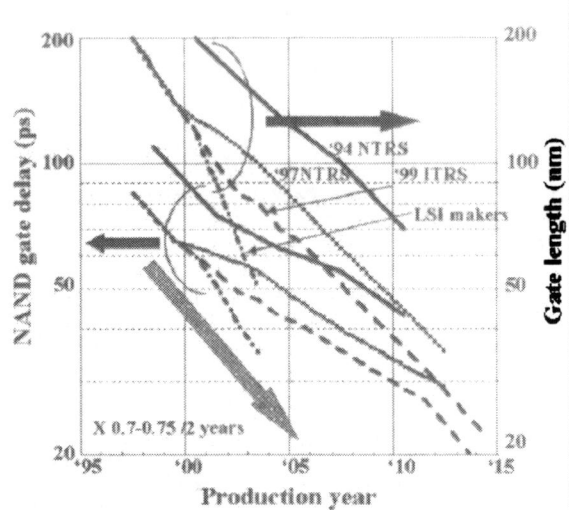

Figure 1. Roadmap for gate length and predicted NAND gate delay.

DEVICE ISSUES FOR MOS SCALING BEYOND 0.1 μm

In Figure 2, various issues related to device scaling beyond the 0.1-μm generation are shown. To keep pace with the acceleration in gate-length miniaturization, we have to suppress short-channel effects first. Since the scaling theory requires that channel-impurity concentration be increased and supply voltage be reduced generation by generation, we have to avoid the trade-off between immunity in short-channel effects and a low threshold voltage. A channel-impurity profile must be optimized two-dimensionally, not uniformly or one-dimensionally, to realize a low threshold voltage with sufficient immunity to short channel effects.

However, continued scaling is expected to yield devices that have higher speeds, as shown in Figure 1. To realize these higher speed devices, we have to increase drive current (reduction of intrinsic and parasitic resistance) and reduce the intrinsic and parasitic capacitances. Not only a reduction of the parasitic resistance, such as contact and extension resistances must be realized, but a reduction of the carrier mobility due to increased channel impurity and a depletion in a gate electrode must be minimized. Intrinsic capacitance can be reduced by the aggressive scaling of gate length. Other capacitive elements, such as overlapped capacitance and junction capacitance can be only moderately or inversely scaled. The weight of these elements is increasing as gate length is scaled.

For low-power operation, the various leak currents, such as gate oxide tunneling, junction leakage, and GIDL, should be suppressed. As transistor dimensions are scaled, the number of impurity atoms controlling the transistor threshold voltage becomes small. This causes variations in threshold voltage due to the statistical fluctuations of impurity number and position. We must study how channel engineering affects this variation.

The suppression of short-channel effects, reduction in resistive and capacitive elements, and understanding statistical fluctuation caused by channel engineering will be reviewed.

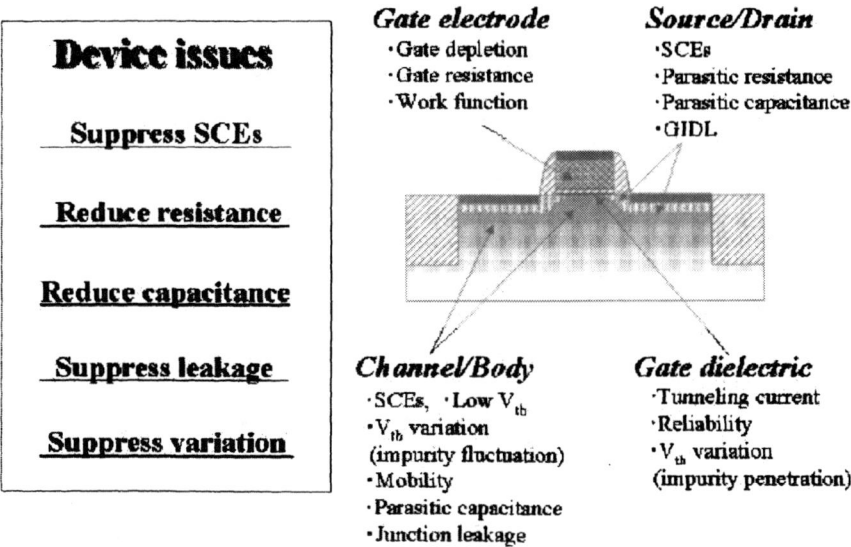

Figure 2. Device issues for MOS scaling beyond 0.1-μm generation.

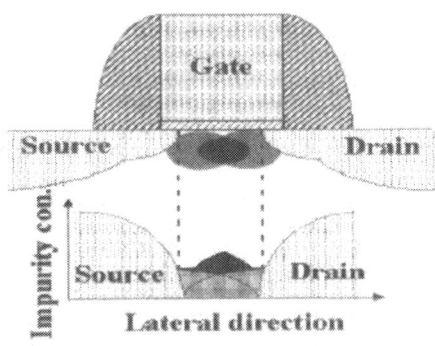

Figure 3. Schematic of proposed channel-impurity profile.

SUPPRESSION OF SHORT-CHANNEL EFFECTS

Channel engineering

A channel-impurity profile must be optimized two-dimensionally, not uniformly or one-dimensionally, to avoid the trade-off between immunity in short-channel effects and a low threshold voltage. At the gate length below which short-channel effects severely occur, the channel-impurity concentration should be increased. Figure 3 shows the schematic of a two-dimensional channel-impurity profile we have proposed [3,4]. This profile can be achieved with tilted channel ion implantation (TCI) after etching a poly-Si gate. Since the overlap of implanted impurity is larger for shorter gate lengths, the concentration of the channel impurities increases and the threshold voltage automatically becomes high. Figure 4 shows simulated depth profiles of channel-impurity concentration at the center of the gate. The channel concentration is low at longer gate lengths and high at shorter ones, which results in weakly dependent characteristics of Vth on the gate length.

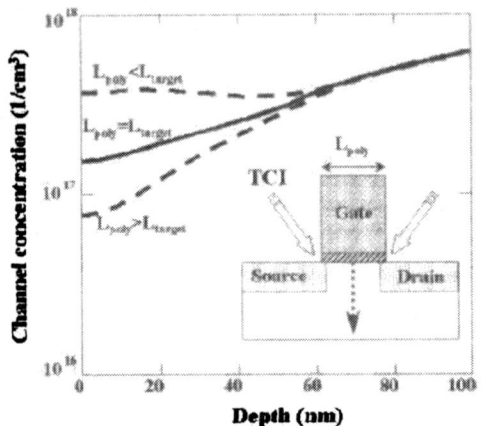

Figure 4. Simulated depth profiles of channel-impurity concentration at the center of the gate.

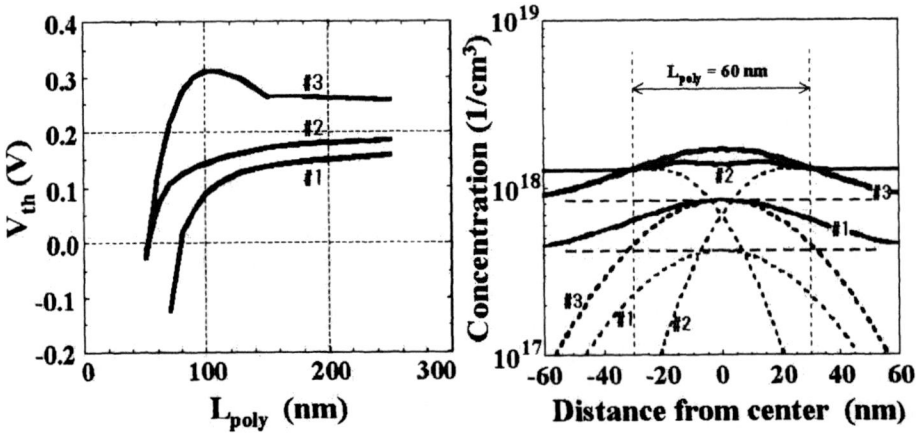

Figure 5-a. Simulated Vth roll-off characteristics of transistors with three types of lateral-channel profiles shown in Figure 5-b.

Figure 5-b. Lateral channel profiles for simulation.

In addition to the self-aligned channel doping by TCI, more precisely controlled channel profiles are going to be required beyond the 0.1-μm generation. This is because the increase in channel-impurity concentration must be minimized to avoid any degradation of carrier mobility and to avoid increases in junction leakage and junction capacitance. It is necessary to make a lateral-channel profile with a high peak concentration and rapid decay. Figure 5-a shows simulated Vth roll-off characteristics of transistors with three types of lateral-channel profiles (Figure 5-b). When a profile is loose, with a low peak concentration, the roll-off becomes poor (#1), but a simple increase in implantation dose causes the large reverse short-channel effects (#3). A high peak concentration and rapid decay such as that indicated by #2, is promising. Therefore, we have to modify the TCI design by using less diffusive atoms with small dispersion in the as-implanted profiles.

Indium is the most attractive p-type impurity because of its smaller diffusivity and heavier mass than boron. Using TCI with indium, nMOSFETs with further suppressed short-channel effects below 0.1 μm have been realized, as shown in Figure 6. Low threshold voltage and strong immunity to short-channel effects have been attained. Due to the reduced short-channel effects, the transistors show low output conductance. These characteristics also improve the RF performance of the transistor. We have applied Indium-TCI to Dynamic Threshold MOS transistors (DTMOS) [5]. We have obtained a 140-GHz f_t and 60-GHz f_{max} at a Vds of 1.5 V and a Vgs of 0.65 V [6]. In addition, we obtained a maximum available gain (MAG) of 6 dB at 30 GHz. This is the highest MAG yet reported for Si-MOSFETs.

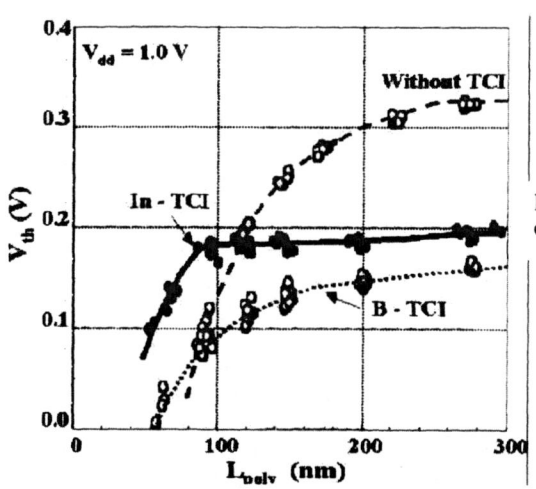

Figure 6. Vth roll-off characteristics of In-TCI, B-TCI, and without TCI.

Variation in Vth caused by channel engineering

Device characteristics are seriously affected by statistical variations in dopant number and position in addition to process variations [7,8]. This is serious because a fluctuation cannot be suppressed by eliminating process variations. The channel-dopant distribution includes not only the number distribution but the position distribution as well. Indium-TCI implantation forms a profile which is characterized by two-dimensionally optimized profiles. Therefore, the TCI profile seems to be susceptible to Vth fluctuations due to impurity-position variations. We studied how the profile affects the Vth fluctuation [9,10].

Figure 7 shows the basis of our measurement for statistical variation in channel-impurity position. We compare only the I-V measurements of one MOSFET between interchanged source/drain bias conditions. As illustrated in the figure, drain voltage, Vd, depletes the channel impurities on the drain edge and Vth is controlled by channel impurities in the local region on the source edge. Asymmetrically placed channel impurities produce different I-V characteristics for the forward (F) and reversed (R) bias conditions. We show an example of measured Id-Vg data at Vd=0.02 V and 1.2 V comparing F and R in Figure 8. While we see no difference between F and R at Vd=0.02 V, a large Vth shift between F and R is found at Vd=1.2 V. This is due to the statistical variation in channel-impurity position. Figure 9 shows the probability distribution of δVth (=Vth(F)-Vth(R)) for transistors with different gate widths. Since the plots fit on linear lines, δVth distribution is regarded as Gaussian-like.

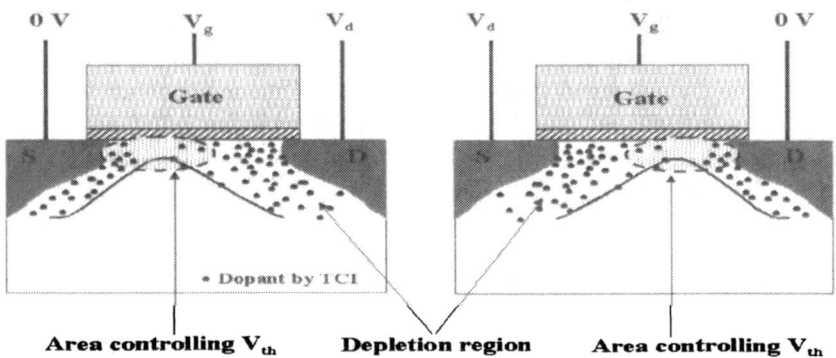

Figure 7. Basis of measurement for statistical variation in channel impurity position.

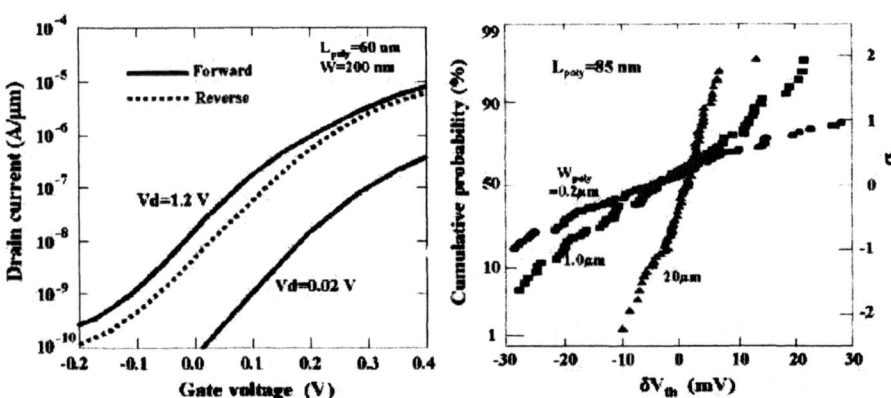

Figure 8. Example of measured Id-Vg data at Vd=0.02 V and 1.2 V comparing F and R.

Figure 9. Probability distribution of δVth (=Vth(F)-Vth(R)) for transistors with different gate widths.

We have investigated the effects of channel engineering on the δVth. We measured δVth with transistors fabricated under a variety of conditions for tilted-channel implantation. Figure 10 shows δVth increases with TCI dose and energy. The large δVth under strong TCI conditions is explained as follows. The region of potential barrier is narrower under a strong TCI condition, resulting in a large dopant fluctuation.

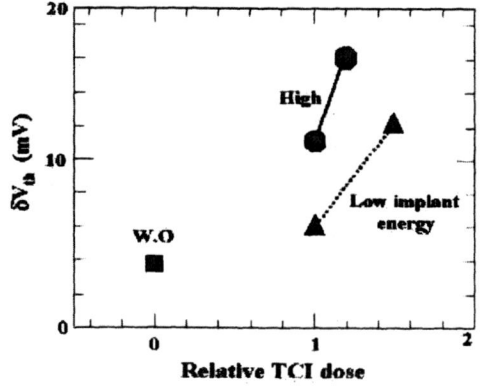

Figure 10. δVth with transistors fabricated under a variety of conditions for tilted-channel implantation.

HIGH ACTIVATION AND SHALLOW JUNCTION

An increase in the drive current is accomplished by reducing the resistive elements in a transistor. The elements are divided into channel and parasitic parts. The reduction in channel resistance means the scaling of gate length and gate-insulator thickness, which is being performed by aggressive scaling, as mentioned before. On the other hand, the parasitic resistance consists of mainly contact resistance between silicide and diffused regions and resistance at extension regions.

Contact resistance increases with scaling. An increase in the electrically active impurity concentration is the most effective method to reduce contact resistance. The traditional approach is to use a higher RTA temperature and a shorter RTA time. The ultimate type of RTA is laser annealing. We have developed a laser annealing process that reduces the contact resistance between a silicide layer and a diffused layer [11]. We achieved a very low contact resistivity of 4 x 10^{-8} $\Omega \cdot cm^2$, which is five times lower than that of conventional RTA. Figure 11 shows the projected contact resistance of laser annealing and RTA devices as a function of technology node, L_D. It is assumed that the distance of the silicided source/drain region is 2 x L_D, and the required contact resistance (2 x Rc) is 10% of the channel resistance. Performance of devices fabricated using the RTA process will be limited by the contact resistance at the 0.1-μm generation. The laser annealing process extends device scaling by maintaining a low contact resistance down to the sub-50-nm node.

The impurity profile at the extension region is the key to reducing the extension resistance. It must have a high surface concentration and be abrupt laterally [12]. Figure 12 shows the simulated drain current for the devices that have the proposed junction profiles shown in the inset. A normal profile is made by a BF_2-ion implantation with a 1-keV acceleration energy. All simulated transistors have the same effective channel length of 50 nm. From this, the abrupt junction increases the drain current, while the graded junction leads to a decrease. Consequently, to reduce the resistance at an extension region, having an abrupt-junction profile of the extension is important. Studies concerning the application of the laser to anneal the extension regions are being vigorously conducted [13].

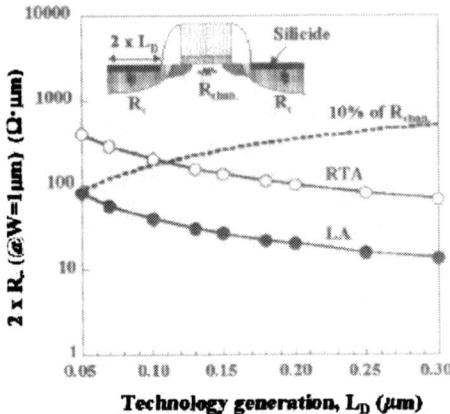

Figure 11. Projected contact resistance of laser annealing and RTA devices as a function of technology generation, L_D.

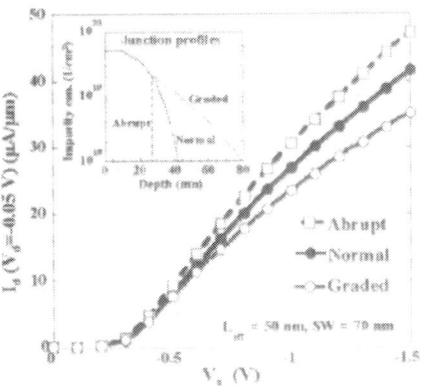

Figure 12. Simulated drain current for devices that have the proposed junction profiles shown in the inset.

DESIGN OF 0.1-μm GENERATION TRANSISTORS

Optimized tilted-channel implantation suppresses short-channel effects, however, the lateral sharpness of extension profiles at source/drain extension regions is insufficient for 0.1-μm generation in which the gate length is around 50 nm. This is due to the limitation of conventional rapid thermal processing. Unless we develop a matured laser-annealing technique, 50-nm transistors will require structures that reduce the gate to source/drain overlap length, such as a double-sidewall structure and a notched-gate structure [14]. The latter structure is also favorable for current drivability since impurities introduced by tilted-channel implantation contaminate the channel region less, as shown in Figure 13. We fabricated a notched-gate transistor where the notch was formed by RIE without introducing any additional process steps.

In Figure 14 we show the threshold voltage roll-off dependence on gate length. The threshold voltage roll-off is well suppressed down to sub-50 nm gate length for both n- and p-channel MOSFETs. In Figure 15 we show transistor gate delay dependence on device off-current. At an off-current of 10 nA/μm, the gate delay is 1.6 psec for an nMOS and 3 psec for a pMOS which is competitive with the published data [14,15].

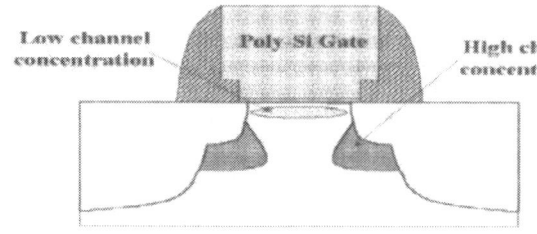

Figure 13. Concept of notched-gate transistor.

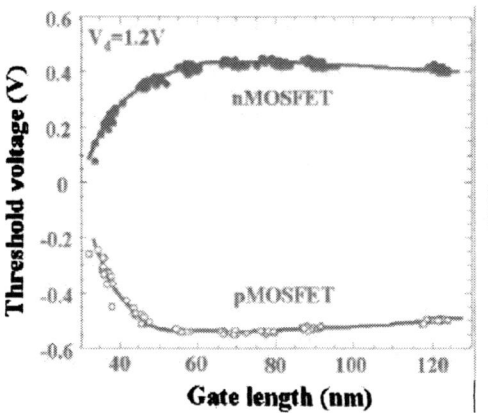

Figure 14. Threshold voltage roll-off dependence on gate length.

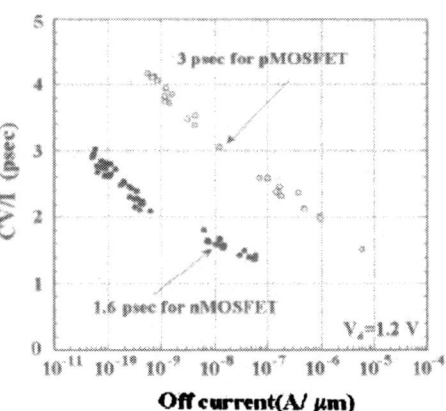

Figure 15. CV/I dependence on device off-current.

CONCLUSIONS

We reviewed the challenges to achieving continued growth in CMOS performance beyond the 0.1-μm generation from the viewpoint of doping related issues. With optimized tilted-channel implantation combined with reduced overlap length by a gate electrode-structure, short-channel effects can be suppressed down to a 50-nm gate length. We studied how the profile affects Vth fluctuation. In order to obtain further improvements of device performance, however, we may need to introduce a revolutionary annealing process for a lateral abrupt junction at an extension region.

ACKNOWLEDGMENTS

We would like to express our gratitude to all members of the ULSI Fabrication Technology Laboratory of Fujitsu Laboratories Ltd. and the Process Development Division of Fujitsu Ltd. for their discussions and the fabrication of the devices.

REFERENCES

1. Semiconductor Industry Association National and International Technology Roadmap for Semiconductors, 1994, 1997, and 1999.
2. T. Sugii, Y. Momiyama, M. Deura, and K. Goto, Abs. Silicon Nanoelectronics Workshop, p. 60 (1999)
3. H. Kurata and T. Sugii, IEEE Trans. on EDL, **45**, 2161 (1998)
4. Y. Momiyama, S. Yamaguchi, S. Ohkubo, and T. Sugii, Technical paper of VLSI Symp. on Tech., p.67 (1999)
5. F. Assaderaghi et al., IEDM Tech. Dig., p. 809 (1994)
6. Y. Momiyama, T. Hirose, H. Kurata, K. Goto, Y. Watanabe, and T. Sugii, IEDM Tech. Dig., p. 451 (2000)
7. H.-S. Wong and Y. Taur, IEDM Tech. Dig. p. 705 (1993)
8. T. Mizuno and A. Toriumi, J. Appl. Phys. **77**, 3538 (1995)
9. T. Tanaka, T. Usuki, Y. Momiyama, and T. Sugii, Technical paper of VLSI Symp. on Tech., p. 136 (2000)
10. T. Tanaka, T. Usuki, T. Futatsugi, Y. Momiyama, and T. Sugii, IEDM Tech. Dig. p. 271 (2000)
11. K. Goto, T. Yamamoto, T. Kubo, M. Kase, Y. Wang, T. Lin, S. Talwar, and T. Sugii, IEDM Tech. Dig. p. 931 (1999)
12. K. Goto, M. Kase, Y. Momiyama, H. Kurata, T. Tanaka, M. Deura, Y. Sanbonsugi, and T. Sugii IEDM Tech. Dig., p. 631 (1998)
13. B. Yu, Y. Wang, H. Wang, C. Riccobene, S. Talwar, and M-R. Lin, IEDM Tech. Dig. p. 509 (1999)
14. T.Ghani, et al., IEDM Tech. Dig., p. 415 (1999)
15. A. Ono, K. Fukasaku, T. Matsuda, T. Fukai, N. Ikezawa, K. Imai, and T. Horiuchi, Technical paper of VLSI Symp. on Tech., p. 136 (2000)

Mat. Res. Soc. Symp. Proc. Vol. 669 © 2001 Materials Research Society

Advanced Ion Implantation Technology for High Performance Transistors.

Kyoichi Suguro, Atsushi Murakoshi, Toshihiko Iinuma, Haruko Akutsu,
Takeshi Shibata, Yoshikazu Sugihara, and Katsuya Okumura*
TOSHIBA CORPORATION
Process & Manufacturing Eng. Center, Semiconductor Company, TOSHIBA Corporation,
8, Shinsugita-cho, Isogo-ku, Yokohama 235-8522, Japan
Phone: +81-45-770-3663,   Fax: +81-45-770-3577
suguro@amc.toshiba.co.jp

ABSTRACT

Cryo-implantation technology is proposed for reducing crystal defects in Si substrates. The substrate temperature was controlled to be below at -160°C during ion implantation.   No dislocation was observed in the implanted layer after rapid thermal annealing.   Pn junction leakage was successfully reduced by one order of magnitude as compared with room temperature implantation.   Precise dose control is indispensable in channel region of high performance MOSFETs.   In order to improve the precision of implanted dose, chip size implantation technology without photoresist mask was developed.   In this technology, chip-by-chip implantation can be carried out by step-and-repeat wafer stage, and different implantation conditions are available in the same wafer independent of wafer size.

INTORDUCTION

Defect control in shallow source/drain and the precise dose control in channel region are important issues in high performance transistors of 0.1-0.13 micron regime.   With shrinkage of junction depth, the thermal budget of annealing after ion implantation becomes smaller in order to suppress impurity diffusion.   On the other hand, it becomes difficult to recover the defects around deep junctions by small thermal budget annealing.   Therefore, the annihilation of defects by annealing with small thermal budget is a key issue for 0.1-0.13 micron regime.   There are several reports concerned with low temperature ion implantation of B, BF$_2$ and P. [1-3]   These papers describe that the residual defect density after furnace annealing for Si substrates implanted at low temperature is lower than that for Si substrates implanted at room temperature.   However, complete annihilation of defects was not reported so far.   We have developed a process module combined with cryo-implantation and rapid thermal annealing in order to annihilate defects in source and drain regions. [4]

Another issue in doping technology is precise implantation in channel region. In order to improve the precision of implanted dose, chip size implantation technology without photoresist mask was developed. In the present technology, chip-by-chip implantation can be carried out by a step-and-repeat wafer stage, and different implantation conditions are available in the same wafer independent of wafer size [5]. In this paper, the experimental results concerned with cryo-implantation and chip-by-chip implantation are reviewed and the some applications are presented.

HOW TO DECREASE DEFECTS

Figure 1 schematically illustrates primary defects and secondary defects for conventional room temperature (RT) implantation and cryo-implantation (Tsub:–135-160°C). Left-hand side is room temperature implantation case. Open circles indicate vacancies and gray solid circles and larger black solid circles indicate interstitial Si atoms and As atoms, respectively. In this case vacancies and interstitial atoms are considered to migrate easily and form clusters. On the other hand, in the case of cryo-implantation, vacancy clustering and interstitial clustering are suppressed as shown in right-hand side.

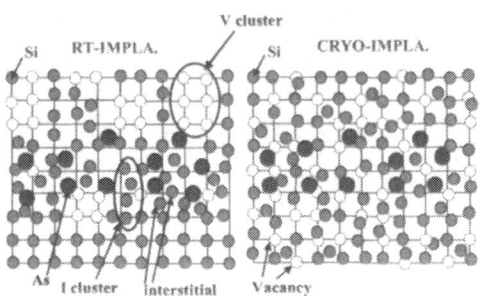

Fig.1 Difference in defect density during implantation.

Figure 2 schematically shows defects in Si after annealing. In the case of RT implantation, vacancy and interstitial atom hardly recombine each other and secondary defects such as dislocation loops and <111> defects are formed. On the other hand, vacancies and interstitial atoms can easily recombine in the case of cryo-implantation and as a result, defects can annihilate after annealing.

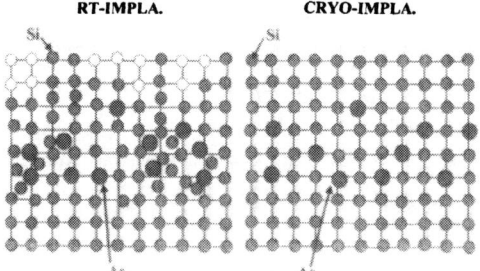

Fig.2 Difference in defect density after annealing.

Figure 3 shows the temperature dependence of probability of various types of vacancies in Si presented by Prof. Watkins in 97 MRS meeting [6]. In usual high current implanters, Si substrate temperature is controlled less than 60°C by cooling the Si substrate with flowing water in the wafer suscepter. However, various types of single vacancies cannot exist at such a high temperature. Therefore, the vacancy binds with another vacancy or impurity atom such as oxygen as shown in Fig. 3.

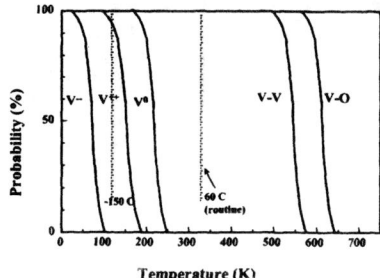

Fig.3 Stability of Various Type of vacancies. (Ref.6)

Positron analysis of As implanted Si substrates showed the clear movement of vacancy distribution profiles towards the Si surface during As ion implantation except cryo-implantation. That is, if the substrate temperature is higher than RT, vacancies at deeper region tend to move to a surface by 5nm, because single vacancies cannot stably exist as described above.

EXPERIMENTAL

Boron or arsenic ions were implanted into (100)Si substrates where the substrate temperature was controlled to be –160°C. Typical acceleration energy of boron and arsenic were 10keV and 20keV, respectively. The implanted dose was in the range from 1E13 to 1E15cm^{-2}. The ion beam current was in the range from 10 to 20 micro A/cm^2. Rapid thermal annealing (RTA) of 900°C for 30sec was carried out for the electrical activation of boron and arsenic. Ramp up rate and ramp down rate were 150°C/sec and 100°C/sec, respectively. Implanted samples were analyzed by SIMS and Rutherford backscattering spectrometry (RBS), transmission electron microscopy (TEM). Pn junction leakage current was measured for some of the implanted samples.

RESULTS & DISCUSSIONS

Figure 4 shows cross section TEM photographs of RT implantation of boron case before and after RTA at 900°C for 30sec. Amorphous layer is not formed in as implanted sample and many dislocations exist even after RTA at 900°C for 30sec.

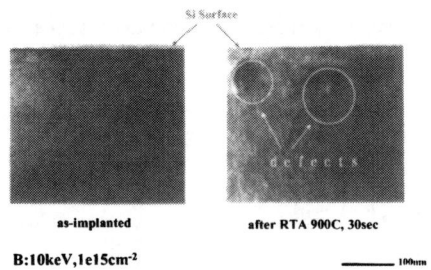

as-implanted   after RTA 900C, 30sec

B:10keV,1e15cm^{-2} ——— 100nm

Fig.4 Cross-section TEM photographs for implanted with B at RT.

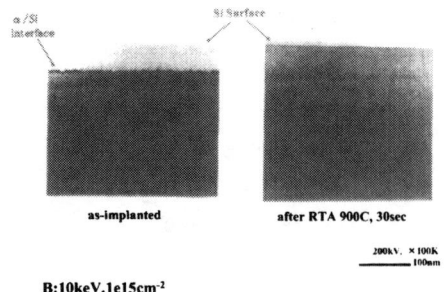

as-implanted   after RTA 900C, 30sec

200kV, ×100K
——— 100nm

B:10keV,1e15cm^{-2}

Fig.5 Cross-section TEM photographs for implanted with B at –160C.

On the other hand, amorphous layer can be clearly observed in as implanted sample in the case of cryo-implantation, and any dislocation cannot be observed after RTA as shown in Fig. 5.

Substrate temperature effect during ion implantation also exists in As implanted case. Figure 6 shows cross section TEM photographs of RT implantation As case before and after RTA at 900°C for 30sec.   Micro defects can be observed below the amorphous/crystal interface in as implanted case.   This result suggests incomplete epitaxial growth during ion implantation due to the substrate heating by energy transfer from implanted ions.   And Micro defects grow to larger defects (dislocations) after RTA as shown in Fig. 6.

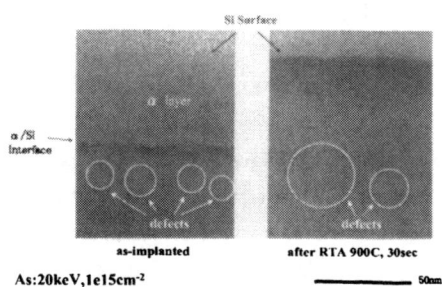

Fig.6 Cross-section TEM photographs for implanted with As at RT.

On the other hand, micro defects cannot be observed in as implanted sample in the case of cryo-implantation, and the amorphous/crystal interface is rather abrupt as shown in Fig. 7.   And no defect is observed after RTA.

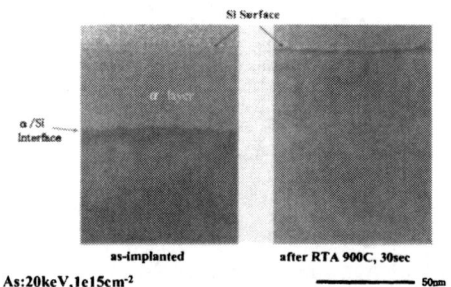

Fig.7 Cross-section TEM photographs for implanted with As at –160C.

Pn junction leakage characteristics show the substrate temperature effect as shown in Fig. 8 and Fig. 9.   Figure 8 shows p^+n junction characteristics for B implanted Si. B dose is 1E13, 1E14, and 1E15 cm^{-2}.   Pn junction leakage effectively decreases by using cryo-implantation.

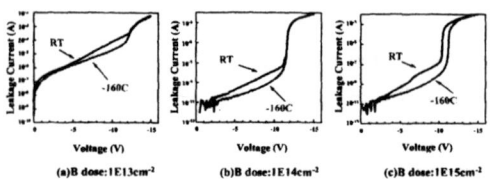

Fig. 8 p⁺/n junction characteristics.

Figure 9 shows n⁺p junction characteristics for As implanted Si. As dose is 1E13, 1E14, and 1E15 cm⁻². In As implanted case, the thermal budget of RTA (900°C, 30sec) is insufficient for RT implantation. Pn junction leakage is effectively reduced by using cryo-implantation as well as B implanted Si. As the implanted dose is increased, the difference in pn junction junction leakage current is more pronounced. There are many advantages in cryo-implantation, however, there is an important issue of difficulty in photoresist mask formation.

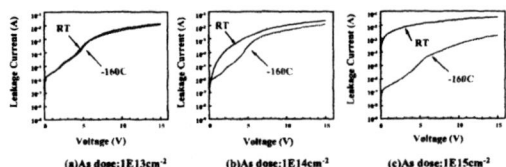

Fig. 9 n⁺/p junction characteristics.

After cryo-implant, many cracks were found in conventional photoresist masks. This is due to the shrinkage of photoresist film during rapid cooling. In order to avoid photoresist cracks, polymer type resist and hard baked resist were developed. Cracks are observed in the film of polymer 1. However, there found no crack in polymer 2 and hard baked photoresist film. These films can be applied to CMOS process by using appropriate resist films.

CONCEPT OF CHIP-BY-CHIP IMPLANTATION

Conventional ion implantation process involves a sequence of steps of photo-resist coating, exposure to light, photo-resist development (photo-resist pattern formation), ion implantation, photo-resist ashing, and wet cleaning using a mixture of H_2SO_4 and H_2O_2.

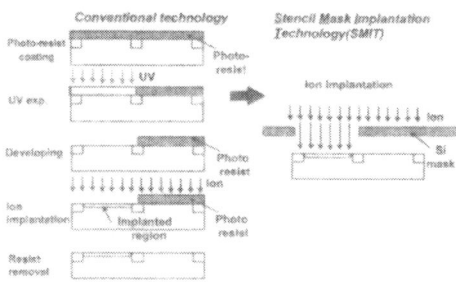

Fig. 10 Comparison of process steps.

In order to simplify ion implantation processes, we newly developed a stencil mask ion implantation technology (SMIT) for eliminating such processes as shown in Fig. 10. [5,7]

Advantage of SMIT (economical merit)

By using SMIT, 6 kinds of processes such as photo-resist coating, exposure to UV, developing, ion implantation, photo-resist ashing and wet cleaning are reduced to 1 process of ion implantation. The reduction of fabrication cost becomes very large if SMIT is used in fabrication of system LSIs where ion implantation process steps exceeds 30. Therefore, the investment for fabricating LSIs can be much reduced as shown in Fig. 11.

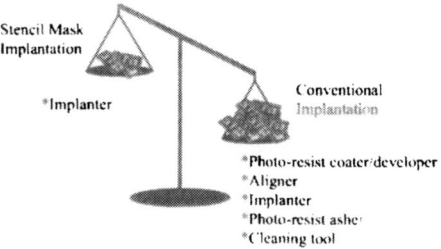

Fig. 11 Comparison of process cost

There is another economical advantage in SMIT. Occupied space of equipments can be effectively reduced as shown in Fig. 12.

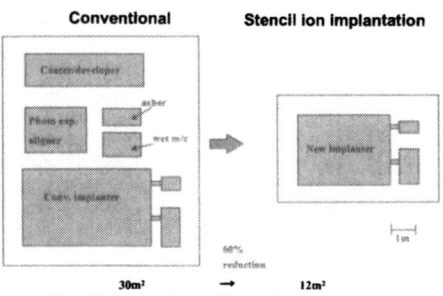

Fig. 12 Comparison of foot print.

Raw process time is reduced from 4 hours in conventional implantation technology to 1 hour in SMIT. The cost of ownership can be much reduced by using SMIT. Advantages of SMIT is summarized in Table 1.

Table 1 Comparison of RPT, COO, Foot print.

	RPT (Hour / Lot)	COO (Yen / Waf)	Foot print (m²)
Conventional	3.9	1150	30.4
SMIT	1.0	550	12.2

Advantage of SMIT (system performance merit)

One of the advantage of stencil mask implantation is that different implanted conditions can be easily achieved on chip by chip. Figure 13 shows subthreshold characteristics of p-MOSFET by using SMIT. Channel region of PMOS is implanted with different kinds of phosphorus dose in chip by chip. Vth of PMOS changes according to phosphorus dose as shown in Fig. 13. It was found that Vth could be precisely controlled by implanted dose without photo-lithography.

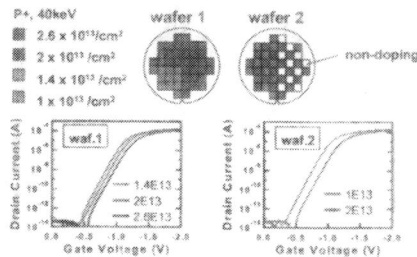

Fig. 13 Subthreshold characteristics of MOSFET in a wafer. Channel dose is changed chip by chip.

Equipment of SMIT

Figure 14 illustrates the schematic diagram of mask and Si substrate wafer stage. Ions come through holes opened in the stencil mask to the substrate wafer. The mask thickness is about 5-10 micron. The distance between the mask and the wafer is ranged from 10 to 100 micron.

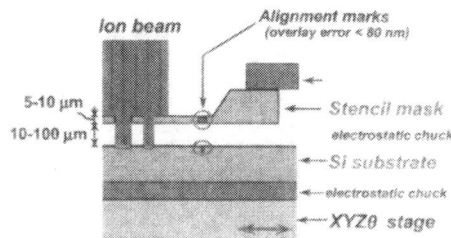

Fig. 14 Magnified schematic diagram of mask and wafer stage.

End station chamber consists of X-Y-Z-theta stage as shown in Fig. 15. Electrostatic chucks for wafer and stencil mask, and optical mask/wafer alignment system. Acceleration energy is ranged from 1keV to 700keV.

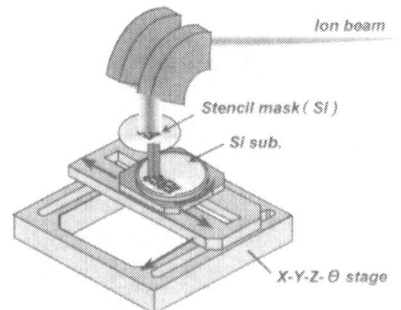

Fig. 15 Schematic diagram of implanter.

Since the scan width of 1 ion beam shot is 10 to 30 mm, the incident angle distribution of ion beam to a Si substrate is always constant.   Therefore, each chip or each shot area can be implanted like a copy of one implanted area and the uniformity is improved as compared with conventional technology where ion beam is scanned from wafer edge to wafer edge as shown in Fig.16.

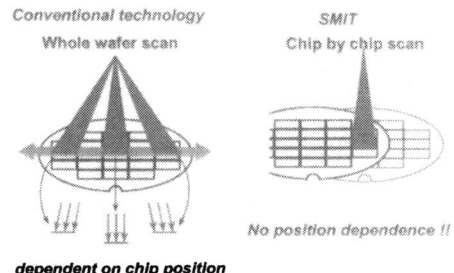

Fig. 16 Comparison of ion beam incident angle distribution.

Phosphorous atoms were implanted to bare (100)Si substrates or SiO_2 films on Si substrates at 40 keV to a dose of 5E14 cm^2.   Implanted area was delineated by defect etching by using the blended acid, which attacks defect rich region.   Used mask is 1 micron L/S pattern as shown in Fig. 17.

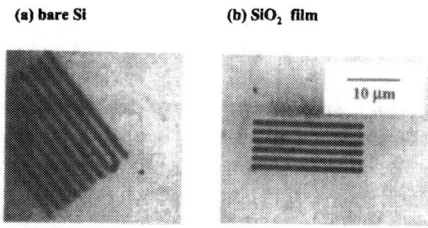

(a) bare Si   **(b) SiO₂ film**

Fig. 17 Ion implanted line pattern region delineated by defect etching.

One of issues of stencil mask implantation is the deformation of the stencil mask (membrane) during ion implantation. The stencil mask consists of Si membrane with 5–10 micron in the thickness surrounded by 500-micron thick Si substrate. Since the distance between the stencil mask and Si substrate is set at less than 100 micron before ion implantation, the strength of the membrane is very important. If non doped Si membrane is used in this experiment, the membrane deform and the distance between the stencil mask and Si substrate changes with implantation dose as shown in Fig. 18. The strength of Si membrane can be improved by doping impurities such as nitrogen, carbon, etc. which increases Young's modulus of Si.

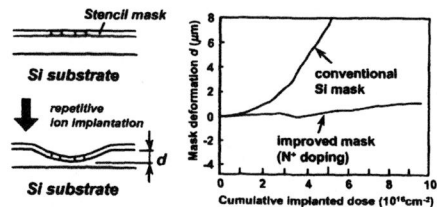

Fig. 18 Mechanical stability of stencil mask .

By improving the strength of the stencil mask, $1E18cm^{-2}$ dose can be achieved without detectable deformation. This result means 1000 wafers can be implanted when 100 shots are implanted in each wafer. The stencil mask life can be elongated by more improvement.

CONCLUSION

Defect free pn junction technology is proposed. By combining cryo-implantation and rapid thermal annealing, lower leakage pn junctions can be realized both in boron and in arsenic implanted silicon substrates and pn junction leakage current was reduced by one to two order of magnitude. The present technology will be very useful in realizing high performance transistors. Stencil mask implantation technology (SMIT) is very effective in reducing investment cost, raw process time, cost of ownership and foot print of equipments in fabricating LSIs, especially system LSIs where ion implantation process steps exceeds 30. The threshold voltage values can be precisely controlled by ion implanted dose.

ACHNOWLEDGMENTS

The authors thank Mr. M. Sugitani and Mr. F. Sato in Sumitomo Eaton Nova, Japan, and Dr. M. Foad and Dr. S. Moffatt in AMAT for the experiments of cryo-implantation and also thank Mr. T. Nishihashi in ULVAC Japan for the experiments of stencil mask ion implanter, and Drs. Y. Sakurada, J. Fujiyama, and K. Kashimoto in ULVAC Japan Ltd. for supporting the development of SMIT and fruitful discussions. They also thank Prof. Uedono in Tsukuba University for positron annihilation analysis of ion implanted samples.

REFERENCES

1. T. Suzuki, H. Yamaguchi, S. Ohzono, N. Natsuaki, Ext.Abst.of the 22nd Int.Conf.on Solid State Device and Materials (1990) pp.1163-1164.
2. M. Takakura, T. Kinoshita, T. Uranishi, S. Miyazaki, M. Koyanagi, and M. Hirose, Ext.Abst.of the 1991 Int.Conf. on Solid State Device and Materials (1991) pp.219-221.
3. M. Kase, Y. Kikuchi M. Kimura, H. Mori, and R. B. Liebert, J. Appl. Lett., 75 (1994) pp.3358-3364
4. A. Murakoshi, K. Suguro, M. Iwase, M. Tomita, and K, Okumura, Mat. Res. Soc. Symp. Proc. Vol. 610 (MRS, Pittsburg, 2001) B3.8.
5. T. Shibata, K. Suguro, K. Sugihara, H. Mizuno, A. Yagishita, T. Saito, and K. Okumura, IEDM Tech. Dig. (IEEE, New York, 2000) p.869.
6. G. D. Watkins, Mat. Res. Soc. Symp. Proc. Vol. 469 (MRS, Pittsburg, 1997) pp.139-150.
7. T. Nishihashi, K. Kashimoto, J. Fujiyama, Y. Sakurada, T. Shibata, K. Suguro, K. Sugihara, and K. Okumura, Abstract of EIPBN 2001.

* present address: Tokyo University, 4-6-1 Komaba, Meguro-ku, Tokyo 153-8904, Japan

Advances in Dopant Profiling

Mat. Res. Soc. Symp. Proc. Vol. 669 © 2001 Materials Research Society

SSRM and SCM observation of modified lateral diffusion of As, BF2 and Sb induced by nitride spacers.

P. Eyben [1], N. Duhayon, C. Stuer, I. De Wolf, R. Rooyackers, T. Clarysse, W. Vandervorst and G. Badenes

IMEC vzw, Kapeldreef 75, 3001 Leuven, Belgium

ABSTRACT

Initial studies (using Scanning Spreading Resistance Microscopy) on the lateral diffusion of B and As have shown an important influence of the thickness of oxy/nitride spacers. The latter phenomenon was tentatively ascribed to stress enhanced diffusion under the spacer region [1]. These studies have been complemented with Scanning Capacitance Microscopy (SCM) measurements, which confirm the SSRM-data. In fact both techniques shows a similar increase in lateral diffusion with increasing spacer thickness (~ 0.2 nm/nm spacer thickness), whereby no effect is observed on the vertical diffusion. When using spacers with or without TEOS-liner, fairly similar enhancements could be seen. Micro-Raman and CBED stress measurements for these cases do however show a large reduction in stress when a TEOS-liner is used, suggesting that the correlation (at least to the final) stress is not really justified. A possible explanation could however be that the lateral diffusion occurs before the stress relaxation within the thermal treatment. In order to elucidate the diffusion mechanism (initial stress, interstitials, hydrogen incorporation, TED,..) we have expanded the experimental matrix with a vacancy diffuser such as Sb and simulated the potential H-incorporation during the nitride deposition by a hydrogen anneal. Moreover we also have studied the impact of TED by splits with RTP-anneals before the nitride deposition.

1. Introduction

The control of the effective channel length is a major issue in the processing of modern CMOS devices. The electrical characteristics of the transistors are drastically affected by minor changes of this length. The study of the effects that may affect this value is thus crucial. Unfortunately, classical 1D dopant profiling techniques may not be used for that issue and the introduction of new characterization techniques that enables dopant profiling in the lateral direction as the AFM-based techniques (SCM, SSRM) has become a necessity.

The lateral diffusion study presented in this paper originates from observations realized during two-dimensional dopant profiling of LDD/HDD structures in NMOS transistors with the SSRM (Scanning Spreading Resistance Microscopy) technique [2]. We have noticed that the effective channel length of the transistors was systematically affected by a change in the oxy/nitride spacer size (Fig. 1). An increase in the spacer size has shown to be responsible for an enhancement of the lateral diffusion of the source and drain n-

---

[1] eyben@imec.be

doped (As) implants and thus for a decrease of the device's effective channel length. In the meantime, no systematic increase of the implants' depths (vertical direction) has been observed. SCM (Scanning Capacitance Microscopy) [3] measurements were realized on the same sample. As illustrated in Fig. 1, they have shown a perfect agreement with the SSRM measurements.

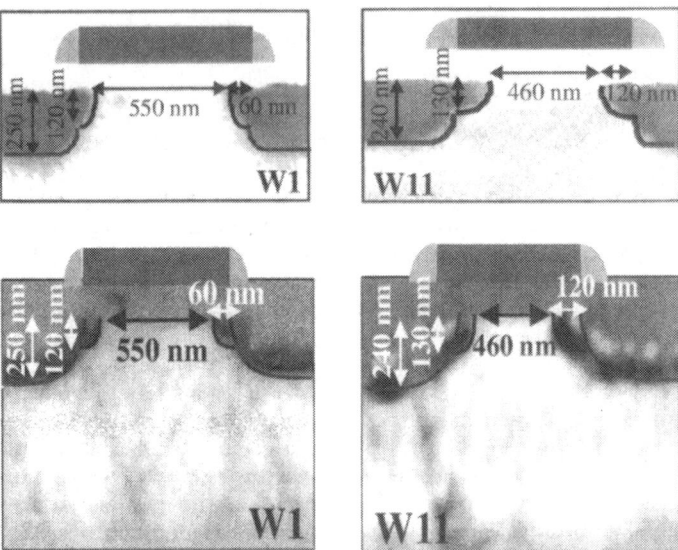

Figure 1. . SCM & SSRM pictures (left: 110nm spacers, right: 160nm spacers) showing an enhanced lateral diffusion for the larger spacers.

2. Lateral diffusion mechanism study

As previously introduced, the change in lateral diffusion appears to be directly linked to a variation in nitride spacers size. The nitride spacers may influence the diffusion by (a) a TED (Transient Enhanced Diffusion) effect occurring during the CVD nitride deposition, (b) the influence of the H_2 incorporated into the silicon bulk during the nitride deposition, or (c) the stress that may exist in the vicinity of the spacers.

2.1. The stress influence

The influence of stress on the diffusion mechanisms has been evidenced many years ago[4] but its importance appears to be more and more pronounced in modern devices due to the shrinking of the dimensions and to the use of various materials. Experimental

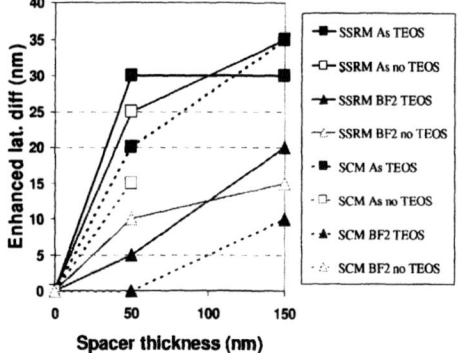

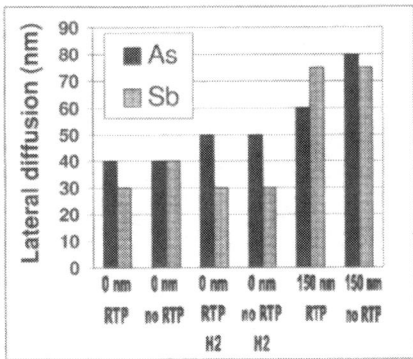

Figure 2. SSRM and SCM observation of the lateral diffusion enhancement with the spacer size increase. The agreement between SCM and SSRM results and the non-influence of the TEOS layer on the enhancement could be observed.

Figure 3. Influence of H_2 incorporation on the lateral diffusion. No influence is observed for Sb. A minor influence is observed for As.

evidence (using SSRM) for stress induced diffusion has, for instance, been obtained in the vicinity of STI (Shallow Trench Isolation) [5].

The initial observations of the lateral diffusion were correlated to the based on the correspondence between the lateral nature of the enhanced diffusion and the results of CBED (Convergent Beam Electron Diffraction) stress measurements [6] showing a large lateral stress (3 times larger than the vertical stress) underneath the nitride spacers increasing rapidly with the spacer size.

For further confirm these conclusions measurements were realized (with SCM and SSRM) on a dedicated set of wafers, where splits on dopants (As vs. BF$_2$), on spacer size (0, 50 and 150nm) and on TEOS-liner (with vs. without) were introduced. A systematic increase of the lateral diffusion with the spacer size for both arsenic and boron implanted samples can be observed (Fig. 2). The enhancement for boron is however rather small and close to the measurement error.

The most interesting result is that the enhancement of the diffusion appears not to be dependent on the presence (or not) of a TEOS-liner (Fig. 2). Knowing that Raman and CBED stress measurements [6] show a drastic decrease of the stress with the introduction of the TEOS-liner, it tends to prove that the observed lateral diffusion mechanism is not driven by the stress. Please note that this conclusion has to be mediated by the fact that measurements of local stress in the near vicinity of the spacers are impossible due to intrinsic characteristics of the stress measurement techniques, and by the fact that the stress is measured on fully processed wafers and thus after the annealing steps. To have an idea of the stress present during the diffusion (which is the important one), in-situ stress measurements would be very necessary. Within the present experimental capabilities it remains thus difficult to definitively conclude on the stress influence.

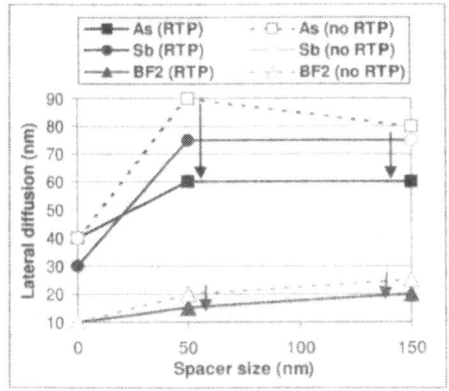
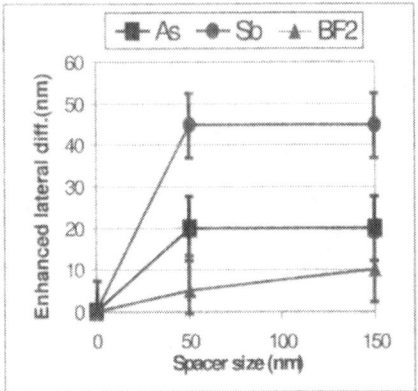

Figure 4. Influence of the RTP anneal on the lateral diffusion. RTP anneal reduces a lot the arsenic enhanced diffusion, a little bit the boron diffusion and seem to have no impact on the antimony diffusion.

Figure 5. Lateral diffusion increase in function of the spacer size for boron, arsenic and antimony implants.

2.2. Influence of the hydrogen incorporation

The chemical reaction involved in the CVD nitride deposition creates hydrogen (H_2), which may enter inside the silicon bulk and influence the diffusion mechanisms. In the present study, we have simulated the effect of the H_2 incorporation during a 150nm nitride deposition by flowing H_2 above the silicon surface with a thermal budget (time and temperature) similar to the 150nm nitride deposition. From measurements realized (with SSRM) on this dedicated set of wafers, with a split on the dopants (As vs. Sb), on the RTP anneal (with vs. without), on the spacer size (0 vs. 150nm) and on the H_2 incorporation (with vs. without), we have observed that the H_2 incorporation has no dominant contribution to the enhancement of the diffusion (Fig. 3). The lateral diffusion values corresponding to the H_2 incorporation without nitride deposition are indeed very close to the samples without nitride deposition (and thus H_2 incorporation). For antimony, the H_2 incorporation seems to have absolutely no influence on the enhancement of the lateral diffusion whereas for arsenic a small effect can be observed.

2.3. Influence of the RTP anneal

In order to eliminate the role of TED during the nitride deposition, a RTP anneal (10'', 970°C) was introduced prior to the nitride deposition. From measurements realized (with SSRM) on a dedicated set of wafers, with a split on the dopants (As, BF$_2$ and Sb), on the spacer size (0, 50nm and 150nm), and on the RTP anneal (with vs. without), we have observed that the RTP anneal was reducing the diffusion enhancement for arsenic and boron (Fig. 4). Moreover, the decrease (in absolute value) appears to be more pronounced for arsenic than for boron implants. Antimony doped implants seem to show no decrease of the enhancement with the RTP anneal process step but, as you may notice in Fig. 4, the Sb study is not completed and more measurements need to be realized to confirm this observation.

Despite the reduction of the TED effect by the RTP anneal, the increase in lateral diffusion remains important (particularly for Sb). This tends to prove that the TED effect is not sufficient to explain the present lateral diffusion effect. Moreover, the lateral nature of the observed diffusion does not fit well with the isotropic aspect of classical TED.

2.4.   Influence of the implanted dopant

The evolution of the lateral diffusion with the spacer size has been studied for different dopants representing different diffusion mechanisms. We have used boron (whose diffusion mechanism is driven by interstitials), arsenic (which uses interstitials and vacancies) and antimony (which uses vacancies). From measurements realized (with SSRM) on a dedicated set of wafers where a RTP anneal step was performed, with a split on the dopants (As, BF_2 and Sb) and on the spacer size (0, 50nm and 150nm), we have observed that the enhancement was minimal (within the error bar) for the boron, intermediate for the arsenic and maximum for the antimony. It tends to prove that the observed lateral diffusion enhancement is strongly dependent on the diffusion mechanism and appears mainly linked to diffusion through vacancies.

3.   Conclusions

We have observed (using SSRM 2D-dopant maps) a systematic enhancement of the lateral diffusion of the source and drain implants in MOS transistors when the spacer size is increased. Under classical processing conditions (with TEOS-liner and RTP anneal processing step), this enhancement is maximum for antimony and minimal for boron. This seem to prove that this diffusion mechanism is mainly related to a diffusion driven by vacancies

Although TED seems to play an important role in the increase of the lateral diffusion, an important enhancement of the diffusion with the spacer size remains present when the TED is drastically reduced by introducing a RTP anneal processing step after the implantation.

The hydrogen (H_2) incorporation in the silicon bulk seems to have no major effect on the diffusion. Antimony, in particular, which represents the largest lateral diffusion enhancement, appears to be completely insensitive to the hydrogen incorporation.

The influence of stress is very difficult to establish. To fully understand the effect of stress, measurements of localized stress underneath the spacers are necessary what remains very difficult today (Raman is not localized and CBED can not measure immediately underneath the spacers). Moreover measurements of the stress present during the diffusion (and thus before and during the annealing processing steps) would be helpful.

In the present state of understanding, the diffusion enhancement does not seem to be caused by one dominant factor but is more likely produced either by a combination of the previously described effects or by an other unknown factor linked to the presence of nitride spacers. For a complete understanding of the nitride spacers induced lateral diffusion, further experiments are however still necessary.

4. Acknowledgements

P. Eyben and N. Duhayon are indebted to the Flemish Institute for Support of Scientific Research IWT for their Ph. D fellowship.

5. References

[1] P. Eyben, N. Duhayon , C. Stuer , I. De Wolf, R. Rooyackers, T. Clarysse, W. Vandervorst and G. Badenes" SSRM and SCM observation of enhanced lateral As- and BF2-diffusion induced by nitride spacers.", Proceedings Material Research Society (MRS) Spring 2000, Symp. B Vol. 610, paper b2.2 (2000)

[2] P. De Wolf, T. Clarysse, W. Vandervorst, J. Snauwaert, L. Hellemans, "1 and 2D carrier profiling in semiconductors by nanoSRP.", J. Vac. Sci. tech B14(1996)380-385

[3] C.C. Williams, J. Slinkman, W.P.Hough, H.K. Wickramasinghe, "Lateral dopant profiling on a 100nm scale by scanning capacitance microscopy", J.Vac.Sci.Technol. A8(2), p. 895, 1990

[4] M.J. Aziz,"Thermodynamics of diffusion under pressure and stress: relation to point defect mechanisms",Appl. Phys. Lett. 70 (21), 26 May 1997.

[5] P. Eyben, M. Xu, N. Duhayon, T. Clarysse, S.Callewaert and W. Vandervorst" "Scanning Spreading Resistance Microscopy and Spectroscopy for routine and quantitative 2D-carrier profiling", Proc. USJ2001 Workshop (to be published in JVST Jan/Feb 2002)

[6] A. Armigliato, R. Balboni, I. De Wolf, S. Frabboni, K.G.F. Janssens and J. Vanhellemont,"Determination of lattice strain in local isolation structures by electron diffraction techniques and micro-Raman spectroscopy",Inst. Phys. Conf. Ser. No 134:section 5, april 1993.

Mat. Res. Soc. Symp. Proc. Vol. 669 © 2001 Materials Research Society

Conductance Imaging of the Depletion Region of Biased Silicon PN Junction Device

Jeong Young Park [1,3], **R. J. Phaneuf** [2,3], **and E. D. Williams**[1,3]

[1] Department of Physics, University of Maryland College Park, Maryland 20742
[2] Department of Materials Science and Engineering, University of Maryland
College Park, Maryland 20742
[3] Laboratory for Physical Sciences, College Park, Maryland 20740

ABSTRACT

Simultaneous conductance imaging and constant current mode STM imaging have been used to delineate Si pn junction arrays over a range of reverse bias conditions. Conductance has been obtained by adding a modulation signal to voltages applied in the p and n regions of a model device, and by measuring the modulation signal of the tunneling current with a lock-in amplifier. Both constant current and conductance imaging of the electrically different regions (n, p, and depletion zone) show a pronounced dependence on applied pn junction bias. The conductance contrast is mainly due to electrically different behaviors of metal-gap-semiconductor junction which are determined by the tip-induced band bending of the oxide-passivated silicon surface.

INTRODUCTION

Semiconductor devices have been widely studied with scanning tunneling microscope (STM) which can probe the surface topographical and electrical structure with high lateral resolution [1-7]. On clean Si surfaces, dopant-dependent contrast is not observed in STM due to surface Fermi-level pinning caused by large density of surface states [8]. However STM images measured on H-terminated or oxide-covered Si devices show variation with dopant type and concentration due to a partial passivation of surface states [9,10]. Recently, Hildner et al.[1] and Yu et al. [2] have used STM and scanning tunneling spectroscopy (STS) to image a depletion zone related feature at a passivated Si pn junction surface. However, the incorporation of both electronic and topographical information in STM constant-current images make it difficult to interpret STM images unambiguously. In this study, we used simultaneous acquisition of STM topographical images and differential conductance images to characterize the electronic features of a lateral Si pn junction. Tunneling spectra were measured across a pn junction to give a qualitative explanation for the variation of conductance.

EXPERIMENTAL DETAILS

The device studied in this experiment has been described elsewhere [1,11,12]. It consists of an array of stripes of p-type doping ($N_A=10^{18}$ /cm^3), within a lightly n-type doped ($N_D = 1.6$ x 10^{14} /cm^3) Si(100) substrate. The surface is terminated by wet chemical oxide, prepared using the Shiraki procedure [13], and placed into the ultrahigh vacuum chamber using a load-lock system within 30 minutes of etching. An STM (a Park Scientific Autoprobe VP) in a UHV chamber with the base pressure of 8×10^{-11} torr was used in this study.

In conductance mapping, the sample bias (U) and the tunneling current (I) are given by following equation (1) and (2),

$$U = U_o + U_1 \cos(\omega_0 t) \tag{1}$$

$$I(U) = I(U_0) + (\frac{\partial I}{\partial U}) U_1 \cos(\omega_0 t) + \frac{1}{2}\left(\frac{\partial^2 I}{\partial U^2}\right)(U_1 \cos(\omega_0 t))^2 + \dots \tag{2}$$

where U_1, ω_0 are modulation voltage and modulation frequency, respectively. The change in the tunneling current corresponding to a small modulated bias voltage, which yields the conductance (dI/dV), is recorded while an STM image is simultaneously acquired. This technique is useful in exploring the spatial variation of the surface electronic structure in a short scanning time. Probing only a limited energy range allows faster data acquisition than measuring the full I-V spectrum [14,15]. A schematic of the system for differential conductance imaging of a biased device is shown in Fig. 1. A small sinusoidal voltage modulation with the height of 85 mV p-p, and the frequency of 4.5 kHz was added to the voltages applied to the n and p regions using two separate summing circuits. The resulting current modulation, which is proportional to dI/dV, is measured with a lock-in amplifier (EG&G 5208). The time constant and the sensitivity of the lock-in amplifier were 5 ms and 5mV.

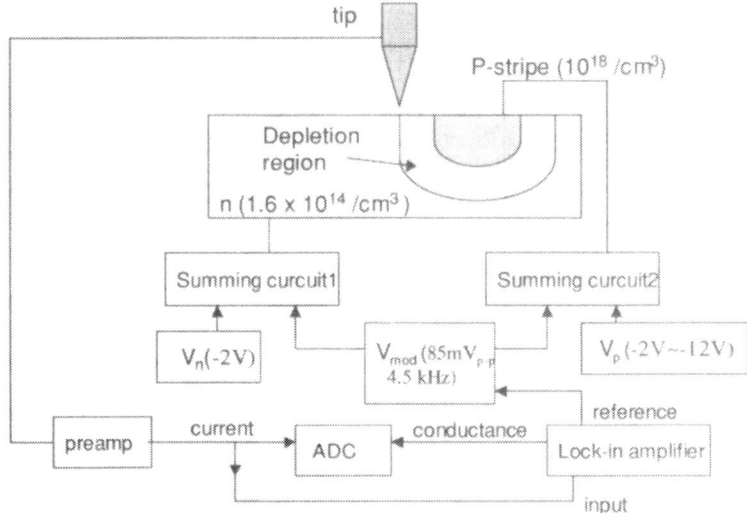

Figure 1. The schematic of a experimental setup for dual scanning (STM topographical imaging and conductance imaging).

RESULTS AND DISCUSSION

Figures 2(a) and 2(b) show an STM topographical image and a differential conductance image, respectively, taken at the same time while a slowly ramping reverse bias (0V ~ -10V) was applied between the p and n region. For both images, the sample voltage (voltage in the n-region) was –2 V and the demanded tunneling current, chosen for attaining a good signal to noise ratio in the conductance measurement was 1nA. A scanning speed of 3.3 seconds per line was used for the acquisition of these images. In Fig 2(a), the bias-dependent features on either side of the p-implanted stripe are attributed to the region of inverted majority carrier ("inverted region") within the depletion zone, consistent with Hildner et al.'s earlier report [1]. Because the metal insulator semiconductor (MIS) junction formed by the tip and the sample is reverse biased, under constant current demand, the tip moves inward and the STM image shows dip-like features in the depletion region. The apparent position of boundary between the depletion region and p region changes with the reverse bias, apparently due to the perturbation of the junction and the change of the effective carrier concentration by the tip-sample bias [3]. The corresponding conductance image of the pn junction is shown in Fig. 2 (b). The contrast is noticeably different from that in the topography map. The conductance image shows (i) the inverted region is always bright and gets wider with reverse bias, (ii) the brightness of p stripe increases with reverse bias, (iii) the brightness in the n region is nearly independent of the reverse bias. The visible left-right asymmetry in the conductance map cannot be explained by the instrumental feedback response time, as measurements of the tunneling current vs lateral position indicate that the feedback responds within 2-3 pixels. Because the asymmetry in the line profiles is scan direction

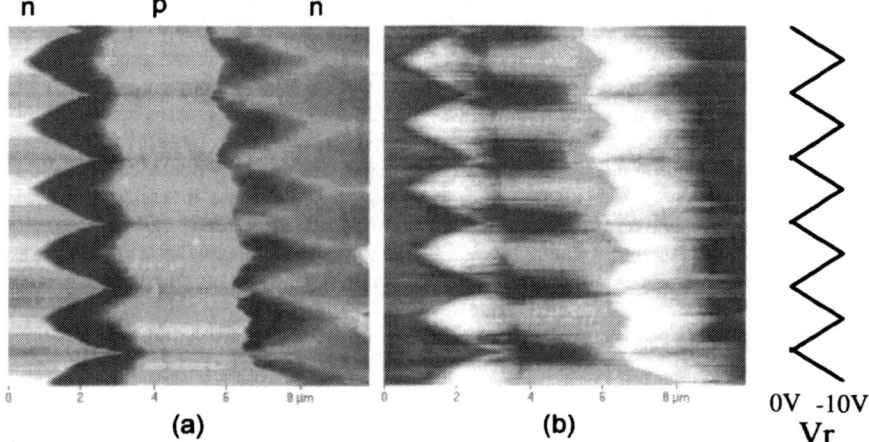

(a)   **(b)**   0V -10V
                                    Vr

Figure 2. (a) A 10 μm x 10 μm constant current mode STM image which shows the p, n, and inverted region (the region of inverted majority carriers). Sample voltage is –2 V and tunneling current is 1 nA. The reverse bias (0 V ~ -10 V) is ramped during the scan with a sweep rate of 0.0032 Hz. (b) The simultaneously measured 10 μm x 10 μm differential conductance image shows clear differences of contrast in electrically different regions (n, p, and inverted region).

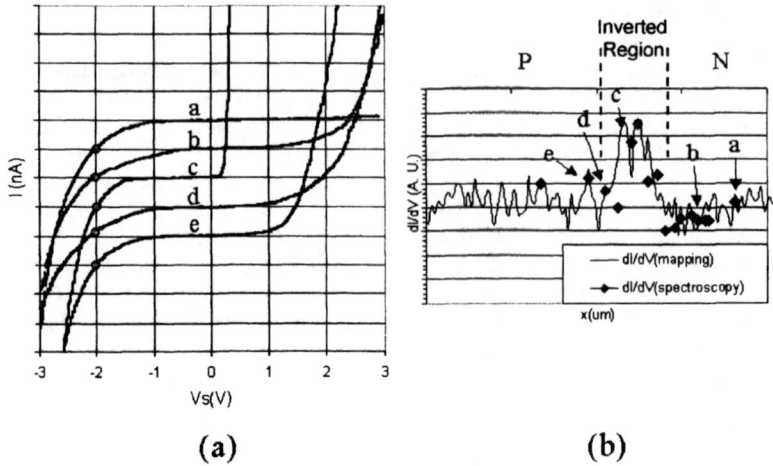

Figure 3. (a) Tunneling spectra measured across pn junction. The starting condition of tip position is determined by a 1 nA demanded current at –2 V sample bias. (b) Conductances (dI/dV) were extracted from the tunneling spectra and mapped across pn junction. The line profile of the conductance mapping at –2 V sample bias is shown as the solid line.

dependent, and reduces with slower scan speed, we attribute it to a slow time response of the inversion charge in the depletion region of the pn junction [16].

We measured tunneling spectroscopy at different points across the pn junction to identify bright and dark regions in the conductance map, as shown in Fig. 3(a). I-V curve a is the tunneling spectrum in the lightly doped n region which shows accumulation at the set-point of – 2V, 1 nA. The tunneling I-V observed at the inverted region (curve c) corresponds to a tip-gap-semiconductor junction electrically in strong inversion, which has large conductance. At the p-stripe, I-V curve e is observed. At the set-point of –2V, 1nA, it shows weak inversion. From these I-V curves, conductance (dI/dV) was extracted and mapped across pn junction as shown in Fig 3(b). The conductance values obtained by two different methods, conductance mapping using lock-in amplifier and measurement of tunneling spectra, shows the rough agreement in the relative contrast between p, n, and inverted zone. The measurements of I-V suggest that the conductance contrast is mainly due to electrical behaviors of metal-gap-semiconductor junction. This in turn identifies the type and concentration of majority charge carriers in the area probed by the STM tip.

CONCLUSIONS

Dual scanning of STM topographical imaging and differential conductance imaging was used to characterize a Si pn junction over a range of the reverse pn-junction bias conditions. conductance maps of semiconductor devices give useful contrast according to the local MIS behavior which is also dependent on the doping type and concentration. The capability of probing electronic features with the high lateral resolution makes conductance imaging potentially quite useful for inspection of operating device.

ACKNOWLEDGEMENT

This work has been supported by the Laboratory for Physical Sciences and in part by the University of Maryland-NSF-MRSEC (NSF-DMR-00-80008).

REFERENCES

1. M. L. Hildner, R. J. Phaneuf, and E. D. Williams, Appl. Phys. Lett. **72**, 3314 (1998).
2. E. T. Yu, K. Barmak, P. Ronsheim, M. B. Johnson, P. McFarland, and J. –M. Halbout, J. Appl. Phys. **79**, 2115 (1996).
3. S. Richter, M. Geva, J. P. Garno, R. N. Kleiman, Appl. Phys. Lett. **77**, 456 (2000).
4. J. V. LaBrasca, R. C. Chapman, G. E. McGuire, R. J. Nemanich, J. Vac. Sci. Technol. **B9**, 752(1991).
5. S. Kordić, E. J. van Loenen, and A. J. Walker, Appl. Phys. Lett. **59**, 3154 (1991).
6. K. –J. Chao, A. R. Smith, A. J. McDonald, D. –L. Kwong, B. G. Streetman, C. –K. Shih, J. Vac. Sci. Technol. **B16**, 453 (1998)
7. S. Hosaka, S. Hosoki, K. Takata. K. Horiuchi, and N. Natsuaki, Appl. Phys. Lett. **53**, 487 (1988).
8. J. A. Stroscio, R. M. Feenstra, and A. P. Fein, Phys. Rev. Lett. **57**, 2579 (1986).
9. L. D. Bell, W. J. Kaiser, M. H. Hecht, and F. J. Grunthaner, Appl. Phys. Lett. **52**, 278 (1988).
10. J. Jahanmir, P. E. West, A. Young, and T. N. Rhodin, J. Vac. Sci. Technol. **A7**, 2741 (1989).
11. R. J. Phaneuf, H. –C. Kan, M. Marsi, L. Gregoratti, S. Günther, and M. Kiskinova, J. Appl. Phys. **88**, 863 (2000).
12. M. Giesen, R. J. Phaneuf, E. D. Williams, T. L. Einstein, and H. Ibach, Appl. Phys. A:Mater. Sci. Process. **64**, 423 (1997).
13. A. Ishizaka, and Y. Shiraki, J. Electrochem. Soc. **133**, 666 (1986).
14. Y. Suganuma, and M. Tomitori, Jpn. J. Appl. Phys. **37**, 3789 (1998).
15. M. P. Everson, R. C. Jaklevic, and W. Shen, J. Vac. Sci. Technol. **A8**, 3662 (1990).
16. J. Y. Park, R. J. Phaneuf, and E. D. Williams (to be published).

Mat. Res. Soc. Symp. Proc. Vol. 669 © 2001 Materials Research Society

Demonstration of the state-of-the-art of formation and characterization of ultra-shallow junctions

P. Borden, A. Al-Bayati[1], J. Madsen, C. Lazik[1], P. Carey[1], L. Bechtler, and A. Mayur[1]
Boxer Cross Inc., 978 Hamilton Court
Menlo Park, CA 94025, USA
[1]Applied Materials, 3050 Bowers Ave.
Santa Clara, CA 95054, USA

ABSTRACT

Doping process windows are becoming very narrow as VLSI technology nodes scale to smaller and smaller dimensions. The time and cost required to develop new doping methods and the desire to re-use equipment will make it likely that current methods will be applied as long as possible. This means that existing process tools will have very tight stability and uniformity requirements, and metrology will be required to drive process control. The paper describes the state-of-the-art of both doping processes involving ion implantation and spike annealing, and new metrology based on Carrier Illumination™ methods that will be required to implement in-line process control for these processes. CI offers depth resolution on the order of 1Å, providing a level of control required to extend existing doping methods. The prospects of new methods such as Laser Thermal Annealing (LTA) are also discussed.

INTRODUCTION

VLSI scaling is approaching the limits of existing doping processes. While these processes – low energy implantation into amorphized layers with spike annealing [1] – can meet the requirements of the 100 nm node, the process window is very narrow. There is a strong motivation to continue existing processes as long as possible. Factors include equipment reuse and the cost, time and risk associated with developing alternate methods. This is analogous to lithography, where the same considerations have motivated extension of conventional exposure methods far beyond what was once thought possible.

The desire to extend processes into narrower and narrower windows makes control of small variations in these complex, multi-step processes critical. This control will come in two ways: improved implanter and annealer performance, and improved metrology. The former will require very tight control of energy, dose, anneal temperature, temperature ramp profile, and uniformity of all processes. The latter will require in-line metrology on product wafers. Measurements on test wafers will not be sufficient, both because of the cost of large diameter substrates and because of the inter-relation between the series of process steps that cannot be captured by monitoring a single step.

Of particular importance is the effect of process variation on device performance, so that measurements and process control tolerances reflect limits based on allowable electrical tolerance. These limits can be determined empirically or through device modeling. Figure 1, for example, shows an example of the relation between drive current and junction depth [2] for 180 nm NMOS transistors, with the junction depth measured on patterned wafers after the extension anneal using Carrier Illumination™ (CI) methods [3,4]. These results show that a 20% variation in junction depth results in a 10% variation in drive current, consistent with transistor models. These models suggest an even greater sensitivity to junction depth for 130 nm transistors. Such correlations can be used to establish process control limits based on circuit tolerance to drive current variation and measurement precision, which is typically better than 2Å based on system response as shown in figure 3.

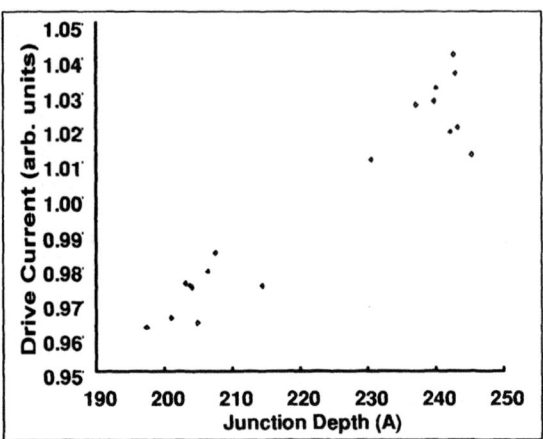

Figure 1: Measured relationship between drive current and junction depth for 180 nm NMOS transistors (from reference 2).

CAPABILITIES OF STATE-OF-THE-ART IMPLANT/ANNEAL PROCESSES

The combination of low energy implant and spike anneal can readily meet the requirements of the 130 nm node. The capability of these processes just touches the edge of the 100 nm node. This is shown in figure 2, which is a plot of the junction depth and sheet resistance space defined by the International Technology Roadmap for Semiconductors (ITRS) overlaid with measured results from various implant/anneal, epitaxial, and laser thermal processing (LTP) processes.

There is some indication that the 100 nm requirements may be excessively aggressive, which suggests conventional extension doping processes will require replacement at the 70 nm node, unless the ITRS requirements for this node are also relaxed.

Nevertheless, it is clear that the process window for these nodes is very narrow. Furthermore, the ITRS roadmap does not indicate the sensitivity of devices to variation in junction depth, which, as indicated above, creates a new set of requirements that can only be met by careful process tool design and implementation of tight process control.

CI MEASUREMENTS APPLIED TO STATE-OF-THE-ART DOPING PROCESSES

CI provides one means to provide in-line control of doping processes. As a non-destructive measurement with a 2 μm spot size, it is readily adapted to the measurement of product wafers. CI has high depth resolution consistent with the requirements of future technology nodes. Figure 3 shows the results of measurements on CVD-grown B-doped layers [5]. These layers have box-like profiles with junction depth set by growth time. Therefore, they are useful as junction depth standards.

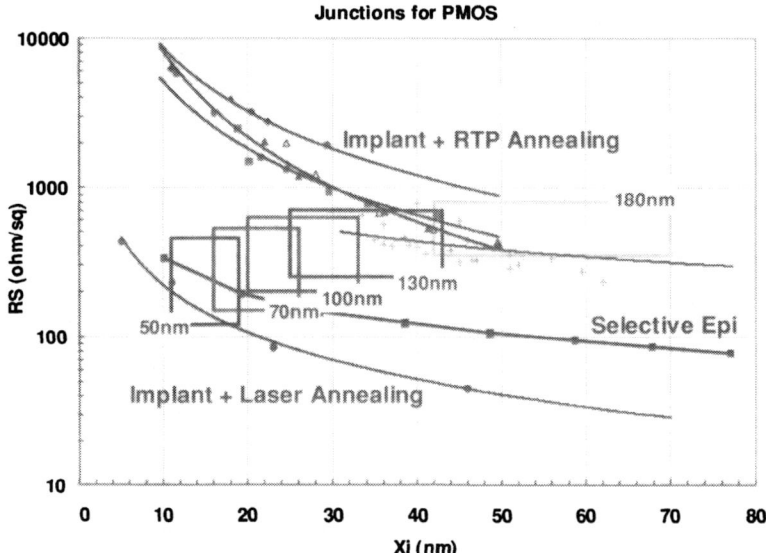

Figure 2: Sheet resistance and junction depth for various ITRS roadmap technology nodes, with best results for implant/anneal and LTA overlaid.

Figure 3 shows the SIMS profiles of the CVD layers and the CI signal as a function of the SIMS-measured depths. CI is an interferometric measurement, and its response follows a cosine function [3]. The measured depth response is about 700 μV/nm. With a noise of about 70 μV, the measurement has a depth resolution of about 0.1 nm (1Å). This provides better than 1% depth resolution at the end of the ITRS roadmap.

Figure 4 shows correlation of CI and SIMS for various low energy implants in the 200 to 500 eV energy and 1 to 5e14/cm^2 dose range followed by spike anneals in the 950 to 1100°C peak temperature range. These are representative of advanced implant/anneal processes, and correspond to various points marked shown in figure 2.

ADVANCED DOPING PROCESSES

While implant/anneal processes will be extended as far as possible, it is clear that new doping methods will at some point be required for the source/drain extensions. The most likely candidates are chemical vapor deposition (CVD) and laser thermal annealing (LTA). Figure 2 shows best results reported with both processes, indicating that they are capable of meeting the sheet resistance and junction depth requirements to the end of the roadmap.

Both processes result in hyper-abrupt junctions. Figure 3 shows a number of SIMS profiles for CVD layers. The abruptness of these layers between 1e18 and 1e19/cm^3 is about 3 nm/decade (well above the SIMS resolution of 0.8 nm/decade). Figure 4 shows correlation to junction depth as measured with SIMS for various low energy boron implants. Figure 5 shows a comparison of a high quality implant/anneal junction and a LTA junction. The LTA junction provides higher doping, shallower depth, and abruptness better than 2.5 nm/decade.

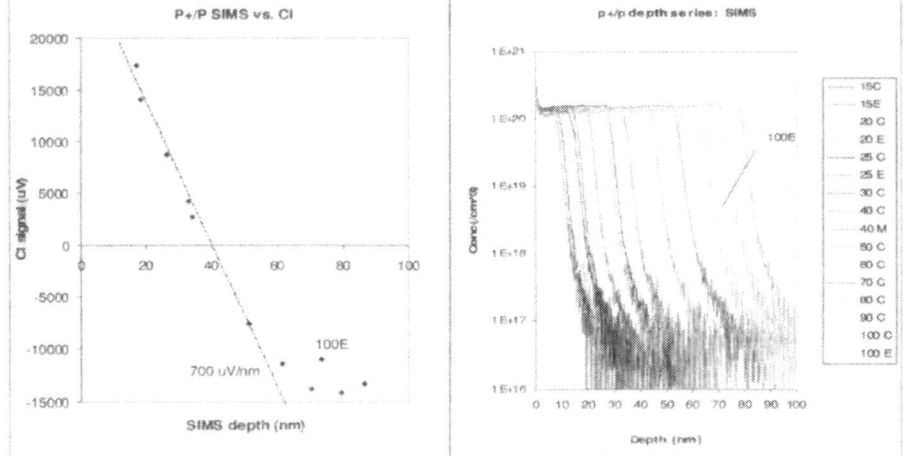

Figure 3: CI response vs. junction depth (left) and SIMS profiles of measured junctions (right) showing cosine response of measurement and depth sensitivity of 700 μV/nm under 50 nm junction depth [5].

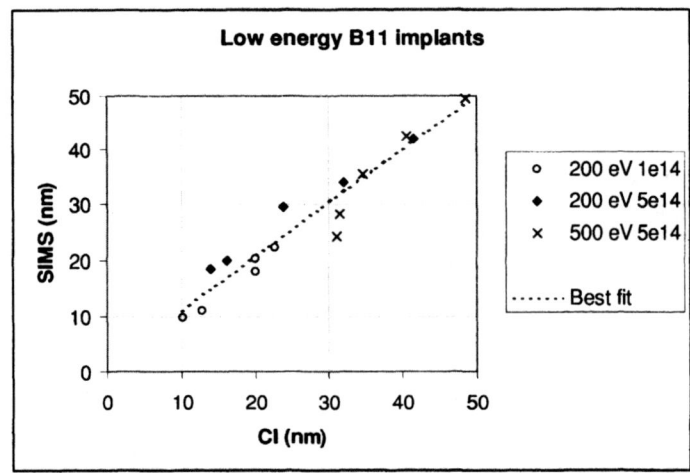

Figure 4: SIMS-CI correlation for various low-energy implants followed by spike anneal.

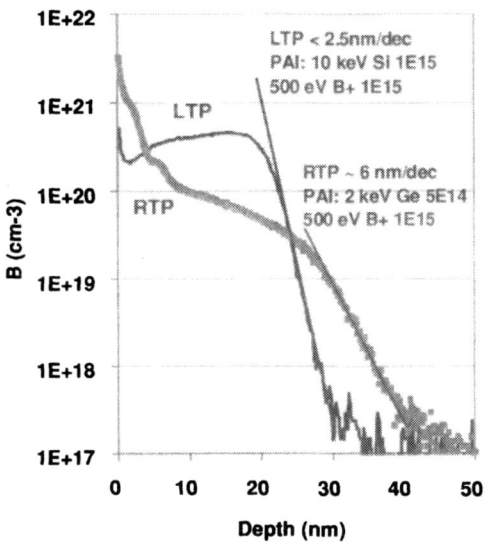

Figure 5: SIMS profiles of RTP and LTA annealed junctions.

CI methods have been used to characterize both types of junctions. Results on CVD layers have been shown above in figure 3, and are summarized in more detail in reference 5. Work characterizing LTA junctions is summarized in detail in reference 6. This work shows that CI can be used to identify the melt transition, the annealing of defects at the junction edge, and movement of the junction beyond the amorphous-crystal interface for higher anneal energy.

Figure 6, excerpted from reference 6, shows an example of CI measurements near the melt transition, and how this data can be used to determine beam uniformity. Without an absorber layer, the transition happens around 0.4 J/cm² anneal energy. This is seen as a sharp drop in the CI signal. Figure 6 is obtained by measuring in a line scan of 1 mm steps across 1 cm wide boxes that have been annealed at progressively higher energies, as indicated in the graph. Note that the left side of the 0.38 J/cm² box anneals first and the right side of the 0.42 J/cm² box anneals last. This movement of the anneal region is due to beam uniformity.

CONCLUSION

While future doping requirements are aggressive, it is likely that implant/anneal can meet the requirements of the roadmap to the 100 nm node. The very narrow process window and uniformity requirements over large diameter substrates means in-line process monitoring tied to statistical process control will be required. CI has shown the depth resolution and throughput required to meet this need, and is a primary candidate to provide SPC measurements for the doping module. This buys time for development of more advanced doping processes such as LTA or CVD that will be needed to reach the end of the ITRS roadmap.

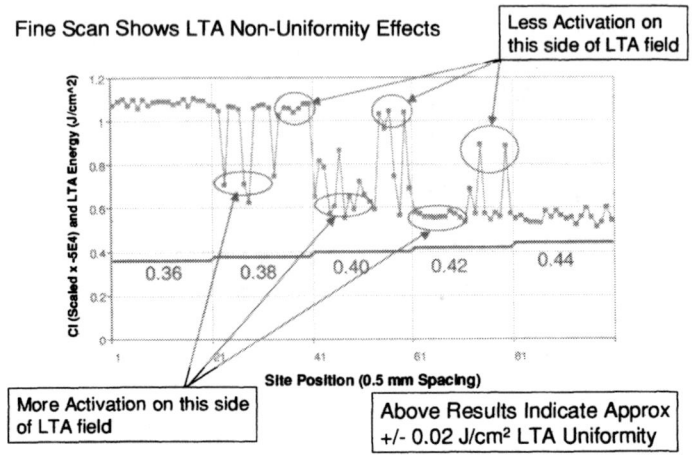

Figure 6: Detail of CI scans across 1 cm² boxes annealed at progressively increasing energies through the melt transition. Irregularity of signal measures annealing laser beam uniformity.

REFERENCES

1. A. Al-Bayati, M. Foad, A. Mayur, S. Tandon, D. Wagner, R. Murto, C. Ferguson, L. Larson, T.S. Wang and A. Cullis, "Exploring the Limits of Pre-Amorphization Implants on Controlling Channeling and Diffusion of Low Energy B Implants and Ultra Shallow Junction Formation," XIII International Conference on Ion Implantation Technology IIT-2000, Alpbach Austria, Sept. 17-22, 2000.
2. G. J. Kluth, L. Bechtler, P. Borden, and J. Mi, "Non-destructive, in-line characterization of shallow junction processes," Spring 2000 MRS meeting, April 24-28, 2000, San Francisco, CA.
3. P. Borden in "Handbook of Semiconductor Metrology," A. Diebold (ed), pp. 97-116, Marcel-Dekker, to be published 4/2001.
4. P. Borden, "Junction depth measurement using Carrier Illumination," 2000 International Conference on Characterization and Metrology for ULSI, Gaithersburg MD, June 26-29, 2000.
5. P. Borden, L. Bechtler, B. Klemme, R. Nijmeijer, E. Judge, A. Diebold, J. Bennett, W. Vandervorst, T. Clarysse, M. Caymax and Y. Peytier, "Progress towards an electrically active, USJ depth reference for carrier illumination, SRP and SIMS", to be presented at the Ultra Shallow Junctions 2001 meeting, April 22-26, 2001, Napa, CA.
6. D. Sing, P. Borden, L. Bechtler, R. Murto, and S. Talwar "Boxer Cross Measurements of Laser Annealed Shallow Junctions," XIII International Conference on Ion Implantation Technology IIT-2000, Alpbach Austria, Sept. 17-22, 2000.

Mat. Res. Soc. Symp. Proc. Vol. 669 © 2001 Materials Research Society

SPV Monitoring of Near Surface Doping – Role of Boron-Hydrogen Interaction; Boron Passivation and Reactivation

D. Marinskiy and J. Lagowski
Semiconductor Diagnostics, Inc., 3650 Spectrum blvd., Ste 130
Tampa, FL 33612, U.S.A.

ABSTRACT

Hydrogen is known to cause the passivation of boron acceptors after such processing steps as wet etching, reactive ion etching, sputter deposition of metal contacts, and Ar ion beam etching. Previous studies of this effect employed CV profiling, spreading resistance profiling, and SIMS measurements on samples diffused with deuterium. These methods are either destructive to the Si surface or require deposition of metal contact. In the present study we used a non-contact small signal ac-surface photovoltage technique, currently available in commercial diagnostic tools. Simultaneous measurements of the semiconductor surface barrier, V_{sb}, and the capacitance of the surface depletion layer, C_D, give the concentration of boron acceptors in a submicron distance from the Si surface or Si/SiO$_2$ interface. The technique has proven very successful in monitoring low dose implants and also near surface doping in oxidized wafers. In bare silicon wafers the method occasionally indicated surface boron concentration noticeably below the bulk value. We found such behavior in wafers after the chemical cleaning, used to prepare a hydrogen terminated surface. Thermal annealing at temperatures from 150°C to 200°C reactivates the boron dopant. We will discuss the effect of various cleaning and annealing conditions on passivation and reactivation of boron acceptors in the near surface region. The results obtained with the non-contact SPV technique show excellent agreement with previous studies. They also provide a basis for reliable measurement of the boron concentration free of interference from hydrogen passivation.

INTRODUCTION

Hydrogen is known to cause impurity passivation in silicon [1,2] and other crystalline semiconductors. It can diffuse into Si and SiO$_2$ during such processing steps as wet chemical cleaning, reactive ion etching, Ar ion beam etching, and sputter deposition of metal contacts. Hydrogen can be released from Al-H, AlO-H, Si-H, SiO-H sites in the Al gate, the SiO$_2$ film, and Al/SiO$_2$ and SiO$_2$/Si interfaces [3,4]. The released hydrogen can diffuse into Si and can cause unintentional changes in the electrically active dopant profile near the surface or Si/SiO$_2$ interface.

The passivation of boron by hydrogen was studied previously using spreading resistance, C-V profiling, and SIMS measurements on samples diffused with deuterium [5,6,7]. In this work we study the passivation of boron by hydrogen introduced during typical surface treatments used in silicon IC fabrication. The electrically active boron concentration is monitored with a non-contact small signal ac-SPV technique, to be referred to as Near Surface Doping (NSD) [8,9,10]. In this method the simultaneous measurements of the semiconductor surface barrier, V_{sb}, and the capacitance of the surface depletion layer, C_D, measured by ac-SPV, give the concentration of boron acceptors, N_A, in the proximity of the Si surface or Si/SiO$_2$ interface.

$$NSD = N_A = \frac{2 \cdot (V_{sb} - \dfrac{kT}{q})}{q \cdot \varepsilon} \cdot C_D{}^2 \tag{1}$$

where ε is the Si dielectric constant, q is the elementary charge, and kT is the thermal energy.

A depletion capacitance is determined from the small ac-SPV signal, V_{SPV}, as originally proposed by Nakhmanson [11] and later on by Kamieniecki [8].

$$C_D = \frac{const \cdot Ieff}{\varpi \cdot V_{SPV}} \tag{2}$$

where $Ieff$ is the effective photon flux, ϖ is the light modulation frequency, and $const$ is the calibration constant.

The silicon surface barrier is determined by measuring the contact potential difference, V_{CPD}, in the dark and under strong illumination [12]. The effective depth of near surface doping measurements corresponds to the depletion layer width, W, which is a function of V_{sb} and the doping N_A:

$$W = \sqrt{\frac{2 \cdot \varepsilon}{q \cdot N_A} \cdot (V_{sb} - \frac{kT}{q})} \tag{3}$$

The NSD technique has proven very successful in measurement of oxidized wafers [9,10,13,14]. In bare silicon wafers measured after typical surface cleaning, the method occasionally indicated boron acceptor concentrations noticeably below the bulk value or the target epi-processing value. The present study explores these effects in terms of boron passivation by hydrogen introduced during surface cleaning. Thermal annealing at 170°C results in complete boron reactivation. Thus, the measurement after annealing provides a reliable means for determining the boron concentration free of interference from hydrogen passivation.

EXPERIMENT

Several types of silicon wafers were used in this study, including boron-doped epitaxial wafers (p/p- and p/p+) and bulk CZ wafers with a thermal oxide. The epitaxial layer doping was in the range from 1e15 to 1e17 cm^{-3}. The bulk CZ wafers had dopant concentrations in the low e15 cm^{-3} range. The following treatments of the Si surface were investigated: 1) dipping in a dilute HF solution followed by a DI water rinse and drying with nitrogen (to be referred to as HF); 2) HF followed by cleaning in a SC1 solution (H_2O_2:NH_4OH:H_2O = 1:1:5) at 65°C for 10 minutes followed by HF (to be referred to as SC1/HF treatment); 3) SC1/HF followed by cleaning in a SC2 solution (HCl:H_2O_2:H_2O = 1:1:6) at 70°C for 10 minutes and followed by HF (to be referred to as SC1/SC2/HF treatment); 4) heating the wafer on the hot plate at 170°C for 5 minutes (to be referred to as thermal stress, TS).

Doping in the near surface region was measured with the FAaST 230 tool manufactured by Semiconductor Diagnostics, Inc.

RESULTS AND DISCUSSION

Boron passivation by SC1/HF and reactivation by TS treatment.

Epitaxial wafers measured after a SC1/HF treatment showed a near surface doping concentration consistently lower than the manufacture target specifications. As shown in Figure 1 this behavior was observed for the entire concentration range from 10^{15} to 10^{17} cm^{-3}. On average the NSD after SC1/HF was about 4 times lower than the epi-layer target doping. Thermal stress significantly increases NSD values, bringing them very close to the target doping values. Longer annealing time or higher temperature has not resulted in any further increase of the dopant concentration.

The probing depth in the NSD measurement is determined by the width of the surface depletion layer. SC1/HF treatment increases the depletion layer width for a p-type substrate or epi-layer, while the thermal treatment reduces it. A major contribution to this is brought about by a corresponding decrease and increase in the acceptor concentration, while the surface barrier, V_{sb}, remains practically unchanged. The values of W calculated using expression (3) are shown in Table 1 for p/p- epitaxial wafers. It is seen that W extends from 0.06 um (for wafer #14 with the highest doping after thermal stress) up to about 1 um (for wafer #12 with lowest doping after SC1/HF treatment). These W values are within the range of hydrogen passivation depth reported in reference 14 from boron dopant profiling measurements on MS diodes prepared on silicon wafers pre-cleaned with a SC1/HF treatment. They are also well within the several microns hydrogen diffusion depth into silicon at room temperature.

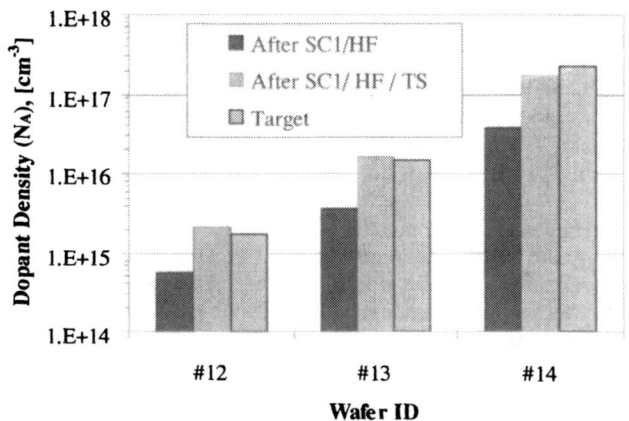

Figure 1. NSD measured on p/p- wafers after SC1/HF treatment and after TS.

Table 1. NSD on p/p- samples after SC1/HF and TS treatments.

Wafer ID	SC1 / HF treatment		After TS		NSD$_2$/NSD$_1$ dopant ratio
	NSD$_1$ [e15 cm^{-3}]	W$_1$ [μm]	NSD$_2$ [e15 cm^{-3}]	W$_2$ [μm]	
#12	0.54	**0.99**	2.0	0.51	3.7
#13	3.76	0.37	15.0	0.19	4.0
#14	39.50	0.12	175.4	**0.06**	4.4

Based on this data we suggest that NSD reduction after SC1/HF treatment is caused by hydrogen diffusion into the silicon and formation of B-H complexes, which passivate the boron acceptors. Referring to Sullivan et al [7] we conclude that elimination of this effect by 170°C annealing is due to thermal dissociation of the B-H complexes, which restores the boron acceptors to an electrically active state.

It is worth to note that hydrogen passivates the same fraction of boron rather then the same amount of B. This is in agreement with observation of Pankove et al [5] that the diffusivity of hydrogen depends on the boron concentration, and also with the SIMS results of Mikkelsen [6] who demonstrated that that the near surface deuterium concentration is nearly equal to the boron concentration on samples treated in atomic deuterium plasma.

Effects of different surface treatments.

The effect of surface treatment on NSD was studied using a uniform oxidized wafer that was cleaved into 4 pieces as shown on figure 2.

Before the wafer was cleaved, NSD was measured on the entire wafer and it was found to be uniform with the average value of 1.13 e15 cm^{-3} with a standard deviation of 0.023 e15 cm^{-3}. Each quarter was subjected to a different treatment: quarter 1 – no treatment; quarter 2 – HF oxide strip only; quarter 3 – HF oxide strip followed by SC1 / HF treatment; quarter 4 – HF oxide strip followed by SC1 / SC2 / HF treatment. After the measurements all quarters were annealed at 170°C for 5 minutes and were measured again. The results are summarized in Table 2.

Very little difference is observed between the reference segment and the one after HF oxide strip. However about 3 times reduction in NSD is observed when SC1 or a combination of SC1 / SC2 cleanings is incorporated into the treatment. For SC1 and SC1/SC2 treated samples NSD increases back to initial level after thermal stress, indicating boron reactivation. No change is observed for reference and HF treated samples.

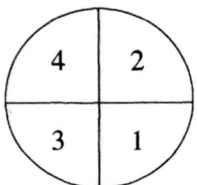

Figure 2. Initially oxidized wafer with a 100A thermal oxide was cleaved into 4 samples. They were treated as follows: 1 – reference, no treatment; 2 – HF oxide strip; 3 – HF oxide strip followed by SC1 / HF; 4 – HF oxide strip followed by SC1 / SC2 / HF.

Table 2. * Effect of different surface treatments on NSD.

Wafer Segment	Reference			HF oxide strip			HF / SC1 / HF			HF / SC1 / SC2 / HF			After TS		
	NSD	Vsb	W	NSD	Vsb	W	NSD	Vsb	W	NSD	Vsb	W	NSD	Vsb	W
1	1.13	0.32	0.61										1.12	0.32	0.61
2				1.20	0.40	0.69							1.11	0.41	0.70
3							0.35	0.41	1.24				1.12	0.39	0.68
4										0.31	0.39	1.29	1.13	0.41	0.69

* NSD in e15 cm^{-3}; Vsb in Volt, W in μm.

Consecutive HF and SC1/HF treatments.

The results of the experiment with different surface treatments were rather unexpected since hydrogen is present in the chemistry of both HF and SC1 solutions [15]. Why was no boron passivation observed during the HF treatment? To answer this question a series of consecutive SC1/HF and HF treatments and measurements was performed for two selected samples. The results (Figure 3) were normalized to the initial measurement when all boron is in the electrically active state. The experiment clearly indicates a significant difference between SC1/HF and HF. When SC1 cleaning is done prior to HF, efficient boron passivation takes place. This is possibly due to the longer time and higher temperature of the SC1 treatment. The boron passivation kept increasing even after 4 SC1/HF treatments. This is due in part to increase of the NSD probing depth with each treatment. In our experiment we observed about 80% passivation of boron at the depth of 0.57 μm.

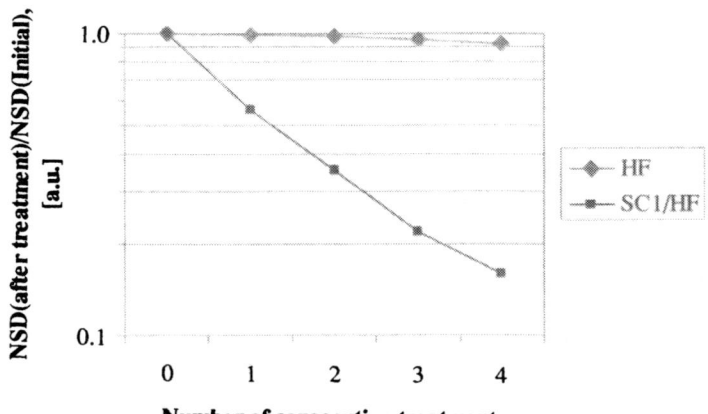

Figure 3. Ratio of NSD after HF and SC1/HF treatments (boron passivation occurs) to initial NSD (all boron is electrically active).

CONCLUSIONS

We studied the passivation of boron by hydrogen introduced into Si during typical surface treatments used in IC fabrication.

1. During HF and SC1 surface treatments hydrogen diffuses into Si to form H-B pairs, passivating boron acceptors. Such behavior was observed for the samples with a boron concentration in the range from 10^{15} to 10^{17} cm^{-3}.

2. Thermal annealing at 170°C for 5 minutes reactivates boron acceptors. Longer annealing times or higher temperatures did not result in any further increase of doping density.

3. Thermal annealing provides a basis for reliable measurement of boron concentrations free of interference from hydrogen passivation.

REFERENCES

1. S.J. Pearton, J.W. Corbett, T.S. Shi, Appl. Phys. A, 43, 153-195 (1987).
2. S.J. Pearton, J.W. Corbett, M. Stavola, *Hydrogen in Crystalline Semiconductors*, Springer, Berlin, 1992.
3. C.T. Sah, J.Y.C. Sun, J.J.T. Tzou, and S.C.S. Pan, Appl. Phys. Lett. 43 (10), 962-964 (1983).
4. C.T. Sah, J.Y.C. Sun, J.J.T. Tzou, Appl. Phys. Lett. 43 (2), 204-206 (1983).
5. J. Pankove, R. Wance, and J. Berkeyheiser, Appl. Phys. Lett. 45 (10), 1100-1102 (1984).
6. J.C. Mikkelsen, Appl. Phys. Lett. 46 (9), 882-884 (1985).
7. J. Sullivan, w. Graham, R. Tung, F. Schrey, Appl. Phys. Lett. 62 (22), 2804-2806 (1993).
8. E. Kamieniecki, J.Vac.Sci.Technol. **20**, 811 (1981).
9. D. Marinskiy, J. Lagowski, M. Wilson, A. Savtchouk, L. Jastrzebski, D. DeBusk, Mat. Res. Soc. Symp. Proc. Vol 591, 225-230 (2000).
10. D. Marinskiy, J. Lagowski, M. Wilson, L. Jastrzebski, R. Santiesteban, K. Elshot, Proceedings of SPIE Vol. 4182, 72-77 (2000).
11. R. Nakhmanson, Solid State Electron. **18**, 617 (1975).
12. M. Wilson, J. Lagowski, A. Savtchouk, L. Jastrzebski, and J. D'Amico, ASTM STP 1382, (1999).
13. D. DeBusk, A. Hoff, Solid State Technology, **42**, April, 1999.
14. R. Santiesteban, D. DeBusk, D. Ramappa, W. Moller, Presented at the Ion Implantation Technology 2000 Conference in Alpbach, Austria. To be published at the Conference proceedings.
15. W. Kern, *Silicon Wafer Cleaning: a Basic Review*, short course at the SCP Global Technologies 7th International Symposium, (2000).

Dopant Diffusion Issues

Mat. Res. Soc. Symp. Proc. Vol. 669 © 2001 Materials Research Society

SELF-DIFFUSION IN INTRINSIC AND EXTRINSIC SILICON USING ISOTOPICALLY PURE ^{30}SILICON LAYER

Yukio Nakabayashi, Hirman I. Osman, Toru Segawa, Kazunari Toyonaga, Satoru Matsumoto,
Keio Univ., Dept. of Electronics and Electrical Engineering, Yokohama JAPAN
Junichi Murota,
Tohoku Univ. Res. Inst. of Electrical Communication, Sendai, JAPAN
Kazumi Wada,
Massachusetts Institute of Technology, Dept. of MS&E, Cambridge
Takao Abe,
Sin-Etsu Handootai, Isobe R&D Center, Gunma, JAPAN

ABSTRACT

Silicon self–diffusion coefficients were measured in intrinsic and extrinsic silicon from 870 to 1070°C using isotopically pure ^{30}Si layer. ^{30}Si diffusion profiles are determined by secondary ion mass spectrometry. The temperature dependence of intrinsic diffusion coefficient in bulk Si is obtained. Comparing it in heavily As-doped or B-doped Si, it is found that Si self-diffusion is entirely mediated by interstitialcy mechanism at lower temperatures below 870°C.

INTRODUCTION

As device dimension shrinks with increasing degree of integrations, accurate prediction and precise control of dopant profiles such as junction depths become very important, and many parameters are necessary for simulation models. Since dopant atoms diffuse by interaction with point defects such as vacancies and self-interstitials, and most Si-LSI fabrication processes such as oxidation and ion-implantation perturb the equilibrium point-defect concentrations, understanding the behavior of point defects and derivation of accurate Si self-diffusivity are essential for device technology.

Self-diffusion in Si can be viewed as a limiting case of dopant diffusion, in which diffusion atoms carry no excess charge and introduce no distortion in the lattice. Thus the study of self-diffusion is very important for the understanding of diffusion mechanism of dopant atoms. To date, Si self-diffusion has been investigated by many researchers [1-4] on account of its scientific and technological importance. However, the results on Si self-diffusion reported by them are less consistent compared with Ge self-diffusion, due to the difficulty of Si self-diffusion experiment caused by the very short half-life of the usually used radioactive tracer ^{31}Si, high ^{30}Si background concentration (3.1%) and ^{30}Si implantation-induced radiation damage. In such situation, there is a common recognition that combining the lower and higher temperature data, a kink exists at about 1000°C in the Arrhenius plot of Si self-diffusion and it indicates a change in the mechanism. Seeger and Chik [6] suggested that both vacancy and interstitialcy mechanisms contribute to the nonlinearity of the Arrhenius plot. More specifically they proposed that in Si at lower temperatures self-diffusion mainly occurs via vacancies, whereas at higher temperatures it is dominated by the interstitialcy mechanism. Further Gösele *et al.* [7] calculated these contributions from diffusion data of Au or Ni in Si. According to the calculation, the interstitialcy component is larger than vacancy component at higher temperatures, but at lower temperatures below 1000°C, its relation is reversed.

Recently, Si self-diffusion has been reevaluated using isotopically controlled Si heterostructures. Bracht *et al.* [5] measured Si self-diffusion coefficients in isotopically enriched ^{28}Si epitaxial layer over a wide temperature range (855-1388°C). Using correlation factors calculated from metal diffusion in Si, they determined the self-diffusion coefficient over seven orders of magnitude with one diffusion enthalpy and described that self-interstitials dominate self-diffusion. Ural *et al.* [8] also studied self-, P and Sb diffusion under inert, oxidizing and nitridizing ambients using the enriched ^{28}Si epilayer. They found that self-diffusion was enhanced in both oxidation and nitridation conditions, indicating direct evidence of dual mechanism, i.e., self-diffusion is mediated by both vacancy and self-interstitial.

In this article, we reevaluate the Si self-diffusion in intrinsic bulk silicon using ^{30}Si/natural Si heterostructures. Then utilizing the advantage of ^{30}Si/natural Si heterostructures, we investigated the extrinsic Si self-diffusion. Isotopic diffusion profiles were measured by secondary ion mass spectrometry (SIMS), and self-diffusion coefficients were determined numerically by fitting calculated profiles to the experimental profiles.

EXPERIMENTAL DETAILS

A ^{30}Si epitaxial layer was grown by gas-source molecular beam epitaxy (GS-MBE) with a base pressure of 4×10^{-10} Torr at 700°C on three different substrates in this experiment. Two of these substrates were heavily doped with B or As ([B] $= 2\times10^{19}$ cm^{-3}, [As] $= 3\times10^{19}$ cm^{-3}) and the other was lightly doped with B ([B] $= 1\times10^{16}$ cm^{-3}). The epitaxial growth of ^{30}Si has been described in detail elsewhere [9].

These samples were annealed simultaneously in a resistance heating furnace in pure Ar (99.999%) ambient from 870 to 1070°C. The temperature was monitored with an accuracy of ±3 K using a calibrated Pt/PtRh thermocouple. The concentrations of the respective Si isotopes and dopant atoms in the Si substrate were measured with SIMS. SIMS analysis was performed on a CAMECA IMS-4f instrument using a 5.5 keV Cs$^+$ primary beam. Each diffusivity was obtained using a numerical fitting process by solving Fick's equation. In this process, the as-grown profile was used as an initial condition. The best fit was determined by minimizing the root-mean-square error.

RESULT and DISCUSSION

The concentration profiles of the respective Si isotopes near the interface between the ^{30}Si epitaxial layer and natural Si substrate are shown in Fig. 1. The ^{30}Si profile was broadened by diffusion during epitaxial growth, but the ^{30}Si epitaxial layer is isotopically well-separated in spite of its high concentration as shown in Fig. 1. In the ^{30}Si epitaxial layer, ^{28}Si and ^{29}Si isotopes are heavily depleted and the isotope composition is ^{28}Si:^{29}Si:^{30}Si = 0.05:0.07:99.88. Thus almost pure ^{30}Si epitaxial layer was grown.

Since carbon influences dopant diffusion in Si, it is desirable that no carbon peak is present at the interface between ^{30}Si/natural Si. Figure 2 shows carbon and oxygen concentration SIMS profiles in the ^{30}Si epitaxial layer and the substrate. At interface, no carbon and oxygen peak are observed. The concentration of carbon and oxygen in the ^{30}Si epitaxial layer are not much different from those of the substrate. Thus this heterostructure of ^{30}Si/natural Si is ideal for the investigation of Si self-diffusion.

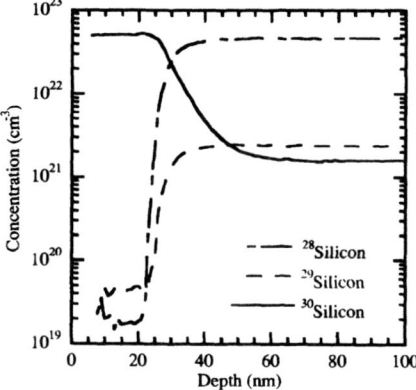

Figure 1. Enriched ^{30}Si as-grown profile.

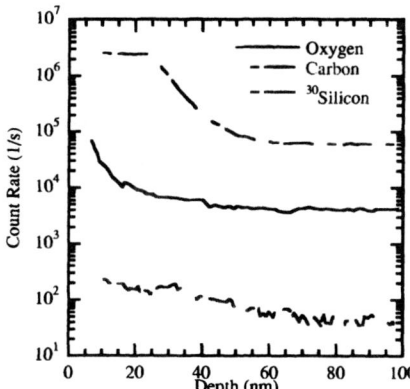

Figure 2. Carbon and oxygen profiles near the interface between ^{30}Si layer and natural Si.

Figure 3 and 4 show the SIMS profiles of ^{30}Si self-diffusion in the bulk silicon for each sample. In these figures, for clarity only every tenth measured SIMS data are plotted, and solid lines show best simulation fits used for extracting diffusion coefficients. The as-grown profile is also given for reference. Since deviations of ^{30}Si profile due to matrix effect are not observed from Figs. 3 and 4, it may be considered that surface oxidation caused by the long diffusion time dose not take place. The diffusivities obtained by simulation fitting are tabulated in Table I. Since the B concentration of a lightly B-doped sample ([B] = 1×10^{16} cm^{-3}) is smaller than the intrinsic carrier density n_i (2.78×10^{18}-9.44×10^{18} cm^{-3}) [10] over the entire temperature range investigated, the sample is referred to as intrinsic silicon. It is found that from Figs. 3 and 4 there is a different tendency at lower or higher temperatures. That is, at 870°C the ^{30}Si diffusion in heavily B-doped Si is found to be enhanced, comparing ^{30}Si profiles with that of intrinsic Si, but in heavily As-doped Si, the diffusion profile is hardly different from that of intrinsic Si. On the other hand, at 1070°C both in As-doped and B-doped samples, the ^{30}Si diffusion is enhanced, and further its enhanced ratio in As-doped silicon is larger than that of in B-doped sample. This different tendency suggests that a complementary mechanism (interstitialcy or vacancy mechanism) is dominant at lower or higher temperatures, respectively.

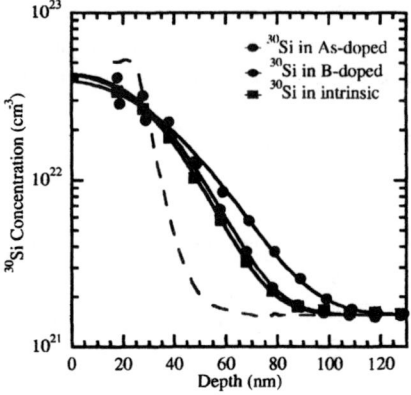

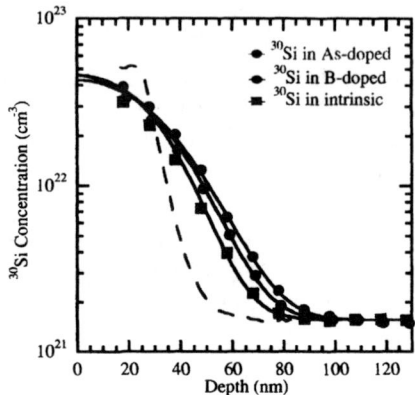

Figure 3. ^{30}Si self-diffusion profiles in bulk Si at 870°C.

Figure 4. ^{30}Si self-diffusion profiles in bulk Si at 1070°C.

Table I. ^{30}Si diffusion coefficients for As, B-doped and intrinsic condition.

Substrate	Concentration (cm^{-3})	$D(^{30}Si)$ ($\times 10^{-19}$ cm^2/s)				
		870°C	920°C	970°C	1020°C	1070°C
As-doped	3×10^{19}	9.54	46.4	271	1340	6090
B-doped	2×10^{19}	16.2	80.5	366	1460	4810
B-doped (intrinsic)	1×10^{16}	8.68	41.9	169	1010	3390

Temperature dependence of intrinsic Si self-diffusion coefficient is shown in Fig. 5. In this figure, the symbols and solid line are obtained from our self-diffusion experiment; dashed lines are literature data as reference. The self-diffusivities of ^{30}Si in intrinsic Si at each temperature almost agree with those of reference data and the activation energy calculated from a gradient of Arrhenius equation is 4.05 eV. Moreover, observing carefully, there is a possibility that the Arrhenius plot is bent near 1000°C as assumed previously. The activation energy of lower or higher temperatures is 3.70 or 4.42 eV, respectively.

Figure 6 shows temperature dependence of extrinsic Si self-diffusion coefficient. As described above, the Arrhenius equation behavior indicates a dissimilar trend in As-doped or B-doped sample. Each activation energy is 4.35 eV in As-doped or 3.83 eV in B-doped silicon, i.e., it is larger or smaller than intrinsic one. This property means that the higher temperatures are, the larger n-type doping effect is. On the contrary, the p-type doping effect much dominates at lower temperatures. Fairfield and Masters [11] investigated Si self-diffusion using radioactive [31]Si in intrinsic and extrinsic silicon at relatively high temperature range (1086-1197°C). They reported that the self-diffusion coefficient was found to increase with n-type ([As] $=7\times10^{19}$-1.8×10^{20} cm^{-3}) and, to a lesser extent, p-type doping ([B] $=8\times10^{19}$-2.2×10^{20} cm^{-3}) above the intrinsic level. Since the concentration of the dopant used in their experiment were much larger than those in the present study, the absolute values of Si self-diffusivities in As or B-doped sample are different from those of their experiment, but the diffusivity ratios of doped to intrinsic near 1080°C are consistent with their results.

Taking into account the Fermi level effect, heavily doping with n-type dopants is expected to raise the equilibrium concentration of vacancy and hence to enhance indirect diffusion via the vacancy mechanism, whereas p-doping should have the opposite effect.

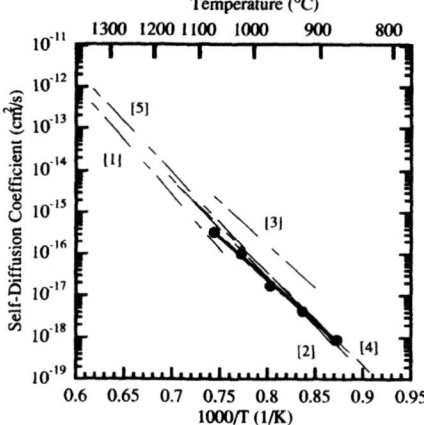

Figure 5. Temperature dependence of intrinsic Si self-diffusion coefficients.

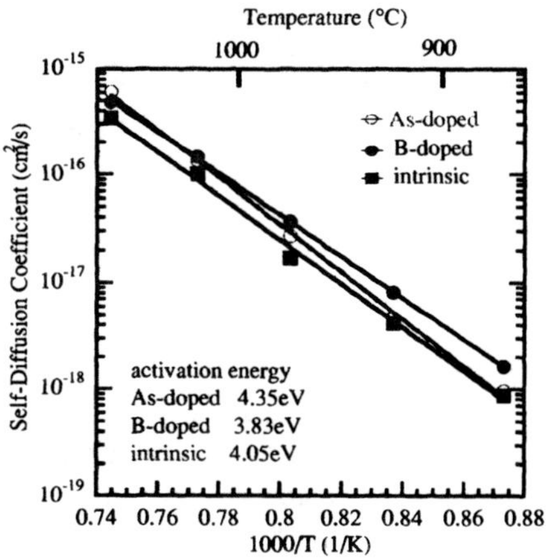

Figure 6. Temperature dependence of extrinsic Si self-diffusion coefficients.

Since As radius is nearly equal to Si, only Fermi level effect can be considered in the As-doped sample. In the case of vacancy, it has an accepter level (V$^-$) in the bandgap. The relation between vacancy concentration C_V and intrinsic vacancy concentration C_{Vi}, using electron density n and intrinsic carry density n_i, can be expressed by

$$\frac{C_V}{C_{Vi}} = \frac{1+(n/n_i)\exp(E_i - E_V/kT)}{1+\exp(E_i - E_V/kT)}$$

where E_i and E_V are intrinsic Fermi level and vacancy accepter level, respectively. If the accepter level is located in the lower half of the bandgap (E_i-E_V >> kT), the ratio of vacancies (C_V/C_{Vi}) becomes nearly equal to the ratio of n to n_i. At lower temperatures below 870°C, n/n_i is larger than 10, i.e., if Si self-diffusion is mediated by vacancy mechanism, [30]Si diffusion in As-doped Si is enhanced by factor 10. However, it is almost equal with that in intrinsic Si; the fractional contribution of vacancy mechanism f_v is smaller than 0.1. This result indicates that interstitialcy mechanism is dominant in Si self-diffusion at lower temperatures.

CONCLUSION

In conclusion, we studied Si self-diffusion using isotopically pure ^{30}Si layer grown on natural Si substrate. The temperature dependence of intrinsic diffusion coefficient in bulk Si was obtained and the present result almost agree with previous data. The doping effects on self-diffusion also were investigated from 870 to 1070°C. The different property is observed in heavily As or B-doped Si. Taking into account the Fermi level effect, Si self-diffusion is entirely mediated by interstitialcy mechanism at lower temperatures below 870°C.

ACKNOWLEDGMENTS

We would like to thank M. Suzuki and A. Takano of NTT-AT for measuring SIMS profiles. We also thank T. Ikeda and Y. Hirose of Nippon Sanso for gas chromatography analysis, and K. Saito and N. Fujiwara for their experimental assistance. This work was partly supported by a Grant-in-Aid for Scientific Research (A) (10305030) from the Ministry of Education, Science, Sports and Culture, and by the Foundation for the Promotion of Material Science and Technology of Japan.

REFERENCE

1. H. J. Mayer, H. Mehrer and K. Maier, *Radiation Effects in Semiconductors 1976*, (Institute of Physics, London, 1977) Con. Ser. 31, p. 186.

2. L. Kalinowski and R. Seguin, *Appl. Phys. Lett.* **35**, 211 (1979).

3. J. Hirvonen and A. Anttila, *Appl. Phys. Lett.* **35**, 703 (1979).

4. F. J. Demond, S. Kalbitzer, H. Mannsperger and H. Damjantschitsch, *Phys. Lett.* **93A,** 503 (1983).

5. H. Bracht, E. E. Haller and R. Clark-Phelps, Phys. Rev. Lett. **81**, 393 (1998).

6. A. Seeger and K. P. Chik, *Phys. Status Solidi* **29**,455 (1968).

7. W. Frank, U. Gösele, H. Mehrer and A. Seeger, in *Diffusion in Crystalline Solids*, edited by G. E. Murch and A. S. Nowick (Academic, New York, 1984).

8. A. Ural, P. B. Griffin and J. D. Plummer, Appl. Phys. Lett. **73**, 1706 (1998).

9. Y. Nakabayashi, T. Segawa, Hirman I. Osman, K. Saito, S. Matsumoto, J. Murota, K. Wada and T. Abe, *Jpn. J. Appl. Phys.* **39**, 1133 (2000).

10. M. E. Levinshtein and S. L. Rumyantsev, *Handbook Series on Semiconductor Parameters*, edited by M.. Levinshtein, S. Rumyantsev and M. Shur (World Scientific Inc., Singapore, 1996), Vol. 1, Chap. 1, p. 1.

11. J. M. Fairfield and B. J. Masters, J. Appl. Phys. 38, 3148 (1967).

Mat. Res. Soc. Symp. Proc. Vol. 669 © 2001 Materials Research Society

A new model for Boron diffusion retardation in SiGe-strained layers accounting for the mechanism of Boron trapping/detrapping by Ge atoms.

Victor I. Kol'dyaev

PDF/Solutions, Inc (333 San Carlos, San Jose, 95110, CA, USA)

ABSTRACT

The main drawbacks of the known models of the B diffusion in strained SiGe layers are summarized. A mechanism is suggested to self-consistently explain the main experimental features and original experimental data which considers the trapping of B atoms by Ge atoms during B diffusion in the Si lattice resulting in the retarded B diffusivity. Fluctuations of Ge atom numbers in a nearest B atom environment result in percolation mechanism of B transport through dilatation centers of random size. A new solid state transport model is generalized by considering dispersion transport of positive and negative point dilatation defects.

INTRODUCTION

Physics based TCAD models are vital in developing a new HBT technology and device architecture. The ready-made TCAD tools for the technology simulation are not accurate for the modern SiGe HBT ("SiGe" is used standing for $Si_{1-x}Ge_x$ where x is the atomic percentage of Ge in a layer). In contrast to known B diffusion physics in Si [1] new principal physical phenomena of a strong B diffusion retardation in strained SiGe, less retardation in relaxed SiGe and B segregation into SiGe are observed [2-15] in SiGe. B in-diffusion into SiGe layer is suggested as a new method for an experimental estimation of B segregation into SiGe. Experiments based on the method are carried out to estimate the segregation coefficient which is higher than observed in [9-11].

There are 2 different models for B diffusion in SiGe [6,9-11]. Model [6] deals with a strain effect coefficient $Q_{as} = dE_A / ds$ being characterized for Si(Ge) and SiGe(B) systems. One has: $D_A = D_{A0} \exp(-E_A / kT) * \exp(+Q_{as}s / kT)$, where s is the coherent local strain due to the Ge presence. In [6] Q_{Ges} =40±5 eV/unit strain (u.s.), E_A =5.3325 eV and D_{A0} =1.03 10^5 cm^2/sec are found for the low content Ge diffusion in Si. Also Q_{Bs} =-17±3 eV/u.s. is extracted for B in SiGe [6]. This is a simple, accurate and useful model describing a lot of features of B and Ge diffusion in SiGe. The model predicts a monotonic dependence of B diffusivity vs. strain and monotonic B profiles at the Si-SiGe interface resulting from the condition: $J_B = -D_B(Si)\nabla C_B(Si) = -D_B(SiGe)\nabla C_B(SiGe) = const(x_b)$ which is not observed in the experiments. The model does not explain the "segregation" phenomena [9-11] and diffusion against the B gradient [4-5,12].

In model [9-11] the segregation phenomena was attributed to the B electrochemical potential difference between Si and SiGe depending on the DOS, the work function, the binding energy, and the lattice contraction energy difference between Si and SiGe. The lattice contraction energy E_{ca} of B in SiGe is taken: $E_{ca} = \theta \beta_B \beta_{Ge} C_{Ge}$, where θ is a constant, β_B and β_{Ge} are the B and Ge lattice contraction coefficients in Si, correspondingly, and C_{Ge} is a Ge concentration. The term clearly indicates that the local interaction of B originated tensile strain and Ge originated compressive strain is neglected in the theory. Also [9-11] does not explain reduction of electrical activity of B in SiGe compared to Si at the same conditions [15].

In [13] and independently in [14] a hypothesis was suggested for the first time that the B diffusion retardation in SiGe as well as the segregation phenomenon could be self-consistently explained assuming that the interaction of local B originated tensile strain and Ge originated compressive strain is important. This possible interaction was out of attention before.

EXPERIMENTAL APPROACH AND RESULTS.

In order to observe the segregation phenomena more accurately a B in-diffusion into SiGe is explored (fig.1). The idea is similar to an experimental structure Si-SiO2 used to observe the segregation mechanism. To fabricate such a structure the following processing steps are done using a relatively high T CVD epitaxial growth process. (i) A thin Si buffer layer is deposited on a lightly p-doped Si substrate. (ii) Then Ge carrying gas is turned on and SiGe layer starts growing (Ge content was determined by a partial pressure of Ge carrying gas). (iii) At some moment B carrying gas is turned on for a short controllable time to deposit a B delta-doped layer (BL2). (iv) After that SiGe layer is continued to grow and at some moment Ge carrying gas is turned off and only Si cap layer is grown. (v) During growing the Si cap layer, B carrying gas is turned on to make another B delta-doped layer (BL1) in absolutely the same manner as for BL2. (vi) A non-doped Si cap layer is grown above BL1. (vii) Anneal of the structures at different T and time is done.

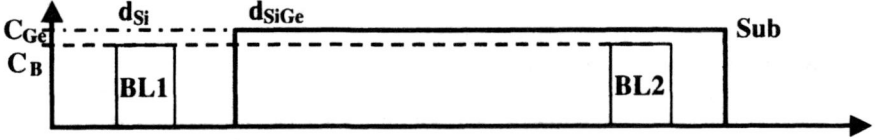

Fig. 1. The basic structure for studying B in-diffusion into SiGe.

Before doing the final experiment a lot of calibration experiments was done to provide reproducibility of the profiles. SIMS measurements are also carefully calibrated especially with respect to the etching rate and absolute concentration estimation. Profiles in fig.2 indicate: (i) Ge profiles are very reproducible and there is little diffusion of Ge for such a small thermal budget (TB) applied. (ii) The TB is enough to redistribute B significantly. (iii) The peak position of B profile inside SiGe layer is almost at the same place indicating a good reproducibility of placing BL2. (iv) The right hand sides of all the profiles from their peak positions are needed to be deconvoluted to account for a SIMS stretching of a profile into depth. Since a rigorous quantitative treatment of the profiles is not done here the profiles are not deconvoluted. Notice that as grown B SIMS profile has some noise for the concentration range below 8e16 cm-3 resulting in error of 2% for the total B dose. The noise does not affect annealed profiles as B concentration is higher than 8e16 cm-3.

After the deconvolution of "BL2" B profiles within SiGe layer and Ge profiles the diffusivity of B is described by model [6] very well. But it is impossible to describe the B in-diffusion from the "BL1" layer into SiGe layer with the same model. The details of BL1 profiles are given in fig.3. Since the segregation coefficient is assumed not to be very high there were no expectations that the difference in concentrations to deal with is going to be very high. A lot of precautions were done to obtain an acceptable resolution. To confirm the measurement accuracy we have integrated the profiles to obtain the total layer concentration for every profile. The total layer concentrations were within 3% of variation from a profile to a profile. In fig.3 the following can be observed. (i) The peak position is moving towards the SiGe layer and penetrating into the layer during anneals. (ii) At some moment the B diffusion seems to be "against gradient". To explain these features a decomposition of the total profiles into 2 profiles is drawn in fig.4. In fig.4 the total profiles are similar to what is observed in fig.3. If there was no B trapping but just a difference in B diffusivity the total profiles should be described by curve-2a and curve-3a (properly scaled). But it is not the case. While these 2 profiles describe the mobile B behavior it is the trapped component that makes the total profiles looking like in experiment: the diffusion against the gradient occurs when only total concentration is considered.

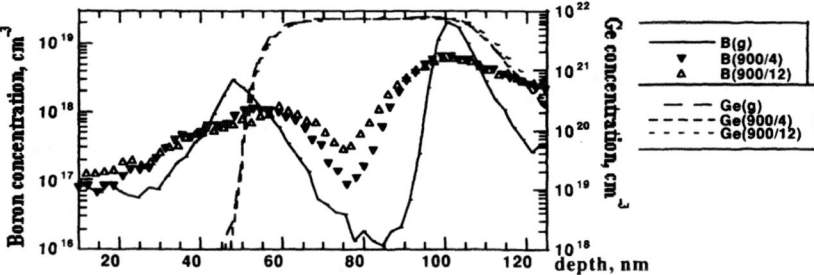

Fig.2 SIMS B and Ge profiles of "17%" samples as grown B(g) and Ge(g) and those after anneal at 900 C for 4 min [B(900/4) and Ge(900/4)] and at T=900 C for 4 min [B(900/12) and Ge(900/12)].

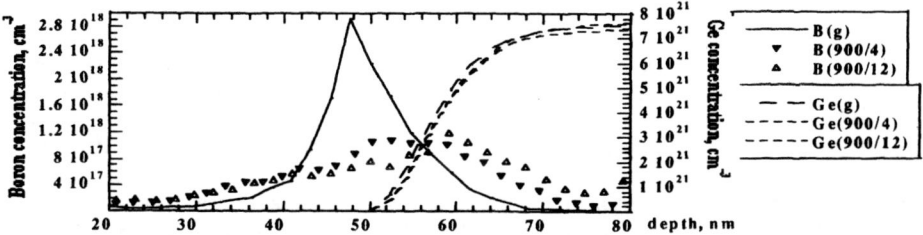

Fig.3. SIMS B and Ge profiles of "17%" samples as grown B(g) and Ge(g) and those after anneal at T=900 C for 4 min [B(900/4) and Ge(900/4)] and 12 min [B(900/12) and Ge(900/12)]. Diffusion against the gradient occurs when comparing profile B(g) with B(900/4) and B(900/12). Note that only BL1 layer is shown here

Note that the profile peak position is moving towards and eventually into the SiGe layer whereas if only diffusivity drops down in SiGe the peak would stop at the interface. From fig.3 one can estimate the segregation coefficient as a ratio of the maximum B concentration at the peak position for sample B(900/12) inside SiGe layer to the B concentration at the "interface" at the Si side. It is difficult to define the interface position very accurately, but picking up a B concentration at around x=48 nm to x=52 nm one can come up with an estimation for the segregation coefficient to be $k_s = 1.7 \pm 0.2$. This value is certainly higher than that found in [9-11] which was 1.3 at almost quasi equilibrium conditions when B penetrates through SiGe layer and goes much further than a SiGe layer thickness. In this experiment an interfacial transition layer, where Ge concentration is changing from 0 to the maximum, occurs which is much wider than the interfacial layer for $Si - SiO_2$ system. The Ge gradient in this transition region results in an electric field. The electric field E in the layer pulls B atoms from Si into SiGe and makes an interpretation more difficult. The field is partially responsible for the segregation process kinetics not impacting the final concentration difference. Moreover the effective B diffusivity is linearly proportional to E as a first order effect whereas the B diffusion retardation effect is exponentially dependent on Ge concentration being the zero order effect.

(a)

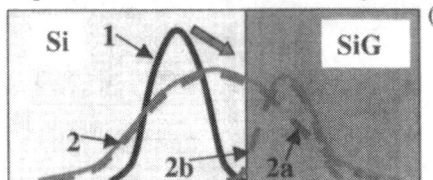 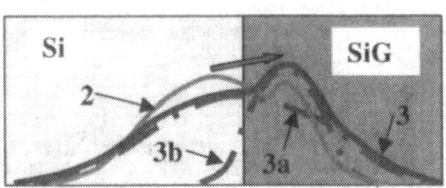

Fig.4. Qualitative model for B redistribution from Si into SiGe layer. (a) Curve-1 (black) is the initial profile (as grown). Curve-2 (green) represents the total B profile as a mobile component (Curve-2a) and the trapped component (Curve-2b) after 4 minutes of anneal. (b) Curve-3 (brown) represents the B distribution after 12 minutes of anneal where Curve-3a is a mobile component and Curve-3b is the trapped component. Note that only BL1 layer is shown here.

A MODEL FOR THE B AND Ge DIFFUSION IN SiGe LAYER.

An assumption that the local interaction of B and Ge due to opposite signs of strain results in a diffusion process with trapping explains roughly the observed features of B transport in SiGe. Also some other experimental observations which are observed [5,12] can be explained with no contradictions by this model. As mentioned there are 2 most important observations. (i) The same magnitude of the strain in a SiGe layer with different Ge contents fabricated using a partial relaxation technique results in different B diffusivity [5]. (ii) The B diffusion in SiGe layer can go against the boron density gradient [12] indicating the segregation process. It is the assumption of [9-11] that the local strain is not important results in the contradiction of the model predictions with what was observed later and in this study. Due to the smaller size of the B atom compared to Si the local tensile strain around a B atom occurs (a positive point dilatation center - PDC). Whereas, the Ge atom size is larger than that of Si and it provides the local compressive strain around a Ge atom (a negative point dilatation center - NDC). A PDC and a NDC should attract each other to create a complex in order to release the stress energy. Such a complex should be much more stable than B-I, B-V, Ge-V and Ge-I complexes implicating that even if I comes to B-Ge pair it should overcome a higher barrier for B-I pair to go into the lattice leaving Ge atom alone. It is very likely that the $Ge - B \equiv Si_3$ complex has also a less cross section for interaction with a mobile I than substitutional B in Si lattice environment ($Si - B \equiv Si_3$). This strain dipole might be still dissolved by interaction with I or V or can attract another B atom to create a cluster/precipitate. The last process should be enhanced by the biaxial strain in SiGe layer.

Energetics and cross-section comparison is schematically presented in fig.5. An electrostatic analogy support the model since the strain is described with a Poisson equation with specific boundary conditions and a screening mechanism. The analogy says that B-Ge interaction looks like a dipole interaction resulting in more localized strain distribution that for a point defect being like an isolated point charge. The local strain interaction of point defects (PD) as a significant driving force is discussed many times [1]. For example it explains why B is prone to interact with I while Sb to V. This is a very bright indication of the importance of the local strain interaction. Another effect in PD interaction is called the "chemical" interaction [16] which confuses the validity of the local strain interaction concept. In fig.6 the "chemical" effect is shown as a couple of Si dangling bonds (DB's) for I and one DB for B-I pair. Thus chemical effect is the extra energy which should be provided into the lattice to keep these DB's. In B-I case there is only 1 extra dangling bond compared to I meaning that B-I pair should be more stable than I during migration. Another chemical component comes from the size and strength of the B-Si bonding which partially reflected in the local

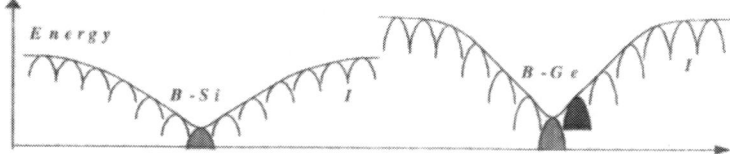

Fig.5 A schematic presentation of a potential relief for a B atom in Si lattice and that when a Ge atom is present in a closest position to the B atom (as seen by I). An energy barrier for a B-I pair to leave the substitutional position and the cross-section for I interaction with B are perturbed by Ge atom on the atomic scale. The potential energy perturbation around I is not

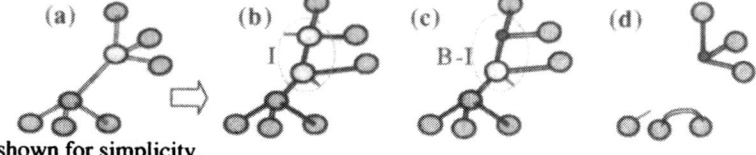

shown for simplicity.

Fig.6 A schematic presentation of the ideal Si network (a), Si interstitial (I) PD (b), B in interstitial position PD (c), and possible PD of B atom interacting with V (d). Sticks (red) are the DB's, small (red) circles are B atom and gray circles are Si atoms.

strain. Therefore both the strain and chemical effect are in favor of B interaction with I. In B-V pair the B atom being in the substitutional position should form 3 fold coordinated structure to the V after the lattice relaxation occurred requesting for an additional energy to be taken from Si lattice for DB's. V is PDC as well as B resulting in repulsion between them. Thus both the strain and the chemical interaction are against B-V pair formation which is proven to be the case. The similar arguments are valid for donors (D) interaction with I and V. But for Sb the local negative strain is so high that it does not interact with I at all. Again both the local strain and chemical structure of a PD are critical in the PD interaction with each other and with the lattice. Now main PD's can be classified as follows. B in substitutional position, V and Ge-V are PDC. I, Ge in substitutional position, Ge-I, B-I are NDC. What sign of strain is around the "dipole" of Ge-B? According to the compensation theory [17] the total strain f_T is a superposition of the B strain and the Ge strain: $f_T = 0.0425c_{Ge} - 0.274c_B$ which indicates that $f_T = 0$ (the complete strain compensation) occurs at the condition $c_{Ge} = 6.45c_B$. Thus the strain developed by a B is much higher than a Ge originated strain resulting likely in the PDC for B-Ge pair. This means that B atom can attract a few Ge atoms creating a cluster to compensate the local stress around it. The more Ge atoms next to B the higher activation energy for migrating should occur. The pre-exponential factor for B diffusivity should be affected according to

$D_{BIeff} = -D_{BI} * [1/(1 + C_{Ge}/C_{Ge0})]$, where C_{Ge0} is a characteristic concentration. The activation energy is affected by C_{Ge} through the local strain dependent on Ge atom numbers around B resulting in the percolation transport mechanism.

How the coherent strain across the structure impacts the PDC and NDC migration barrier and interaction regardless of the chemical component of the total coupling energy for the PD's? Neglecting by the chemical energy change due to the coherent strain means neglecting the changes in the local strain around PDC and NDC due to plunging them into differently strained lattice. It is shown that Ge diffusion is mainly mediated by V (70% [3]) and Ge diffusion along the in-growth direction of a SiGe layer is enhanced. Thus it can be concluded that Ge-V pair as a PDC is randomly walking with an effectively less migration barrier in the tensile strained direction. According to this assumption a B-I pair as the main mediating pair for B diffusion and as a NDC is going to have less diffusivity in SiGe layer in the growth direction. And this is consistent with the observations. It is very likely that the proportions between the coherent strain retardation f_s of B diffusion and the retardation due to trapping f_t are in the same order of magnitude to be $f_s + f_t = 1$ and f_s is in range of 0.3 to 0.7 or so. Otherwise it would be very difficult to observe these both mechanisms. A more sophisticated study needs to be done to clarify this issue based on a profound theory from first principal investigations [16-20].

CONCLUSIONS.

The main drawbacks of the known models are surveyed and represented by the summarization of the experimental facts which are in contradiction with predictions of the models. New model is proposed explaining these experimental facts which assumes that B atoms can be trapped by Ge atoms resulting in the decrease of B diffusivity. The model can be formulated as a set of reactions to be written as a boundary value problem for further study.

ACKNOWLEDGEMENTS.

The author is very thankful to Dr.'s L. Deferm and S. Decoutere for many useful discussions at IMEC where this study was started and to Prof. N. Cowern for valuable e-mail discussion.

References:

1 P.M. Fahey, P.B. Griffin, J.D. Plummer, Rev.Mod.Phys., vol.62, No.2, p.289, 1989.

2. P. Dorner, W. Gust, B. Predel, et.al., , Phil.Mag., vol.49 (No.4), p.557, 1984.

3. P. Fahey, S.S. Iyer, G.J. Scila, Appl.Phys.Lett., vol.54, No.9, p.843, 1989.

4. P. Kuo, J.L. Hoyt, J.F. Gibbons, et.al., Appl.Phys.Lett., vol.62, No.6, p.612, 1993.

5. P. Kuo, J.L. Hoyt, J.F. Gibbons, et.al., Appl.Phys.Lett., vol.66, No.5, p.580, 1995.

6. N.E.B. Cowern, P.C. Zalm, P. van der Sluis, Phys. Rev. Lett., vol.72 (No.16), p.2585, 1994.

7. G.H. Loechelt, G. Tam, J.W. Steele, et.al., J.Appl.Phys., vol.74, No.9, p.5520, 1997.

8. D. Krüger, P. Gaworzewski, R. Kurps, J.Vac.Sci.Tech., vol.B14, No.1, p.341, 1996.

9. S.M. Hu, Phys. Rev. Lett., vol.63, No.22, p.2492, 1989.

10. S.M. Hu, D.C. Ahlgren, P.A. Ronsheim, Phys. Rev. Lett., vol.67, No.11, p.1450, 1991.

11. S.M. Hu, Phys. Rev. (B), vol.45, No.8, p.4498, 1992.

12 T.T. Fang, W.T.C. Fang, P.B. Griffin, Appl.Phys.Lett., vol.68, No.6, p.791, 1996.

13. V.I. Kol'dyaev, IMEC Internal Report of 11th of March, 1998, (unpublished).

14. R.F. Lever, J.M. Bonnar, and F.W. Willoughby, J. Appl. Phys., vol.83, No.4, p.1988, 1998.
15. P. Gaworzewski, D. Krüger, R. Kurps, et.al., J.Appl.Phys., vol.75, No.12, p.7869, 1994.
16. C.G. Van de Walle, and J. Neugebauer, Phys. Rev., (B), vol.52, No.20, p.R14320, 1995.
17. W.P. Maszada, T. Thompson, J.Appl.Phys., vol.72, No.9, p.4477, 1992.
18. M.J.Aziz, E. Nygren, W.H. Christie, Mater.Res.Soc. Symp. Proc., vol.36, p.101, 1985.
19. G. Subramanian, K.S. Jones, M.E. Law, Mat.Res.Soc.Symp., vol.610, p.10.1, 2000.
20. H. Park, K.S. Jones, J.A. Slinkman, and M.E. Law, J.Appl.Phys., vol.78, No.6, p.3664, 1995.

Mat. Res. Soc. Symp. Proc. Vol. 669 © 2001 Materials Research Society

SHALLOW JUNCTIONS FOR SUB-100 NM CMOS TECHNOLOGY

Veerle Meyssen,[1] Peter Stolk,[2] Jeroen van Zijl,[1] Jurgen van Berkum,[3] Willem van de Wijgert,[3]
Richard Lindsay,[4] Charles Dachs,[2] Giovanni Mannino,[5] and Nick Cowern[1,2]
[1]Philips Research Laboratories, [3]Philips CFT; Eindhoven, THE NETHERLANDS
[2]Philips Research Leuven, [4]IMEC; Kapeldreef 75, B-3001 Leuven, BELGIUM
[5]CNR-IMETEM, Catania, ITALY

ABSTRACT

This paper studies the use of ion implantation and rapid thermal annealing for the fabrication of shallow junctions in sub-100 nm CMOS technology. Spike annealing recipes were optimized on the basis of delta-doping diffusion experiments and shallow junction characteristics. In addition, using GeF_2 pre-amorphization implants in combination with low-energy BF_2 and spike annealing, p-type junctions depths of 30 nm were obtained with sheet resistances as low as 390 Ω/sq. The combined finetuning of implantation and annealing conditions is expected to enable junction scaling into the 70-nm CMOS technology node.

I. INTRODUCTION

The semiconductor industry is characterized by an incessant drive for ever smaller and faster electronic devices and circuits. The achievement of shallow, low-resistance junctions is seen as one of the important prerequisites for this evolution. In CMOS technology, the fabrication of shallow source/drain extensions at the edge of the polysilicon gate is the most critical doping step. Figure 1 shows an example of two-dimensional doping profiles that were obtained on 0.18 μm transistors, using selective etching and transmission microscopy for the NMOS (Fig. 1a, Ref. 1) and voltage contrast scanning electron microscopy for the PMOS (Fig. 1b, Ref. 2).

The contrast fringes in Fig. 1(a) clearly reveal the two As implantation steps by which the junction areas have been fabricated. The drain extension region underneath the spacer is clearly separated from the deeper, highly-doped drain (HDD), as desired. In comparison, the voltage contrast image in Fig. 1(b) indicates that there is no clear distinction between the extension and HDD junctions in the PMOS device. This arises from the fact that boron undergoes much faster diffusion than arsenic due to the phenomenon of transient enhanced diffusion (TED) [3]: ion-generated silicon self-interstitials couple with B atoms during thermal activation, leading to a temporary enhancement in the boron diffusivity. In addition, the excess interstitials drive clustering of B atoms below the solid solubility limit [3], thereby limiting the activation efficiency. These physical effects associated with ion implantation jeopardize the suitability of implanted B junctions for sub-100 nm CMOS.

It has been demonstrated [4] that the coupling between silicon self-interstitials and B (and P) decreases with increasing temperature, thereby reducing the driving force for TED. This implies that rapid thermal annealing (RTA), which uses high temperatures and short annealing times, is the most suitable industrial approach for the formation of shallow, electrically active, junctions. The need to achieve good electrical activation of the implanted dopants and full removal of implant damage has led to the development of so-called "spike" annealing cycles [5-7], with heating ramps as high as a few hundred degrees per second (up to ~400 °C/s), and high peak temperatures at short soaking times (<1 s). In addition to RTA optimization, it is known that pre-amorphization of the Si surface prior to B implantation leads to a reduction of the channeling tail, thereby allowing for shallower, more abrupt junctions [6,8,9]. In addition, it has been found that implanting BF_2 instead of B (at the same equivalent B energy) results in shallower junctions after annealing, a result which is tentatively ascribed to the co-implanted fluorine [8]. In an approach to combine these two benefits [10], we have used pre-amorphization with GeF_2 ions to fabricate shallow p-type junctions.

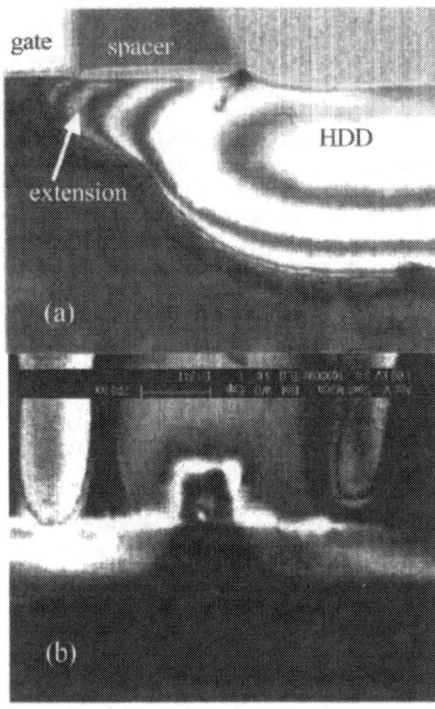

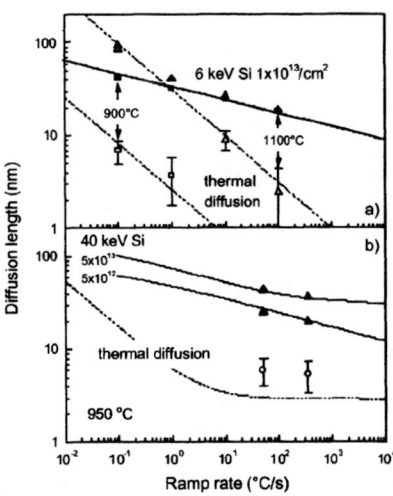

Figure 1 (left). 2D doping profiles for (a) NMOS and (b) PMOS devices fabricated in 0.18 μm CMOS technology.

Figure 2 (right). Diffusion length versus ramp rate. Annealing conditions used were (a) "spike" annealing at 900 and 1100 °C; (b) "soak" annealing for 10 seconds at 950 °C. Open symbols/dashed lines: data and simulations of diffusion in unimplanted samples. Solid symbols/thin solid lines: data and simulations for the implanted case. Doses and energies of the Si damage implants are indicated. Thick solid lines: simulated contribution of TED [11].

II. RESULTS AND DISCUSSION

In a recent publication by Mannino et al. [11], the influence of heating ramp variations on TED has been investigated. For this purpose, δ-doped B marker layers were used to probe accurately the dependence of interstitial-enhanced diffusion on heating ramp rates. The results in Fig. 2 show that thermal diffusion can be substantially reduced by increasing the heating ramp rate, a trend which reflects the decrease in overall thermal budget. At the same time, increasing the ramp rate also reduces the effects of transient enhanced diffusion (TED). For samples implanted with 6 keV Si, the difference in diffusion lengths for 900 and 1100 °C spikes becomes small at ramp rates above ~10°C/s. Apparently, damage annealing and thus TED, is entirely completed during the ramp-up phase to 900 °C, while the high ramp rates assure that the thermal component of diffusion is negligibly small above 900 °C. These results demonstrate the success of spike annealing. When using sufficiently high ramp rates, temperatures can be raised to as high as 1100 °C for improving dopant activation without inducing significant *thermal* dopant diffusion after TED has completed. At the same time, high ramp rates ensure that TED itself is minimized.

In an attempt to improve the activation of implanted B junctions while minimizing junction depths, various annealing strategies were explored on a conventional lamp-based RTA system. In brief, four recipes were studied, all with heating and cooling rates of ~100 and ~80 °C/s, respectively. In the reference spike anneal, a 5 sec stabilization at 700 °C was applied before ramping to the final plateau (1 second soak at 1050-1100 °C). In the second variant, the soak time was minimized to "0 seconds" while keeping the same stabilization during the heating phase. In the remaining 2 variants, the soak time was kept minimal while taking out or extending the stabilization phase.

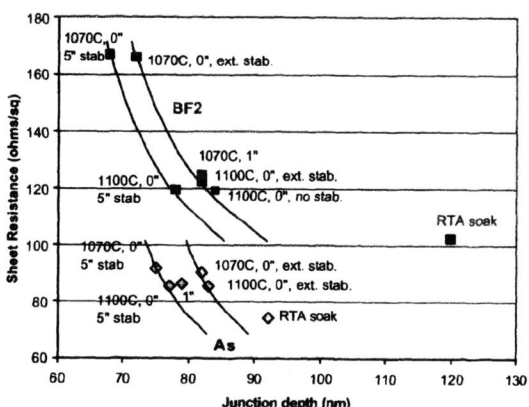

Figure 3. Junction parameters for n-type (30 keV As, $4x10^{15}/cm^2$) and p-type (12 keV BF_2, $4x10^{15}/cm^2$) junctions annealed with various spike recipes (see text). Data points for conventional RTA are also shown ("soak"). Junction depths were derived from SIMS profiles at concentrations of $10^{19}/cm^3$ (As) and $10^{18}/cm^3$ (B).

Figure 3 summarizes the measurements of junction depth and sheet resistance for both n- and p-type junctions as obtained with the different annealing strategies. For both junction types, minimizing the soak time leads to an improved trade-off between junction depth and sheet resistance. Measurements of sheet resistance across the 8" wafers indicate that non-zero stabilization steps improve within-wafer uniformity. The apparent need for temperature stabilization illustrates one of the practical limitations of lamp-based system for reaching true "spike conditions". Furthermore, wafer cooling in these systems is mostly controlled by radiation, imposing an upper limit on the maximum cooling rate that can be achieved (typically 110 °C/s). Alternative annealing systems based on, for example, mechanical handling to control the wafer temperature excursion may be needed to realize the optimal spike annealing conditions. In the near future, RTA is expected to evolve to true spike anneals at peak temperatures ranging from 1100-1200 °C, with heating and cooling rates approaching or even exceeding 1000 °C/s.

Although spike annealing has proven beneficial for shallow junction formation, full CMOS integration calls for efficient dopant activation in the polysilicon gate electrode as well. When avoiding gate pre-doping in the CMOS process, the gate dopant level is predominantly determined by the HDD implant conditions and the subsequent spike anneal. First studies indicate that efficient gate doping requires soak times of ~1 sec to reduce gate depletion to below the required 10% [12]. Further experiments are needed to co-optimize spike annealing for gate activation, while avoiding the problem of B penetration in PMOS transistors.

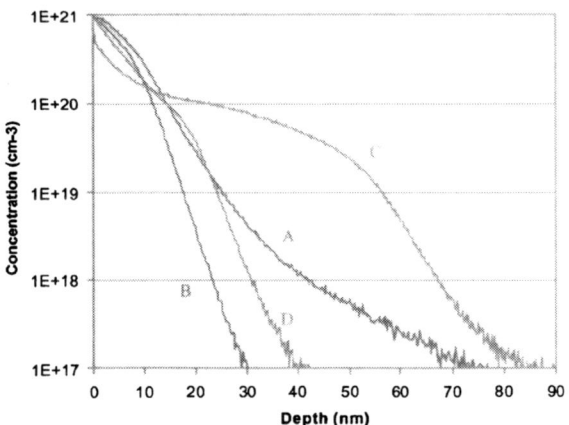

*Figure 4. Secondary ion mass spectrometry profiles of B junctions after implantation (A - 1 keV B, $1x10^{15}/cm^2$ into crystalline Si; **B** - 4.5 keV BF_2, $1x10^{15}/cm^2$ into $1x10^{15}/cm^2$ GeF_2-pre-amorphized Si) and after spike annealing (A→C; B→D - 1100°C, 1 sec).*

In addition to optimizing RTA procedures, the ion implantation conditions can be varied to improve junction parameters. Here, we have used GeF_2 implants to simultaneously amorphize the near surface region to suppress B channeling, while at the same time introducing fluorine to reduce TED during annealing. It should be noted that GeF_4 is generally used as a gas in ion sources to generate Ge ions. It is therefore straightforward to switch to PAI by GeF_2 instead of Ge by tuning the mass selection magnet of the ion implanter system.

Figure 4 shows B concentration profiles after implantation and after spike annealing (1100 °C, 1 sec). When changing from a 1 keV B implant into crystalline Si to a 4.5 keV BF_2 implant into GeF_2-pre-amorphized Si, a significant reduction of the channeling tail is seen, as expected. A spectacular difference in junction profile is observed for these two samples after spike annealing: the junction depth X_J is decreased from 65 nm for the reference sample to 30 nm for the GeF_2/BF_2 sample. Changing the concentrations of co-implanted fluorine by using other implant variants (always with the same effective Ge and B implant energies) leads to a proportional scaling of X_J in between 30 and 65 nm. This suggests that the junction depth reduction is strongly related to the co-implanted F. Indeed, SIMS measurements after RTA show that going from BF_2 implantation to GeF_2/BF_2 leads to a significantly stronger incorporation of F into the near surface region. This suggests that fluorine may lead to the trapping of excess Si interstitials thereby reducing the driving force for TED. Alternatively, the reduced junction may result from a chemical interaction between F and B. Clearly, this subject needs further investigation, and first studies on this subject have been presented at this conference [13].

The sheet resistance ρ_{sh} for the shallowest junction (X_J=30 nm) is 390 Ω/sq. To our knowledge, these junction parameters are the best obtained until today using ion implantation and rapid thermal annealing, breaking through the trade-off curve between X_j and ρ_{sh} that is generally observed for p-type junctions. It is expected that further optimization of the pre-amorphization and implantation conditions (e.g., <1 keV B) together with more aggressive spike anneals will lead to additional improvements in junction performance.

The approach of Pre-Amorphization with co-implanted Fluorine (PAF) was used for forming extensions in a 0.13 μm PMOS process flow, featuring a 2.3 nm pure gate oxide. From electrical analysis (not shown) it is found that the underdiffusion of the extension decreases with increasing fluorine concentration. For the implant/anneal variants shown in Fig. 4, the overall source/drain underdiffusion is reduced by 25 nm when using GeF_2/BF_2, consistent with the measured reduction in X_j. Furthermore, the measured source/drain series resistance was ~500 $\Omega\mu$m, with a 10% improvement for the PAF approach. It can therefore be concluded that PAF is suitable for the fabrication of sub-100 nm PMOS transistors. One drawback, however, is that the co-implanted fluorine is observed to enhance the penetration of B through the thin gate oxides, leading to undesired shifts in the threshold voltage. This implies that a successful implementation of PAF may require the utilization of heavily nitrided gate oxides (or even high-k dielectrics?) to increase the resistance against B penetration. Alternatively, the CMOS processing scheme could be modified to prevent that fluorine atoms are introduced into the gate electrode during extension implantation.

IV. CONCLUSION

In this paper, we have studied the scalability of ion-implanted shallow junctions for future silicon technology. It is concluded that the concomitant optimization of implant and annealing conditions will enable the fabrication of both n- and p-type junctions that are compatible with sub-100 nm CMOS technology. For further device scaling, novel approaches such as elevated source/drains or laser-annealing may become indispensable.

ACKNOWLEDGMENTS

We would like to acknowledge Anne Lauwers (IMEC) and Marcel Verheijen, Monja Kaiser, Sjoerd Mentink, Fred Roozeboom, Pierre Woerlee, Jurriaan Schmitz, and Youri Ponomarev (Philips) for their contributions to this work.

REFERENCES

[1] C.J.J. Dachs, M.A. Verheijen, M. Kaiser, P.A. Stolk, and Y.V. Ponomarev, Proc. ESSDERC-2000, pp. 360-363 (2000).
[2] S.A.M. Mentink et al., 12th European Congress on Electron Microscopy (2000).
[3] P.A. Stolk H.-J. Gossmann, D.J. Eaglesham, D.C. Jacobson, C.S. Rafferty, G.H. Gilmer, M. Jaraiz, J.M. Poate, H.S. Luftman, and T.E. Haynes, J. Appl. Phys. **81**, 6031 (1997).
[4] A.E. Michel, W. Rausch, P.A. Ronsheim and R.H. Kastl, Appl. Phys. Lett. **50**, 416 (1987).

[5] A. Agarwal, A.T. Fiory, H.-J. Gossmann, C.S. Rafferty and P. Frisella, Mat. Sci. Semi. Proc. 1, 237 (1998).

[6] S. Saito, S. Shishiguchi, K. Hamada and T. Hayashi, Mat. Chem. and Phys. **54**, 49 (1998).

[7] M. Mehrotra, J.C. Hu, A. Jain, W. Shiau, S. Hattangady, V. Reddy, S. Aur, and M. Rodder, Tech. Digest IEDM, 419 (1999).

[8] D.F. Downey, C.M Osburn, S.D. Marcus, Solid State Tech., 71 (Dec. 1997).

[9] A. Al-Bayati, S. Tandon, A. Mayur, M. Foad, D. Wagner, R. Murto, D. Sing, C. Ferguson, L. Larson, Proceedings of the Ion Implantation Technology conference (2000).

[10] T.H. Huang, H. Kinoshita, and D.L. Kwong, Appl. Phys. Lett. **65**, 1829 (1994).

[11] G. Mannino, P.A. Stolk, N.E.B. Cowern, W.B. de Boer, A.G. Dirks, F. Roozeboom, J.G.M. van Berkum, P.H. Woerlee, and N.N. Toan, Appl. Phys. Lett. **78**, 889 (2001).

[12] P.A. Stolk et al., Mat. Res. Soc. Symp. Proc. Vol. 610 (2000).

[13] L.S. Robertson, K.A. Gable and co-workers.

Mat. Res. Soc. Symp. Proc. Vol. 669 © 2001 Materials Research Society

The Role of Ion Mass on End-of-Range Damage in Shallow Preamorphizing Silicon

Mark H. Clark[1], Kevin S. Jones[2], Tony E. Haynes[3], Charles J. Barbour[4], Kenneth G. Minor[5] and Ebrahim Andideh[6]
[1,2]University of Florida, Dept of Materials Science and Engineering,
Gainesville, FL 32611-6130, U.S.A.
[3]Oak Ridge National Laboratory,
Oak Ridge, TN, 37831-6048, U.S.A.
[4]Sandia National Laboratories,
Albuquerque, NM, 87185-1056, U.S.A.
[5]Sandia National Laboratories,
Albuquerque, NM, 87185-1056, U.S.A.
[6]Intel Corporation,
Portland, OR, 97124, U.S.A.

ABSTRACT

Preamorphization is commonly used to form shallow junction in silicon CMOS devices. The purpose of this experiment was to study the effect of the preamorphizing species' mass on the interstitial concentration at the end-of-range (EOR). Isovalent species of Si, Ge, Sn and Pb were compared. Silicon wafers with a buried boron marker layer (4700 Å deep) were amorphized using implants of 22 keV $^{28}Si^+$, 32 keV $^{73}Ge^+$, 40 keV $^{119}Sn^+$ or 45 keV $^{207}Pb^+$, which resulted in similar amorphous layer depths. All species were implanted at a dose of $5x10^{14}$ /cm^2. Cross-sectional transmission electron microscopy (XTEM) was used to measure amorphous layer depths (approximately 400 Å). Post-implantation anneals were performed at 750 ^0C for 15 minutes. Plan-view transmission electron microscopy (PTEM) was used to observe and quantify the EOR defect population upon annealing. Secondary ion mass spectrometry (SIMS) was used to monitor the transient enhanced diffusion (TED) of the buried boron marker layer resulting from the EOR damage introduced by the amorphizing implants. Based upon the SIMS results Florida Object Oriented Process Simulator (FLOOPS) calculated the resulting time average diffusivity enhancements. Results showed that increasing the ion mass over a significant range (28 to 207 AMU) not only affects the quantity and type of damage that occurs at the EOR, but results in a reduced diffusivity enhancement.

INTRODUCTION

Increasing the packing density, speed, and power efficiency of future devices requires the vertical and lateral scaling down of device dimensions. One of the most challenging problems of device scaling is forming shallow source and drain junctions. With each successive generation, the junction depth for the source and drain is slated to shrink with the gate length. A 70 nm gate in 2008 will require a junction depth as shallow as 20 nm [1].

Currently, ion implantation is the most commonly used technique for forming shallow junctions, due to its superior dose control. Shallow n^+-p junctions are formed relatively easily by arsenic implantation. Since arsenic is a heavy ion, it has a small projected range and does not experience serious ion channeling in crystalline silicon. On the other hand, shallow p^+-n junctions are difficult to form due to the small atomic mass of boron, which is commonly used as the p-type dopant. When boron is implanted into crystalline silicon, channeling of the implanted boron ion occurs, resulting in a dopant profile channeling tail, and consequently a much deeper junction [2].

Preamorphization of single-crystal silicon by ion implantation effectively eliminates the channeling of subsequent dopant implants [2, 3] and increases activation, thus lending itself to shallow junction formation [4]. The primary drawback to the technique of preamorphization is the EOR damage (e.g., {311}'s and dislocation loops), which remains beyond the original amorphous crystalline (α/c) interface after the amorphous layer has been regrown by solid phase epitaxial (SPE). EOR damage can have several adverse affects; for example, if the resulting dislocation loops are located in the depletion region of the device, a large leakage current can result [5]. Additionally, the dissolution of EOR defects upon subsequent anneals causes enhanced diffusion [6, 7] of interstitial diffusers (e.g., boron) [8], resulting in a deeper junction.

Ideally, a preamorphizing implant should produce minimal EOR damage so as to have minimal deleterious affects. The implant should also be isovalent (i.e., for silicon group IV of the periodic table) so as not to affect the electrical characteristics of the subsequent dopant implant. In this study amorphizing implants of silicon [9], germanium [10], tin [4] and lead are studied to determine the effect of atomic mass on the resulting EOR damage and subsequent diffusion enhancements.

EXPERIMENTAL DETAILS

To monitor the diffusion enhancement created by the release of interstitials from the EOR damage, all implants were performed into a silicon wafer with a single boron marker layer [11] [12]. The wafer was constructed by epitaxially growing a buried boron marker layer (9E19 /cm^3 boron spike) and subsequent undoped layer of silicon (4700 Å thick) by chemical vapor deposition on an existing Czochralski grown p-type <100> silicon substrate. Four samples of the wafer were independently preamorphized by implants of 22 keV ^{28}Si$^+$, 32 keV ^{73}Ge$^+$, 40 keV ^{119}Sn$^+$ and 45 keV ^{207}Pb$^+$, all at a dose of 5E14 /cm^2. The dose rate of all the preamorphizing implants was maintained at 0.23 μA/cm^2, while the temperature was maintained at approximately 25$\pm$1 $^\circ$C. Implant energies were chosen based on TRIM [13] simulations to achieve an α/c interface of approximately 400 Å for each implant species.

Depth of the α/c interface was observed by XTEM on a JOEL 200CX, with images taken in bright field near the [110] zone axis. Additionally, XTEM was used to ascertain if the damage remaining after SPE regrowth was confined to the EOR. Regrowth of the amorphous layer and post amorphizing implant annealing was performed in a tube furnace with a nitrogen ambient at 750 $^\circ$C for a total duration of 15 minutes. After annealing, PTEM images of the EOR damage were taken using the g$_{220}$ reflection and weak beam dark field conditions. The two main types of defects observed by PTEM were {311} defects and dislocation loops. Quantification of the concentration of interstitials trapped by the EOR defects was determined by following an adapted procedure set forth by Bharatan [14].

Measurement of the time average diffusivity enhancement ($<D_B>/D_B*$) began with depth profiling of the boron marker layer by dynamic SIMS [12]. SIMS depth profiles were performed on a Cameca IMS-3f with a magnetic sector detector using an O$_2^+$ primary ion source, a 10 keV primary accelerating voltage, a 250 μm raster with 1 turn down, a field aperture of 60 μm and a contrast aperture of 50 μm. The resulting boron marker layer profiles are normalized to the same depth and dose. Based on boron depth profiles obtained by SIMS, FLOOPS [15] was used to calculate the $<D_B>/D_B*$.

DISCUSSION

XTEM of the amorphized samples showed that all implant masses produced an α/c interface of approximately 400 Å in depth.

End-of-range defects

Damage resulting from amorphizing implants of $^{28}Si^+$, $^{73}Ge^+$, $^{119}Sn^+$ and $^{207}Pb^+$ after annealing at 750 °C for 15 minute is shown in figures 1a-d. It is evident from these PTEM micrographs that the atomic mass of the amorphizing ion species has a substantial effect on the resulting damage. In general, increasing the atomic mass of the preamorphizing implant decreases the size and density of the resulting defects.

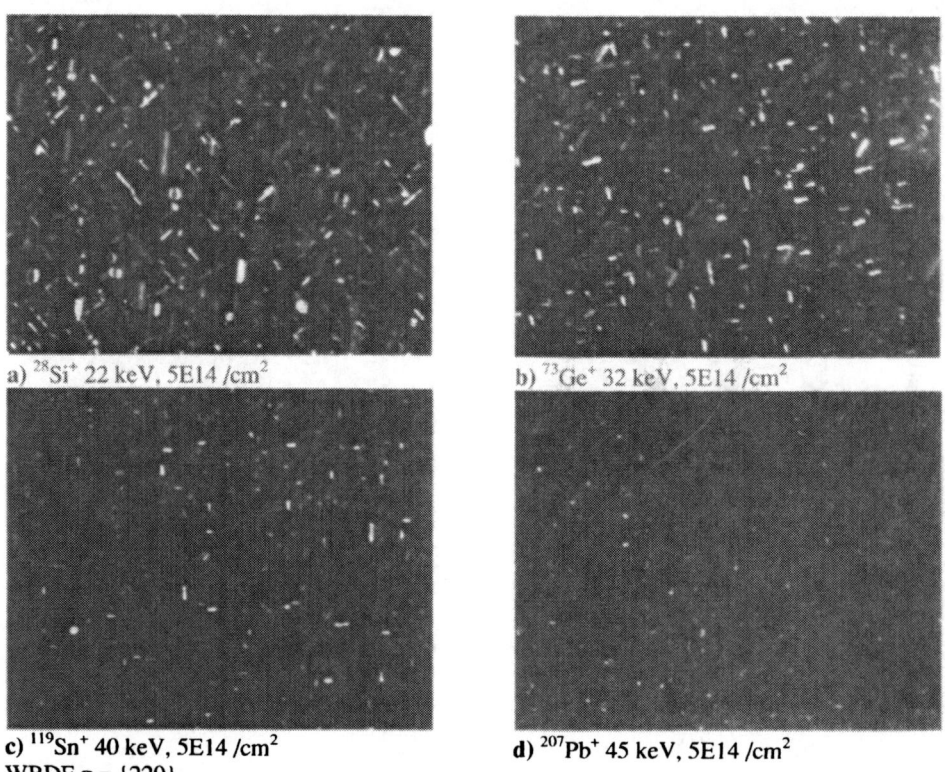

a) $^{28}Si^+$ 22 keV, 5E14 /cm^2

b) $^{73}Ge^+$ 32 keV, 5E14 /cm^2

c) $^{119}Sn^+$ 40 keV, 5E14 /cm^2
WBDF **g** = {220}

d) $^{207}Pb^+$ 45 keV, 5E14 /cm^2

Figures 1a-d. Plan-view transmission electron micrographs of the damage created by amorphizing implants of increasing atomic mass after annealing at 750 °C for 15 minutes.

As caveat, XTEM analysis of these samples revealed that in the case of the Pb implant not all the damage observed by PTEM was confined to the EOR, but a noticeable portion was littered across the regrown amorphous region. The existence of defects in the regrown amorphous region coupled with the negligible solubility of Pb in Si [16] suggests the existence of Pb precipitates. For the Si, Ge and Sn implants the damage observed by XTEM was predominately confined to the EOR.

Trapped interstitial concentration

EOR defects observed in the PTEM micrographs consist of {311}'s (rod-like) and dislocation loops. Both defect morphologies are composed of interstitials resulting from the damage introduced by the amorphizing implant. The effect of amorphizing implant mass on the concentration of interstitials trapped in the EOR defects is shown in figure 2 (annealed at 750 °C for 15 minutes). The bar graph quantitatively shows that increasing the atomic mass of the amorphizing implant species decreases the number of interstitials trapped in both {311}'s and loops.

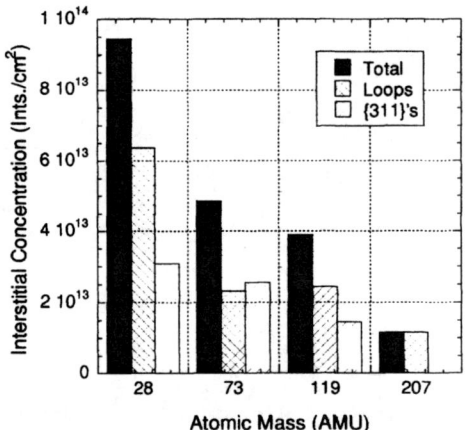

Figure 2. Stack bar graph of the concentration of interstitials trapped in the EOR defects as a function of implant atomic mass after annealing at 750 °C for 15 minutes.

Assuming that TED [17] is dominated by the release of interstitials from the EOR damage [7], figure 2 implies that since the total number of interstitials trapped in the EOR damage decreases with increasing atomic mass so should the diffusion enhancement.

Diffusivity enhancement

The time average diffusivity enhancement experienced by the boron marker layer due to damage introduced by preamorphizing implants of $^{28}Si^+$, $^{73}Ge^+$, $^{119}Sn^+$ and $^{207}Pb^+$ is shown in figure 3, after annealing at 750 °C for 15 minutes. In agreement with the mass effect on the concentration of interstitials trapped in the EOR, figure 3 clearly shows that increasing the atomic mass of the amorphizing implant reduces the $<D_B>/D_B^*$.

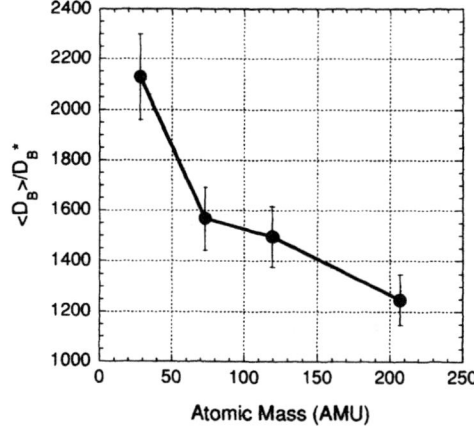

Figure 3. Time average diffusion enhancement as a function of atomic mass of the preamorphizing ion for a 15 minute anneal at 750 °C.

CONCLUSIONS

This study shows that changing the atomic mass of the preamorphizing species affects the EOR damage and the subsequent diffusion enhancement. Specifically, increasing the atomic mass of the preamorphizing ion reduces the size and density of the resulting EOR defects, which lowers the trapped interstitial concentration. Decreasing the trapped interstitial concentration limits the number of interstitials that can contribute to TED, thus reducing the resulting diffusion enhancement. In applications where enhanced diffusion of the implanted dopant profile can not be tolerated, increasing the atomic mass of the preamorphizing species offers a lower diffusion enhancement alternative.

REFERENCES

1. SIA, "International Technology Roadmap for Semiconductors," Semiconductor Industry Assoc., San Jose, CA 1999.
2. T. M. Liu and W. G. Oldham, "Channeling Effect of Low Energy Boron Implant in (100) Silicon," *IEEE Electron Device Lett.*, vol. EDL-4, pp. 59-62, 1983.

3. R. B. Simonton, "The Use of Self-implanted Amorphized Silicon Substrates to Eliminate Channeling Effects in Low Energy Boron Implants," *Nucl. Inst. and Methods in Phys. Res. B*, vol. 21, pp. 490-492, 1987.

4. M. Delfino, D. k. Sadana, and A. E. Morgan, "Shallow Junction Formation by Preamorphization with Tin Implantation," *Appl. Phys. Lett.*, vol. 49, pp. 575-577, 1986.

5. S. N. Hong, G. A. Ruggles, J. J. Wortman, and M. C. Ozturk, "Material and Electrical Properties of Ultra-Shallow p^+-n Junctions formed by Low-Energy Ion Implantation and Rapid Thermal Annealing," *IEEE Transactions on Electron Devices*, vol. 38, pp. 476-486, 1991.

6. D. Eaglesham, P. Stolk, J.-Y. Cheng, H.-J. Gossmann, T. Haynes, and J. Poate, "{311} Defects in Ion-implanted Si: the cause of Transient Diffusion and a Mechanism for Dislocated Formation," presented at Microsc. Semicond. Mater. Conf., Oxford, 1995.

7. A. Claverie, L. Laanab, C. Bonafos, and C. Bergaud, "On the Relation Between Dopant Anomalous Diffusion in Si and End-of-Range Defects," *Nuclear Instruments and Methods in Physics Research B*, vol. 96, pp. 202-209, 1995.

8. K. S. Jones, L. H. Zhang, V. Krishnamoorthy, M. Law, D. S. Simmons, P. Chi, L. Rubin, and R. G. Elliman, "Diffusion of Ion Implanted Boron in Preamorphized Silicon," *Appl. Phys. Lett.*, vol. 68, pp. 2672-2674, 1996.

9. M. Y. Tsai and B. G. Streetman, "Recrystallization of Implanted Amorphous Silicon Layers. I. Electrical Properties of Silicon Implanted with BF_2^+ or $Si^+ + B^+$," *J. Appl. Phys.*, vol. 50, pp. 183-187, 1979.

10. A. C. Ajmera and G. A. Rozgonyi, "Elimination of End-of-Range and Mask Edge Lateral Damage in Ge^+ Preamorphized, B^+ Implanted Si," *Appl. Phys. Lett.*, vol. 49, pp. 1269-1271, 1986.

11. H. S. Chao, S. W. Crowder, P. B. Griffin, and J. D. Plummer, "Species and Dose Dependence of Ion Implantation Damage Induced Transient Enhanced Diffusion," *J. Appl. Phys.*, vol. 79, pp. 2352-2363, 1996.

12. H.-J. Gossmann, G. H. Gilmer, C. S. Rafferty, F. C. Unterwald, T. Boone, and J. M. Poate, "Determination of Si Self-interstitial Diffusivities fro the Oxidation-enhanced Diffusion in B Doping-superlattices: The Influence of the Marker Layer," *J. Appl. Phys.*, vol. 77, pp. 1948-1951, 1995.

13. J. F. Ziegler, J. P. Biersack, and U. Littmark, *The Stopping and Range of Ions in Solids*. New York: Pergamon Press, 1999.

14. S. Bharatan, "Transmission Electron Microscopy (TEM)," in *Material and Process Characterization of Ion Implantation*, M. I. Current and C. B. Yarling, Eds.: Ion Beam Press, 1997, pp. 224-243.

15. M. E. Law, G. H. Gilmer, and M. Jaraiz, "Simulation of Defects and Diffusion Phenomena in Silicon," in *MRS Bulletin*, 2000, pp. 45-50.

16. R. W. Olesinski and G. J. Abbaschian, "The Si-Pb Binary," *Metal Progress*, pp. 55,57, 1985.

17. P. A. Packan and J. D. Plummer, "Transient Diffusion of Low-concentration B Due to ^{29}Si Implantation Damage," *Appl. Phys. Lett.*, vol. 56, pp. 1787-1789, 1990.

Mat. Res. Soc. Symp. Proc. Vol. 669 © 2001 Materials Research Society

Use of isotopically pure silicon material to estimate silicon diffusivity in silicon dioxide

D. Tsoukalas, C. Tsamis and P. Normand
Institute of Microelectronics, NCSR 'Demokritos'
15310 Aghia Paraskevi, Greece

ABSTRACT

In this paper we report measurement of the silicon diffusion coefficient in silicon dioxide films using isotopically enriched ^{28}Si silicon dioxide layers. ^{30}Si atoms are introduced in excess in a stoichiometric isotopically pure silicon dioxide layer either by ion implantation or by a predeposition technique and the time evolution of the ^{30}Si concentration profile under various thermal conditions is monitored using SIMS. The estimated diffusivity values are significantly higher than previously reported values for Si diffusion within a stoichiometric oxide and closer to reported values for excess Si diffusion within an oxide.

INTRODUCTION

Diffusion of silicon atoms in silicon dioxide has attracted the research interest for many years because of the technological importance of silicon dioxide and the silicon/oxide interface on device performance. Moreover, knowledge of diffusion mechanisms of silicon atoms within the oxide are very important for recent experimental methods for the formation of nanocrystals within the oxide for microelectronic and optoelectronics application.

Various experiments, both direct and indirect, have been performed in the past in order to estimate silicon diffusivity in SiO_2. Early experiments was performed by Brebec et al.[1]. These investigators have used SiO_2 layers enriched in ^{30}Si deposited on silicon dioxide that contained ^{30}Si in its natural abundance percentage. The diffused Si was measured directly by Secondary Ion Mass Spectroscopy (SIMS) for concentrations ranging from 1.5 10^{21} cm^{-3} and above.

Other experiments have been reported by Nesbit [2] who extracted Si diffusivity after simulating the formation of Si precipitates in Si rich SiO_2 deposited by Chemical Vapor Deposition and subsequently annealed, by Celler et al [3] who have used impurity thermo-migration experiments and by Tsoukalas et al [4] who have used extended defects formed in bulk Si to monitor the diffusion of Si in a thermally grown silicon dioxide layer after generation of interstitial Si atoms at a top Si layer. The diffusivity values estimated from these three indirect experiments [2-4] are about four orders of magnitude higher than the direct experiments of Brebec et al [4]. This discrepancy has been attributed [2,3] to the diffusion of SiO molecules rather than of Si atoms in the SiO_2 layer.

In this work we report results using isotopically enriched ^{28}Si to form an isotopically enriched ^{28}Si silicon dioxide layer after thermal oxidation where ^{30}Si atoms are introduced either by ion implantation or by predeposition techniques. The approach followed here allows for a direct estimation of diffusivity by analysis of SIMS profiles of ^{30}Si atoms with a concentration limit as low as 10^{19} cm^{-3}

EXPERIMENTAL

The starting material used for the current experiments was a 5 μm n-type isotopically pure silicon layer that was grown epitaxialy on top of a float zone Si wafer. The wafer was obtained from a commercial vendor (Isonics Corporation, San Jose). The concentration of ^{30}Si, ^{29}Si and ^{28}Si isotopes in the epitaxial film are 10^{18} cm^{-3}, $3.4 \ 10^{19}$ cm^{-3}, $4.9 \ 10^{22}$ cm^{-3} respectively as compared with the $1.5 \ 10^{21}$ cm^{-3}, $2.3 \ 10^{21}$ cm^{-3}, $4.5 \ 10^{22}$ cm^{-3} corresponding values measured in normal Si wafers. Carbon concentration in the epitaxial films was lower than 10^{16} cm^{-3} and oxygen concentration was $2.2 \ 10^{17}$ cm^{-3}. Two set of experiments were performed, corresponding to two different methods to introduce extra silicon atoms in the oxide, using either ion implantation or predeposition.

In the first set of experiments, the isotopically pure Si wafer was oxidized in dry O$_2$ at 1100 °C to form a 180 nm thick SiO$_2$ layer and then cut in samples. Some of them were implanted with ^{30}Si at 50 keV to a dose of $2 \ 10^{15}$ cm^{-2} dose and some others were used for the predeposition process that we shall describe later. For the selected implantation conditions there is no formation of nanocrystals within the oxide, that could complicate the analysis of our experiments. After implantation the samples were covered with a thin (30 nm) Si$_3$N$_4$ layer deposited by LPCVD and subsequently annealed in inert ambient for different times at temperatures 1050 °C, 1100 °C, 1125 °C and 1150 °C. Distribution profiles were measured with SIMS using a commercially available service (Charles Evans & Ass.).

Fig.1a shows a representative SIMS profile that depicts the distribution of all three Si isotopes and oxygen throughout the structure, while figure 1b shows the results of experiments at 1100 °C. The duration of the thermal treatments performed in N$_2$ were 1h, 3, 6 and 12 days.

From figure 1a -as well as from other similar ones not presented here- it is evident that for the experimental conditions used there is no ^{30}Si diffusion taking place from the Si$_3$N$_4$ layer to the

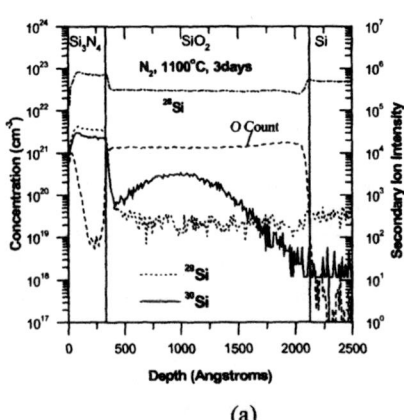

(a)

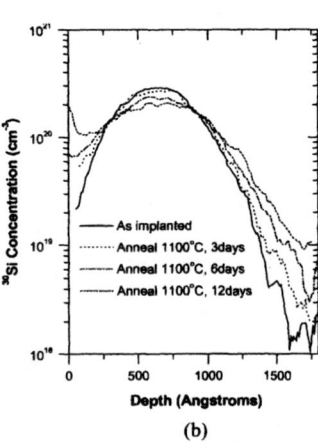

(b)

Figure 1. (a) Representative SIMS profile of a structure composed of 30nm Si$_3$N$_4$ and 180 nm isotopically pure SiO$_2$ thermally grown on an isotopically enriched ^{28}Si substrate after annealing at 1100°C for 3 days. (b) SIMS profile of ^{30}Si for as implanted samples and after annealing at 1100°C for various times.

oxide layer. This is expected due to the strength of the Si-N bonds. After 1h annealing (not shown in the figure) no effect of any transient enhanced diffusion due to ion implantation damage was observed. Experiments performed at 1050 °C for 30 days, 1125 °C for 6 days and 1150 °C for 3 days allow the estimation of the diffusivity as a function of temperature.

For the second set of experiments a predeposition technique was used to study the diffusion of ^{30}Si in the isotopically pure SiO_2 layer. Silicon Wafer Bonding technique was used to bond a commercial SIMOX wafer on top of isotopically pure SiO_2 films grown as described previously. Details of the process have been previously reported in the literature [4]. The final structure, shown in figure 2a, consists of a 200 nm Si crystalline layer that contains the Si isotopes in their natural abundance. The samples were subsequently oxidized in N_2O for 6h and 11 h at 1100 °C to generate interstitial Si atoms in the top Si film. The use of N_2O was necessary to avoid consumption of the silicon layer by the thermally grown oxide as one expects if O_2 is used instead. The resulting profiles of ^{30}Si were monitored by SIMS in the isotopically pure SiO_2 layer. The experimental as well as simulated profiles are shown in fig. 2b. On the same figure a reference sample -that has not seen the oxidation process- is also shown. Alternatively, in one experiment we have deposited by LPCVD a 130 nm polysilicon layer on top of the isotopically pure oxide and then oxidized this layer for 6h. The resulting profile was exactly the same as the one obtained for an isotopically pure SiO_2 layer with crystalline Si on top. We have observed also that the polysilicon surface becomes rough after oxidation and a roughness of 10 nm (rms) was measured by Atomic Force Microscopy. After removal of the polysilicon layer by selective etching the surface roughness measured at the oxide surface was at 4 nm. Both these numbers are higher as compared with what we measured using crystalline Si but evidently they do not affect the diffusion of Si atoms from the polysilicon into the oxide. For the predeposition experiments, accurate SIMS measurements required the selective removal of the Si or polysilicon top layer. This was performed by using an Ethylenediamene-Pyrocatechol (EDP) solution prior to the measurements. This solution etches silicon dioxide at a very low etch rate (less than 0.5 nm/min). If SIMS measurements are performed without removing this Si layer there is still clear evidence of a ^{30}Si diffusion process taking place. However in that case we were not able to correctly

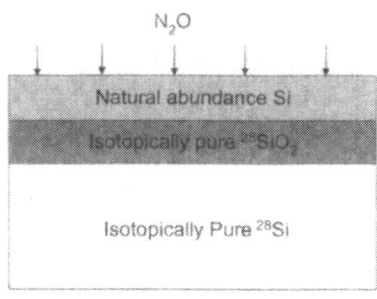
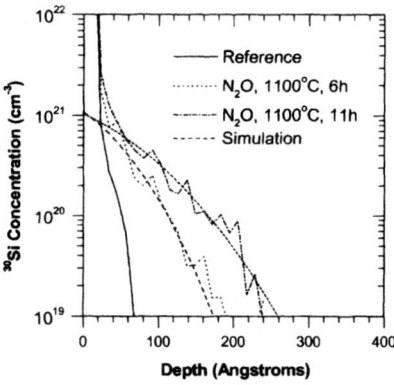

Figure 2. (a) Experimental test structure for the predeposition experiments. (b) Experimental and simulated results of predeposition experiments.

estimate its diffusivity in the oxide because the profile of ^{30}Si is much more deeper due to noise from the Si top layer.

DISCUSSION

For the estimation of the diffusivity values, commercially available software was used (Athena-Silvaco). SIMS measurements of the as-implanted samples were used as the initial profile for all the simulations at all temperatures. The diffusivity within the oxide layer was extracted after simulating the ^{30}Si profile evolution assuming a constant diffusivity value. Especially for the thermal anneals that were performed at 1100°C for various times, ^{30}Si diffusivity was extracted independently for each time and an average diffusivity value was calculated. However, since the diffusivities that were extracted for each time did not differ significantly, this mean value gives very good agreement between the simulated and the experimental profiles for all times.
In figure 3 we show the experimental and simulated results for various temperatures and times. We can see that there is very good agreement between the experimental and the simulated profiles.

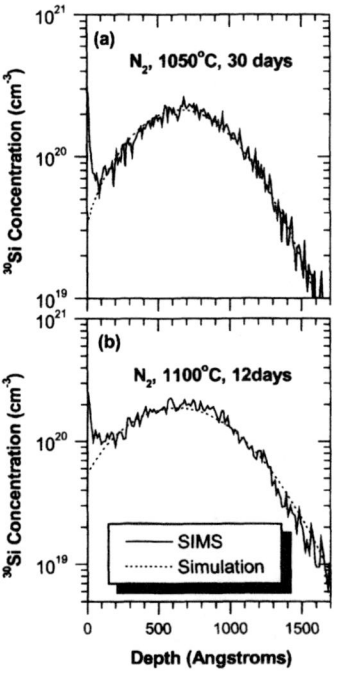
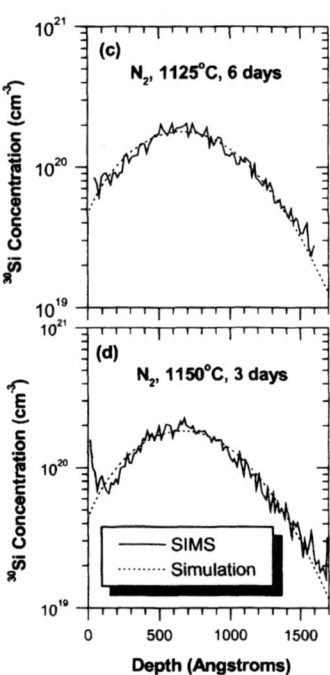

Figure 3. Comparison between experimental and simulation results for various times and temperatures for the implantation experiements.

For the estimation of the diffusivity from the predeposition experiments, simulations were performed by assuming a constant surface value of the ^{30}Si on the surface of the isotopically pure oxide and a constant diffusivity value The simulation shows evidence of a complementary error function (erfc) behavior of ^{30}Si profiles resulting from ^{30}Si diffusion from the top layer in the isotopically pure SiO$_2$ layer A diffusivity value of ~10^{-17} cm^2 sec^{-1} was obtained at 1100°C for all diffusion times(Fig. 2b).

In figure 4 the estimated diffusivity is shown as a function of temperature, for all experiments. In the same figure experimental results from previous works are also included. The diffusivity obtained from the implantation experiments exhibits an Arrhenius type dependence on temperature with an activation energy of 4.74 eV. The data can be best fitted to the expression $D_{Si} = 1.378\ exp\ (-4.74\ eV\ /\ kT)\ cm^2 s^{-1}$. It is worthwhile to remark that relatively small changes in diffusivity lead to an error bar of ±0.25 eV of the estimated activation energy of 4.74 eV.

As we have reported in our introduction direct studies of silicon diffusivity in SiO$_2$ are rare in the literature due to the difficulty of the experiments. The only direct measurement we found is the one performed by Brebec et al [1] who studied self-diffusion of Si in a vitreous oxide and another one by Jaoul et al [5] who have made self-diffusion studies in quartz. Both these research groups performed their diffusion studies using predeposition like experiments from a deposited SiO$_2$ layer that was enriched in ^{30}Si on top of an underlying SiO$_2$ where ^{30}Si was found in its natural abundance and both report much lower values of Si diffusivity than the present work. The significantly higher diffusivity values (~300 times higher on the average) that are reported within this work as compared with those measurements are attributed to the fact that in our experiments silicon is introduced in excess in a stoichiometric oxide, so its diffusion in the SiO$_2$ proceeds most likely through a different mechanism as compared with the case of the experiments performed in [1] and [5].

Our results are closer (~about 50 times lower on the average) to other existing in the literature (ref. 2-4) where Si diffusivity in SiO$_2$ is estimated using indirect techniques –where Si is provided in excess into the SiO$_2$ layer- and the higher measured values of diffusivity compared

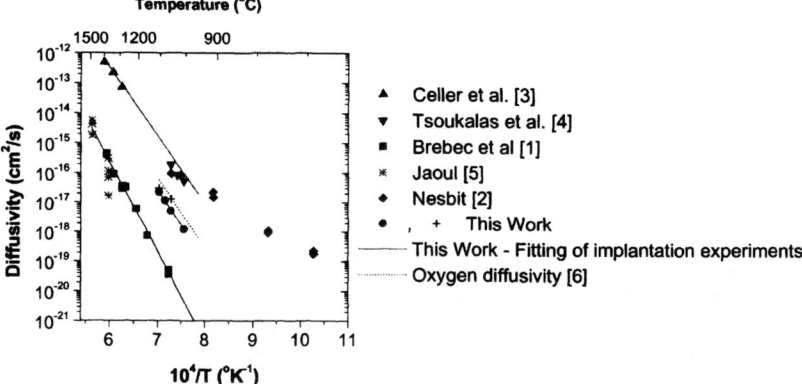

Figure 4. Si diffusivity in silicon dioxide as a function of temperature as estimated from the two types of experiments (^{30}Si implantation or predeposition) as well as data from literature.

with the data by Brebec et al [1] is attributed to SiO diffusion within the oxide. These species are produced from the reaction of Si with SiO_2. It is also worth to notice that our measurements are close to the oxygen diffusivity data in SiO_2 (with an estimated activation energy of 4.7 eV and a pre-exponential term equal to 2.6) as reported by Mikkelsen in his work on 'network oxygen diffusion' [6]. As remarked by this investigator Si and O diffusion in SiO_2 could be correlated through the formation or motion of several network related atoms involved in their diffusion process.

It is possible that diffusivity of silicon in the oxide depends on the oxide formation process (deposition or thermal oxidation) and the method that the Si atoms are introduced in it. For instance, depending on the oxide deposition or growth, the vacancy concentration in the oxide may vary, which in turn would affect the silicon diffusivity. Moreover, during ion implantation the implanted atoms can either replace existing network silicon atoms or/and create vacancies that can trap the diffusing atoms. These effects as well as the possible presence of elements such as hydrogen may result in lower diffusivities that those reported in refs. 2-4. The differences obtained for the extracted diffusivity values between the implanted and the oxidized samples, might indicate this influence, although this difference is not so significant. More experiments are necessary in order to clarify this point.

CONCLUSIONS

In summary, we have presented direct measurements of silicon diffusivity in isotopically pure SiO_2 layers using either ion implantation and annealing studies or predeposition-oxidation experiments. The possibility that the diffusing species is the SiO molecule cannot be excluded by our experiments. The activation energy estimated is 4.74 eV very close to previous data on oxygen network diffusion [6]. The predeposition-oxidation experiments might also prove a versatile tool to probe the recombination properties of interstitials arriving at the Si/SiO_2 interface allowing for quantitative studies to be undertaken.

ACKNOWLEDGMENTS

The authors acknowledge Dr. P. Morfouli for her assistance with AFM measurements, Dr. D. Skarlatos for his help with oxidations and Prof. Y. Stoemenos for enlightening discussions. The work is partially supported by the ESPRIT project RAPID (no 23481) of EU.

REFERENCES

[1] G. Brebec, R. Seguin, C. Sella, J. Bevenot, J.C. Martin, Acta Metallurgica **28**, 327, (1980)
[2] L. A. Nesbit, Appl. Phys. Lett. **46**, 38 (1985)
[3] G. K. Celler and L. E. Trimble, Appl. Phys. Lett. **54**, 1427 (1989)
[4] D. Tsoukalas, C. Tsamis and J. Stoemenos, Appl. Phys. Lett. **63**, 3169 (1993)
[5] O. Jaoul, F. Bejina, F. Elie, F. Abel, Phys. Rev. Lett. **74**, 2038 (1995)
[6] J. C. Mikkelsen Jr., Appl. Phys. Lett. **45**, 1187 (1984)

Mat. Res. Soc. Symp. Proc. Vol. 669 © 2001 Materials Research Society

Phosphorus diffusion in silicon; influence of annealing conditions

J. S. Christensen[1], A. Yu. Kuznetsov[1], H. H. Radamson[1], and B. G. Svensson[1,2]
[1]Royal Institute of Technology (KTH), Department of Electronics, Electrum 229, SE-164 40 Kista-Stockholm, Sweden
[2]University of Oslo, Physics Department/Physical Electronics, P. B. 1048 Blindern, N-0316 Oslo, Norway

ABSTRACT

Phosphorus diffusion has been studied in both pure epitaxially grown silicon and Cz silicon, with a substantial amount of impurities like oxygen and carbon. Anneals have been performed in different atmospheres, N_2 and dry O_2, as well as in vacuum, at temperatures between $810 - 1100$ °C. Diffusion coefficients extracted from these anneals show no difference for the P diffusion in the epitaxially grown or the Cz silicon. The diffusion coefficients follow an Arrhenius dependence with the activation energy $E_a = 2.74 \pm 0.07$ eV and a prefactor $D_0 = (8 \pm 5) \times 10^{-4}$ cm^2/s. These parameters differ considerably from the previously reported and widely accepted values (3.66 eV and 3.84 cm^2/s, respectively). However, vacuum anneals of the same samples result in values close to this 3.6 eV diffusion mode. Furthermore, control anneals of boron doped samples, with similar design as the phosphorus samples, suggest the same trend for boron diffusion in silicon – lower versus higher values of activation energies for nitrogen and vacuum anneals, respectively. These results are discussed in terms of the concentration of Si self-interstitials mediating the diffusion of phosphorus and boron.

INTRODUCTION

The scaling of electronic devices down to nanometer-technology modes makes it important to control how the dopant atoms redistribute with an accuracy of a few monolayers during device manufacturing. Phosphorus is one of the most commonly used impurities, and its diffusion in silicon has been extensively studied. A lot of experiments concerning phosphorus diffusion have been performed during the 70's and 80's [1-5], and these experiments showed that the diffusion of phosphorus in Si, at concentrations lower than 10^{19} P/cm^3, is well described by a diffusion coefficient with an activation energy $E_{old}^P \sim 3.6$ eV[1-5]. However, more recent results for P diffusion in Si yield an activation energy $E_{new}^P \sim 2.8$ eV [6,7]. The difference between E_{old}^P and E_{new}^P is huge (~0.8 eV) and should be studied and interpreted accordingly.

There can of course be several sources for this discrepancy. For example: (i) experimental errors during either "new" or "old" measurements. (ii) different types of silicon used for the different measurements (orientation, phosphorus and other impurity concentrations , etc.), (iii) different or uncontrollable atmospheres during anneals.

In the present paper we have addressed a part of the issues above and the intrinsic phosphorus diffusion have been studied in both pure epitaxially grown silicon and ordinary Cz silicon with oxygen and carbon concentrations in the 10^{17} and 10^{16} cm^{-3} range, respectively. Anneals have been performed in different atmospheres, N_2 and dry O_2, as well as in vacuum. Nitrogen anneals of both epitaxial and Cz samples performed over the range of $810 - 1100$ °C confirm the existence of the 2.8 eV diffusion mode for phosphorus in Si. Surprisingly, the vacuum anneals of the same samples result in values close to those described by the 3.6 eV diffusion mode. Interestingly, control anneals of similarly designed boron doped samples suggest a similar trend for boron diffusion in silicon – lower versus higher values of activation energies for nitrogen and vacuum anneals, respectively.

EXPERIMENTAL

The samples used were manufactured by low pressure chemical vapor deposition (LPCVD) in an Epsilon 2000 ASM CVD reactor. Three types of samples (referred to below as P1, P2, and B1) were used in this work. Sample P1 was used to study P diffusion in pure epitaxial Si and was fabricated as follows. Firstly, an intrinsic silicon buffer layer 3000 Å thick was grown on a Cz silicon (100) n-type substrate. Then a 3000 Å thick Si-layer having a P-concentration of $\sim 10^{18}$ P/cm^3 was deposited. Finally, a 2000 Å cap-layer of intrinsic (undoped) silicon was deposited. A typical as-deposited phosphorus profile for this sample is shown in figure 1. (solid line) Sample P2 was used to study P diffusion in Cz silicon and was prepared similar to that of P1, but without a CVD grown buffer layer, so that the P atoms diffuse directly into the Cz silicon substrate. Sample B1, with a structure similar to P1 but doped with boron instead of phosphorus, was used to compare the P diffusion with B diffusion for similar annealing conditions.

Anneals were performed in two different furnaces. One furnace is a tube furnace with a flow of either N_2 or O_2. In this furnace the temperature is measured by thermocouples inside the tube, and both the temperature and ambient are monitored during annealing. The temperature is controlled with an accuracy of ± 2 °C. Prior to loading into the furnace at ~ 600°C, the samples were cleaned in diluted HF in order to remove the native SiO_2 layer. The furnace was then heated to a desired temperature. At the end of the anneal the furnace was cooled down to ~ 600 °C (keeping the N_2-flow) and the samples were removed. Eight anneals of the P1 samples were performed in N_2 atmosphere in temperature range of 810 °C – 1100 °C. The duration of anneals varied between 50 hours and 30 minutes for the 810 °C and 1100 °C anneals, respectively. A control anneal of the sample P1 was done in an O_2-ambient at 810 °C. Samples P2 and B1 were

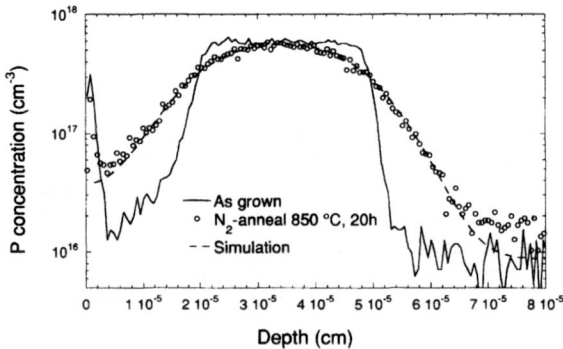

Figure 1. Typical SIMS profiles of sample P1 before (solid line) and after anneal at 850 °C for 20 hours (o). A simulation of the phosphorus diffusion, using $D^P = 4.2 \times 10^{-16}$ cm^2/s is also included. (dashed line).

annealed in this furnace in N_2 flow at four different temperatures in the range of 810°C – 900°C. Another part of the anneals has been performed a vacuum furnace operating at ~10^{-6} Torr. In this case the samples were introduced through a lock-chamber into the furnace kept at the desired temperature. The accuracy of the temperature measurement is somewhat worse in this furnace, ±10 °C, and the composition of residual gases at the pressure of 10^{-6} Torr is not known. All three types of samples were annealed in the vacuum furnace at 850 °C, 900 °C, and 950 °C. For all samples the SiO_2 thickness was measured by spectroscopic ellipsometry before and after annealing to determine if any oxidation took place.

Dopant concentration versus depth profiles were analyzed in as-deposited and annealed samples by secondary ion mass spectrometry (SIMS), using a 13.5 keV Cs^+ beam and counting the secondary $^{31}P^-$ ions or a 8 keV O_2^+ beam and counting the secondary $^{11}B^+$ ions. Time to depth conversion was done using the crater depths measurements obtained with an Alphastep profilometer, and the calibration of concentration was obtained using a reference silicon sample having a known phosphorus concentration. The effect of cascade mixing caused by Cs^+ ion damage was estimated from the abruptness of the P-profile in the as-grown sample, and it is concluded to be negligible relative to the broadening caused by diffusion.

Phosphorus diffusivities were obtained by fitting numerical solutions of Fick's equation to the diffused profiles measured by SIMS. The as-grown profile was used as the initial condition and the diffusivity was treated as a free parameter. A typical SIMS profile after annealing of sample P1 as well as a simulated fit are shown in figure 1. For the anneals at temperatures above 900 °C, where the diffusion time becomes comparable with the furnace heating and cooling times, the temperature ramp was included in the simulations.

RESULTS

Figure 2 shows an Arrhenius plot which summarizes our P diffusion measurements for samples P1 and P2 annealed under different conditions (see the caption in figure 2). Firstly, the absolute values for the P diffusivities in samples P1 and P2 annealed in N_2 atmosphere are almost identical (see the overlap between circles and crosses in figure 2). A fit of these data to a straight line gives the prefactor $D_0 = (8\pm5)\times10^{-4}$ cm^2/s and the activation energy $E_{new}^P = 2.74\pm0.07$ eV. This value for E_{new}^P is consistent with those recently reported by Haddara et al. [6] ($E_a = 2.81$ eV) and by Fage-Pedersen [7] ($E_a = 2.87$ eV), but ~1 eV lower than the "classical" values for E_{old}^P ~3.6 eV reported, for example, by Fahey et al. [8] and shown as a dashed line in figure 2. One annealing of P1 in dry O_2 atmosphere resulted in an enhancement of the P diffusion by a factor of 6 as compared to the N_2 anneals (compare the diffusivities at 810 °C in figure 2). This intentional oxidation resulted in the formation of a 150 Å thick SiO_2 layer. For the N_2-anneals an oxide less than 20 Å was detected. Assuming that the formation of an oxide gives rise to a steady-state injection of interstitials, and comparing the results of the anneals in O_2 and N_2 at 810 °C, the enhancement of the diffusivity in the case of the N_2-anneal, compared to the case where no interstitials were injected, is estimated to be 6 %, which is within the uncertainties of the diffusivities.

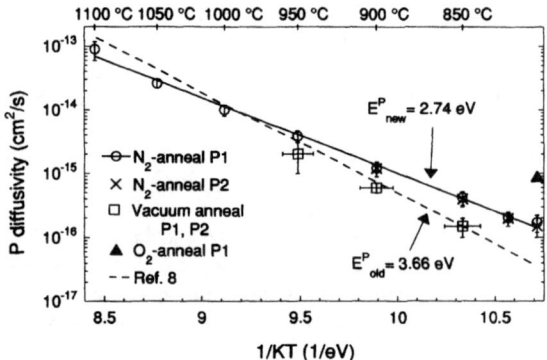

Figure 2. Arrhenius plot for P diffusion. N_2-annealing results for P1 (o) and P2 (×) coincide and correspond to $E^P_{new} = 2.74$ eV (solid line). Results from vacuum anneal ($\square$), however, correspond to $E^P_{old} = 3.66$ eV reported in ref. 8 (broken line). Finally, annealing in O_2 atmosphere at 810 °C ($\blacktriangle$) shows enhanced diffusion.

Surprisingly, the result for the vacuum anneals using the same kind of samples are different from those obtained for the N_2 anneals and consistent with the classical parameters for P diffusion in Si (see how the squares agree with the dashed line in figure 2). In this case we were able to detect formation of an oxide (50-200 Å thick, depending on the annealing temperature).

Since P and B are considered to be governed by the same, interstitialcy, diffusion mechanism in Si [8] we have also performed both N_2 and vacuum anneals for the sample B1, and figure 3 summarizes the diffusivities extracted from this measurement. Interestingly, anneals in N_2 atmosphere give higher B diffusivities than those obtained after vacuum annealing, indicating the same trend as for phosphorus. Moreover, the vacuum anneal data are again in close accordance

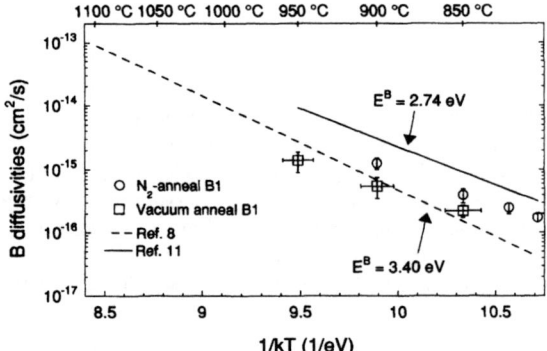

Figure 3. Arrhenius plot for B diffusion. The N_2 anneals (o) result in a higher diffusivity than the vacuum anneals ($\square$). The vacuum anneals yield values in correspondence with $E^B = 3.40$ eV (broken line), reported in ref. 8. However, more recent new boron diffusivities are reported in ref 11 with $E^B = 2.74$ eV (solid line).

with the values given by Fahey *et al.* [8] (dashed lines in figure. 3). The N_2 annealing results are, on the other hand, somewhat closer to those recently reported by Zangenberg *et al.* [9]

DISCUSSION AND CONCLUSIONS

As mentioned in the introduction different reasons can be put forward for the discrepancy between E_{new}^P and E_{old}^P, these will be discussed below. Firstly, we exclude any systematic error extracting either E_{new}^P or E_{old}^P, for instance the measurements of P diffusion in Si samples annealed in N_2 atmosphere (giving E_{new}^P =2.74 eV) were performed over a wide temperature range so that the diffusivity changes almost over three orders of magnitude, which substantiates the reliability of the measurements. We also presently believe that "old" measurement were accurate, though Haddara *et al.* [6] pointed out that the presently used SIMS profiling technique is superior compared with previously employed spreading resistance [1,3,5], radiochemical [1,2], and marker measurements [4]. However, these "old" experiments have been designed in accordance with the limitations of the measurement techniques used. For example, a poorer depth resolution was normally compensated by longer diffusion times, and the data collected by different techniques are in a good agreement with each other. However if both E_{new}^P and E_{old}^P are true observations then the question is – why are two activation modes observed at 810 °C – 1100 °C?

It is known that phosphorus diffusion is mediated by Si self-interstitials (I's), which implies that the P diffusion depends on the I-concentration [8]. The concentration of I's is a function of temperature, annealing conditions, dopant concentration and presence of other impurities, which act as traps/generation centers for the I's. Consequently, the discrepancy between E_{new}^P and E_{old}^P may be due to a different concentration of I's during "old" and "new" measurements. For instance carbon is known to trap interstitials, and influence the diffusion of dopants, governed by the interstitialcy mechanism [10]. However, the comparison of P diffusion in pure epitaxial Si (sample P1) and Cz Si, with oxygen and carbon concentrations in the 10^{16} - 10^{17} cm^{-3} range (sample P2) does not reveal any difference for P diffusivities in these samples.

Instead we have observed a strong influence of the annealing conditions on the phosphorus diffusion, annealing in either N_2-atmosphere or vacuum results in different values of P diffusivities. Moreover, the results for the N_2- and vacuum anneals are consistent with 2.8 eV and 3.6 eV activation energies for P diffusion, respectively. However, some of the "old" measurements were performed in vacuum while others were done in nitrogen, so our observations are not fully consistent with the host of the "old" data. In addition, we presently do not understand why the results of the N_2- and vacuum anneals are different. Considering only the difference in pressure, assuming the same pressure dependence for P as for B, given in ref. 11, we estimate that the effect of external N_2-pressure is negligible, $<10^{-6}$ eV.

More realistic is to consider a difference in point defect balance caused by chemical reactions at the sample surface during annealing. As mentioned above, unintentional oxidation took place during the vacuum anneals. Consequently, enhanced P diffusion may be expected for this case. However, the absolute P diffusivities obtained for the vacuum anneals are below those for the N_2-anneals (where negligible oxidation took place), figure 2. Moreover, intentional oxidation of our phosphorus samples in dry O_2 reveals a strong enhancement of the P diffusion, as expected because of the injection of excess I's [8]. Thus, based on these observations, we conclude that the vacuum anneals are not accompanied by an injection of I's, but rather a reduction of the concentration of I's, at least for the 850 – 950 °C range.

This hypothesis is further supported by the boron results, shown in figure 3. For sample B1 annealed in N_2-ambient the obtained B diffusivities are higher compared to the diffusivities obtained from the same sample annealed in vacuum. Thus, the B diffusivities follow a similar trend on the annealing conditions as the phosphorus diffusion does. Hence, it may be concluded that reactions at the silicon surface play a decisive role for the concentration of point defects mediating diffusion of phosphorus and boron. One such reaction is oxidation-induced injection of I's. It is, however, not the only candidate and further work is in progress to elucidate other mechanisms/reactions.

ACKNOWLEDGEMENTS

Partial financial support of this work was received from the Swedish Foundation for Strategic Research (SSF) and the Nordic Academy for Education and Research Training (NorFA), and the EU commission, contract No. ERBFMRXCT980208 (ENDEASD-THR network). Margareta K. Linnarsson is kindly acknowledged for support on the SIMS-measurements.

REFERENCES

1. R. N. Ghoshtagore, Phys. Rev. B **3**, 389 (1971)
2. J. S. Makris and B. J. Masters, J. Electrochem. Soc. **120**, 1252 (1973)
3. A. M. Lin, D. A. Antoniadis, and R. W. Dutton, J. Electrochem. Soc. **128**, 1131 (1981)
4. C. Hill, in *Semiconductor Silicon 1981*, edited by H. R. Huff, J. R. Kriegler, and Y. Takeishi (Electrochemical Society, New York, 1981) p. 988
5. Y. Ishikawa, Y. Sakina, H. Tanaka, S. Matsumoto, and T. Niimi, J. Electrochem. Soc. **129**, 644 (1982)
6. Y. M. Haddara, B. T. Folmer, M. E. Law, and T. Buyuklimanli, Appl. Phys. Lett **77**, 1976 (2000)
7. J. Fage-Pedersen, PhD. Thesis, university of Aarhus, 2001
8. P. M. Fahey, P. B. Griffin, and J. D. Plummer, Rev. Mod. Phys. **61**, 289 (1989)
9. N. R. Zangenberg, J. Fage-Pedersen, J. Lundsgaard Hansen, and A. Nylandsted Larsen, To be published in Defect Diffus. Forum **194-199** (2001)
10. H. Rücker, B. Heinemann, W. Röpke, R. Kurps, D. Krüger, G. Lippert, and H. J. Osten, Appl. Phys. Lett. **73**, 1682 (1998)
11. Y. Zhao, M. J. Aziz, H.-J. Gossmann, S. Mitha, and D. Schiferl, Appl. Phys. Lett. **74**, 31 (1999)

Mat. Res. Soc. Symp. Proc. Vol. 669 © 2001 Materials Research Society

DIFFUSION OF IMPLANTED NITROGEN IN SILICON AT HIGH DOSES

Lahir Shaik Adam, Lance Robertson, Mark E. Law, Kevin Jones, Kevin Gable, SWAMP Center, Univ. of Florida, Gainesville, FL 32611 USA
Suri Hegde, Omer Dokumaci, SRDC, IBM Corp., East Fishkill, NY 12533

ABSTRACT

Nitrogen implantation is used to retard gate oxide growth thereby making it particularly useful for dual- V_T and System On A chip technologies. This paper discusses the diffusion behavior and the concomitant defect evolution at high doses of implanted nitrogen in silicon. This paper shows that as the nitrogen implant dose is increased, the extent of nitrogen diffusion reduces. This paper also reports based on TEM studies, that upon annealing at 750°C , $5 * 10^{14}$ N_2^+/cm^2, 40 keV implant produces Type I extended defects. However, $2 * 10^{15}$ N_2^+/cm^2, 40 keV implant, produces a continuous amorphous layer to a depth of about 800 to 900 Å from the surface. In addition, upon annealing at 750°C, the $2 * 10^{15}$ N_2^+/cm^2, 40 keV implant produces Type V or solid solubility defects in addition to End of Range or Type II defects.

INTRODUCTION

Nitrogen implantation is used to control gate oxide thickness [1]. Ref 1 shows that upon implantation and subsequent oxidation, nitrogen retards gate oxide growth. Therefore, by varying the dose of the nitrogen implant across the wafer, it is possible to vary the gate oxide thickness across the wafer. This ability to have variable gate oxide thickness across the wafer is of particular importance in dual $-V_T$ and System On a Chip technologies. Previous studies done by us reported the diffusion behavior of implanted nitrogen in silicon at low doses of $5 * 10^{13}$ N_2^+/cm^2, 40 keV [2]. By "low" doses, we mean doses low enough so as not to create extended defects. That study showed that nitrogen diffused towards the surface rapidly with time at 750°C. In this paper, we discuss the diffusion behavior of implanted nitrogen in silicon at doses high enough to create extended defects. Through TEM studies, we also discuss the concomitant extended defect evolution for high doses of nitrogen implants.

DIFFUSION AND DEFECT STUDIES ON $5 * 10^{14}$ N_2^+/cm^2, 40 keV IMPLANTS

Nitrogen at a dose of $5 * 10^{14}$ N_2^+/cm^2 was implanted at 40 keV into <100> Czocharlski silicon through a 50 Å screen oxide. After implantation, the samples were furnace annealed at 750°C for various times. Figure 1 shows the diffusion profiles. As can be observed from Figure 1, the nitrogen profile shrinks with time. To that extent, there is qualitative similarity between the diffusion behavior of nitrogen implanted at the lower dose of $5 * 10^{13}$ N_2^+/cm^2 at 40 keV and the present case. However, we can observe a couple of distinct differences in the diffusion behavior between the lower dose implant of $5 * 10^{13}$ N_2^+/cm^2, 40 keV and the present case.

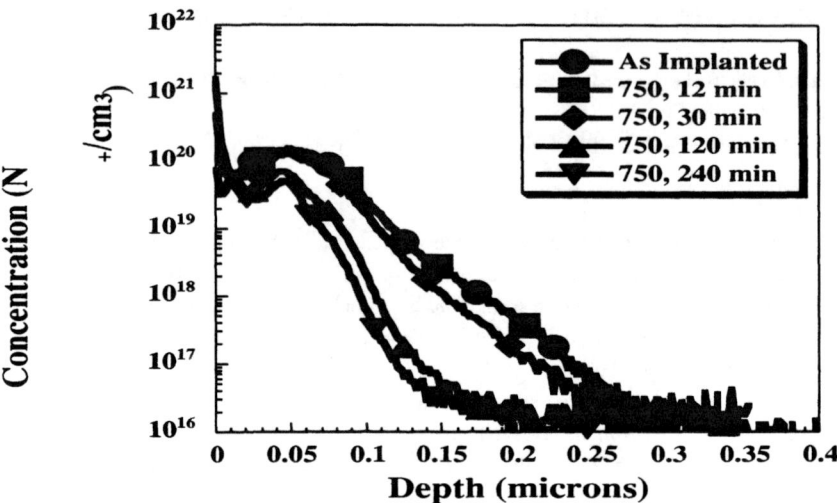

Figure 1: Diffusion of nitrogen implanted into silicon at $5 * 10^{14}$ N_2^+/cm^2, 40 keV

Upon annealing, in the case of the lower dose implant, most of the nitrogen had diffused towards the surface after annealing at 750°C for 120 minutes. However, in the present case, there is still a substantial amount of nitrogen that remains in the bulk even after annealing for 240 minutes at 750°C. Therefore, the motion of the nitrogen profile is drastically reduced in the case of the higher dose implant of $5 * 10^{14}$ N_2^+/cm^2, 40 keV. Furthermore, in the case of the lower dose implant, upon annealing, the nitrogen peaks shifted towards the surface with time. However, in the present case of $5 * 10^{14}$ N_2^+/cm^2, 40 keV, as observed in Figure 1, the peaks are pinned at about the projected range. Both the observations mentioned above indicate that there could possibly be some extended defects that form at the projected range that act as a sink for nitrogen. This could effectively slow down the overall motion of nitrogen towards the surface. Further, using insight from the model in Ref. 2, these extended defects (if they form) can also alter the efficiency of the conversion of nitrogen atoms to mobile nitrogen interstitials. This in turn can affect the gradient of the nitrogen interstitials towards the surface and hence the peaks may not move towards the surface with time.

In order to verify if there are extended defects that form at the projected range, Cross Section Transmission Electron Microscopy (XTEM) was performed. As can be seen from Figure 2, there is a band of extended defects that form at the projected range. This band of defects has been well characterized as Type I defects in the literature [3].

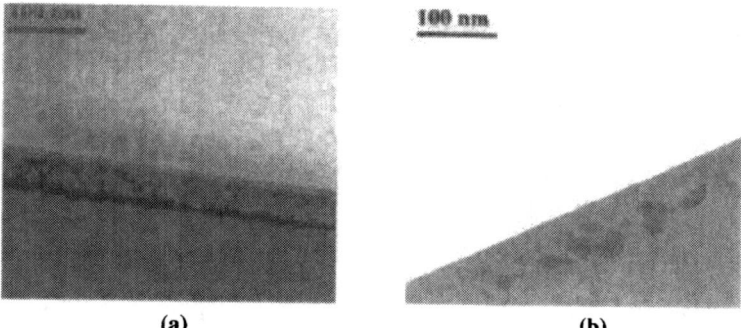

(a) **(b)**

Figure 2: (a) shows the XTEM micrograph of $5 * 10^{14}$ N_2^+/cm^2, 40 keV as-implanted sample. The band of defects at the projected range is clearly visible in (b) which corresponds to the XTEM micrograph of $5 * 10^{14}$ N_2^+/cm^2, 40 keV annealed for 120 min at $750^\circ C$. The imaging axis for (a) is g {220} and for (b) is {110}2A. The image of {110}2A for Figure 2 (b) was chosen for the sake of clarity.

Plan view Transmission Electron Microscopy (PTEM) was performed on the annealed samples. The evolution of the defects with time is as shown in Figure 3. As can be seen from Figure 3, we can see that there is a bimodal distribution of the extended defects in the samples. This is especially observable in the TEM's of the samples annealed for 120 min and 240 min at $750^\circ C$. As the sizes of a good proportion of the extended defects is small and also very close to each other, there is an error in the estimation of both the density of the defects and also in the number of interstitials contained in them. The method employed for counting both the extended defect density and the number of intersititals contained in the defects was as follows: On the PTEM micrographs, the images where we were sure that extended defects formed were first marked and counted. Then the images where we were not sure as to the number of defects in a specific image position or images that were quite dim to identify positively as a defect were then marked. The sum of the "sure" and the "unsure" images then gives an upper bound on the defect density and the number of interstitials contained in them. The difference of the "sure" and "unsure" images then gives a lower bound on the defect density and the number of interstitials contained in them. The density of the Type I defect evolution with time is shown in Figure 4(a) and the density of the interstitials contained in them is shown in Figure 4 (b) along with their respective lower and upper bounds of count errors.

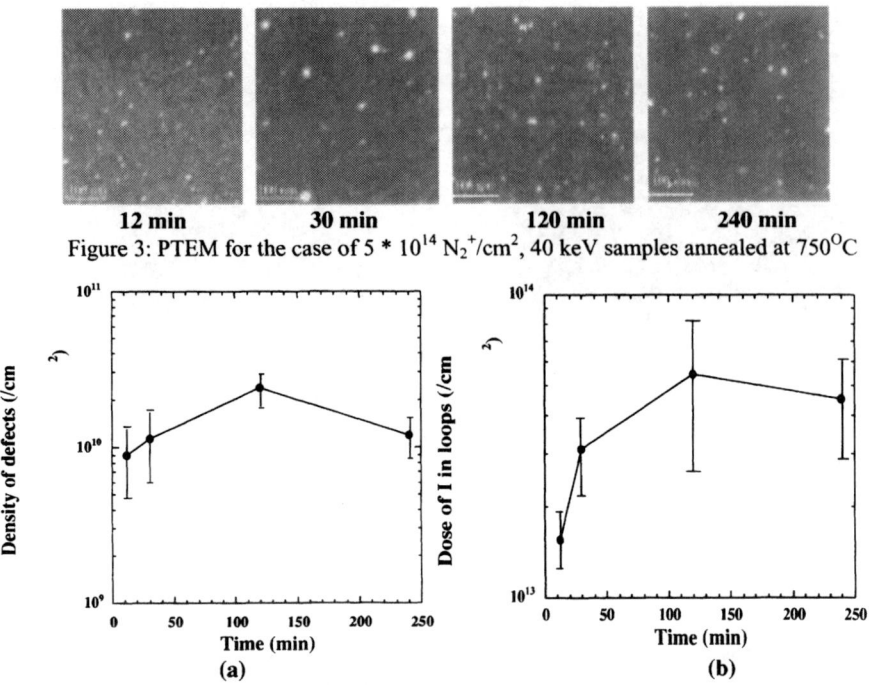

| 12 min | 30 min | 120 min | 240 min |

Figure 3: PTEM for the case of $5 * 10^{14}$ N_2^+/cm^2, 40 keV samples annealed at $750^{\circ}C$

Figure 4: (a) shows the dose of defects with time and (b) shows the dose of I in those defects with time

DIFFUSION AND DEFECT STUDIES ON $2 * 10^{15}$ N_2^+/cm^2, 40 keV

Nitrogen at a dose of $2 * 10^{15}$ N_2^+/cm^2 was implanted at 40 keV into <100> Czocharlski silicon through a 50 Å screen oxide. After implantation, the samples were furnace annealed at $750^{\circ}C$ for various times. The evolution of the nitrogen profile is shown in Figure 5. As seen from Figure 5, there are some similarities between the diffusion behavior of $5 * 10^{14}$ N_2^+/cm^2 and the present case. In the present case, we see that there is accumulation of nitrogen at around 900 Å. This could possibly suggest that we form End of Range (EOR) defects at around that depth as these EOR defects can trap nitrogen. Further, we can also see that the nitrogen profile shrinks with time just like the cases discussed so far. However, the extent of nitrogen diffusion is much less than any of the previous cases discussed so far. We first need to verify if Type II defects form at the end of range. The XTEM micrograph shown in Figure 6(a) indicates that a continuous amorphous layer forms to depths of about 900 Å upon implantation of $2 * 10^{15}$

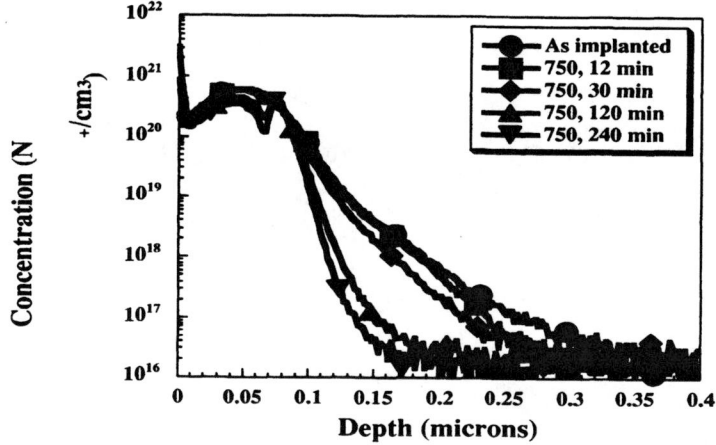

Figure 5: Time evolution of the profiles for a dose of $2 * 10^{15}$ N_2^+/cm^2, implanted at 40 keV and subsequently annealed at 750^oC for various times

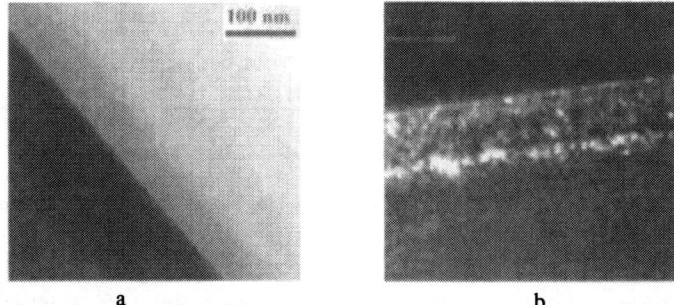

a   b

Figure 6: (a) XTEM micrograph of $2 * 10^{15}$ N_2^+/cm^2, 40 keVas-implanted sample. (b) XTEM micrograph of $2 * 10^{15}$ N_2^+/cm^2, 40 keV sample annealed at 750^oC, 120 min in an inert ambient

N_2^+/cm^2. Therefore upon annealing, Type II defects form at the EOR as observed in Figure 6(b). In addition, Figure 6(b) also indicates the formation of solid solubility related defects (Type V defects). The evolution of the extended defects with time is shown by the PTEM micrographs in Figure 7. As can be seen from Figure 7, we can see that there is a bimodal distribution of the extended defects in the samples. This is similar to the defect distribution seen for the previous case of $5 * 10^{14}$ N_2^+/cm^2. Once again the bimodal distribution of the defects is especially observable in the TEM's of the samples annealed for 120 min and 240 min at 750^oC. The method employed for counting both the extended defect density and the number of intersititals contained in the defects was the same as the previous case of $5 * 10^{14}$ N_2^+/cm^2. The density of the

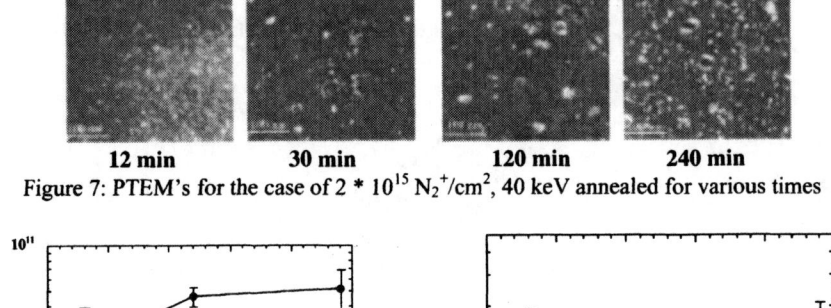

12 min   **30 min**   **120 min**   **240 min**

Figure 7: PTEM's for the case of $2 * 10^{15}$ N_2^+/cm^2, 40 keV annealed for various times

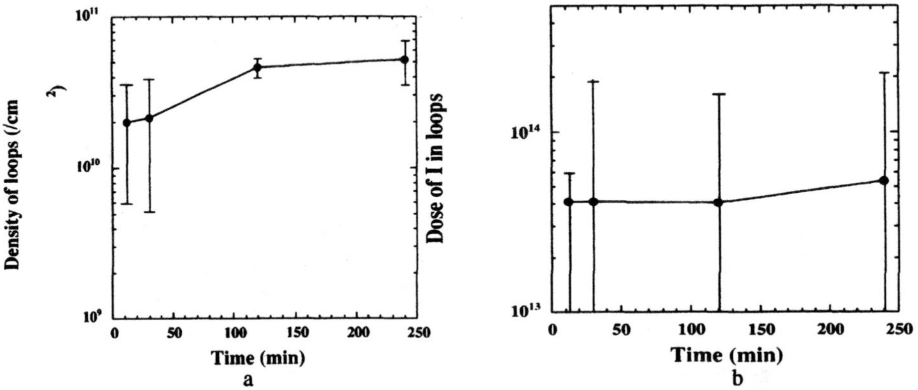

Figure 8: (a) shows the density of loops with time and (b) shows the dose of I in those loops

Type II defect evolution with time is shown in Figure 8(a) and the density of the interstitials contained in them is shown in Figure 8(b) along with their respective lower and upper bounds of count errors. Note that in Figure 8(b) the error bars are quite large due to the small sizes of the loops (bimodal distribution) and also due to the small sized loops being close to each other.

CONCLUSIONS

The diffusion behavior and concomitant defect evolution of $5 * 10^{14}$ N_2^+/cm^2 and $2 * 10^{15}$ N_2^+/cm^2 have been discussed and shown that they are anomalous. It has been shown that as the dose of implanted nitrogen increases, the extent of diffusion reduces.

REFERENCES

1. C.T. Liu, *et. al.*, IEDM 96, pp. 499-502
2. L.S. Adam, *et. al.*, IEDM 2000, pp. 507-510
3. K.S. Jones, *et. al.*, Applied Physics A, v45, pp. 1-34

Poster Session

Mat. Res. Soc. Symp. Proc. Vol. 669 © 2001 Materials Research Society

NONMELT LASER ANNEALING OF 1 KEV BORON IMPLANTED SILICON

Susan Earles[1], Mark Law[1], Kevin Jones[1], Somit Talwar[2], and Sean Corcoran[3]

[1]SWAMP Center, University of Florida, Gainesville USA

[2]Verdant, San Jose, CA USA

[3]Intel Corp, Portland OR USA

ABSTRACT

Heavily-doped, ultra-shallow junctions in boron implanted silicon using pulsed laser annealing have been created. Laser energy in the nonmelt regime has been supplied to the silicon surface at a ramp rate greater than 10^{10}°C/sec. This rapid ramp rate will help decrease dopant diffusion while supplying enough energy to the surface to produce dopant activation. High-dose, non-amorphizing 1 keV, 1e15 ions/cm^2 boron is used. Four-point probe measurements (FPP) show a drop in sheet resistance with nonmelt laser annealing (NLA) alone. Transmission electron microscopy (TEM) shows the NLA dramatically affects the defect nucleation resulting in fewer defects with post annealing. Hall mobility and secondary ion mass spectroscopy (SIMS) results are also shown.

INTRODUCTION

One of the key issues involved in scaling PMOS transistors is reducing the depth of the p-type source/drain extensions. Junction depths less than 30 nm are required for 70 nm gate lengths [1]. The simplest method of producing p-type junctions is to implant boron, a p-type dopant. After the implant, the wafer is rapid thermally annealed (RTA) in an effort to activate the boron and remove damage created by the implant.

Upon annealing, the heating of the lattice and the damage from the implant results in boron diffusion, boron clustering, and defect evolution [2, 3, 4]. This produces deeper junctions, lower boron activation, and reduced mobility. Variations in the implant energy and thermal annealing techniques are thus required to produce shallower junctions.

Previous studies have investigated the use of high power pulsed lasers to melt the implanted layers to achieve high activation and abrupt junctions [5, 6]. Complications arising from melting and regrowth however limit the use of this technique [7, 8, 6].

Experiments show increasing the ramp up rate during thermal processing has been shown to decrease the TED of boron in silicon [9, 10, 11]. Plots of the ramp up rate versus diffusion length show that the ramp up rate would need to be around 10^{10}°C /sec to result in a diffusion length of zero, and hence no TED [12]. Unfortunately, conventional RTA systems have peak ramp up rates of 200-400°C. However, using a laser for thermal processing results in a ramp up ramp which approaches the 10^{10}°C /sec that current data suggests is needed for zero TED. The ramp down or cooling rate is also dramatically higher for the laser annealed sample since only a small surface region of the wafer is heated during the NLA thermal conductivity dominates the cooling down of the wafer instead of radiation. In a continued effort to reduce TED while achieving high dopant activation, the following study investigates the effects of nonmelt laser annealing on silicon heavily-doped with 1 keV boron.

EXPERIMENTAL RESULTS

A 1 keV, 1e15 ions/cm^2 B+ implant into a CZ grown <100> silicon wafer was processed with a 308 nm XeCl excimer laser. The 1 keV implants received one or ten 15 ns pulses at a constant energy density of 0.55 J/cm^2. Following the NLA some samples received an RTA for 5 sec at 1040°C. Control samples received the RTA and no NLA. These samples were then analyzed using SIMS, Hall Effect, FPP, and plan-view TEM. Indium contacts were used for the Hall Effect measurements.

Figure 1 shows the SIMS of the 1 keV samples as-implanted, after 10 laser shots, after the RTA, and after 10 laser shots followed by the RTA. SIMS for the sample receiving one laser shot was nearly identical to the as-implanted profile, while SIMS for the one laser shot plus the RTA was nearly identical to the RTA alone. Figure 1 shows that the NLA of ten laser shots prior to the RTA decreases the boron diffusion while one laser shot shows no noticeable effects.

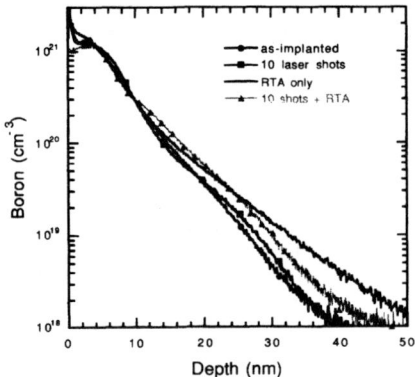

Figure 1. SIMS of 1 keV, 1e15 ion/cm2 B following multiple laser pulses and 1040oC, 5 sec RTA.

Figure 2 shows the change in sheet resistance as the number of laser pulses is increased for samples receiving just the NLA and for those receiving the NLA followed by the RTA. The results show that the NLA alone results in a decrease in sheet resistance equivalent to the drop seen for the samples receiving the RTA. The decrease in sheet resistance following the NLA occurs with little change in the junction depth (Figure 1).

Figure 3 plots the Hall mobility versus number of laser shots following the RTA (squares) and without RTA (the X). Ohmic contacts could not be made to the sample receiving one laser shot therefore no Hall measurements were made. The sample receiving ten laser shots has a Hall mobility of 15 cm^2/V-s while those receiving the RTA had a nearly constant mobility around 30 cm^2/V-s (Figure 3).

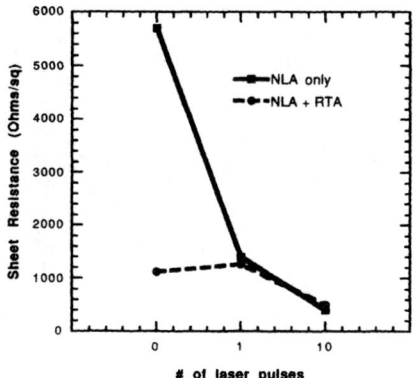

Figure 2. Sheet resistance vs laser pulses for 1 KeV, 1e15 ions/cm2 B following 0, 1, or 10 laser pulses and 1040°C, 5 sec RTA.

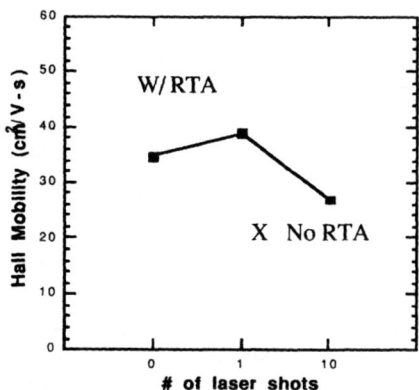

Figure 3. Hall mobility vs laser pulses for 1 keV Boron.

Plan-view TEM pictures of the 1 keV boron after 10 laser shots, after the RTA, and after 10 laser shots plus the RTA are shown in Figure 4. Figure 4 shows the NLA of ten laser shots alone nucleates numerous small loops (left-most picture). Figure 4 also shows that this NLA prior to the RTA reduces the final loop density (right-most picture) compared with the RTA alone (center picture).

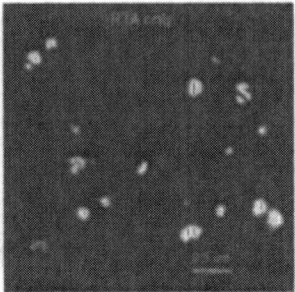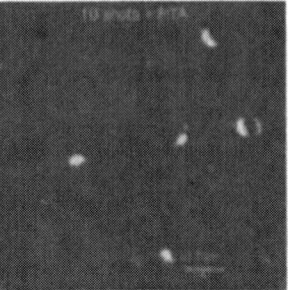

Figure 4. Plan-view TEM of 1 KeV, 1e15/cm2 boron implanted silicon following (from left to right) 10 shots, RTA only, 10 shots plus RTA.

DISCUSSION

Contrary to 5 keV, 1e15/cm^2 SIMS results[14] which show that the boron profile diffuses more when given an NLA prior to the RTA, SIMS results of the 1 keV boron show that the NLA prior to the RTA actually decreases the boron diffusion. Also, the NLA prior to the RTA resulted in an increase in the loop density for the 5 keV samples and a decrease in the loop density for the 1 keV samples.

The variations in loop densities and diffusion can be attributed to the interaction of the laser beam with the damage and boron following the implant. The laser used has an absorption depth around 100 A. The peak of the boron as-implanted profile is around 260 A for the 5 keV implant and 50 A for the 1 keV implant. For the 1 keV implant, the effect of the laser is distributed across the bulk of the damage and dopant profile. While for the 5 keV implant, the laser interacts with less than a fifth of the damage and dopant profile. During the NLA the surface is heated to 1200-1400°C for a few nanoseconds. This allows time for the silicon interstitials to move around, but not the boron. Thus during one laser shot interstitials diffuse to the surface where they recombine while some remain behind in clusters.

For the 5 keV implant, during one laser shot a region around 100 A thick rich in interstitials is heated resulting in the nucleation of numerous small loops. When followed with an RTA these loops grow and act as traps for interstitials which would normally recombine at the surface. This concentration of loops between the surface and the peak of the dopant/interstitial profile results in more interstitials diffusing into the bulk instead of towards the surface. Both cases increase the number of interstitials available to contribute to TED and defect formation during post-annealing.

For the 1 keV, the bulk of the interstitials and boron are within 100 A of the surface. This high concentration of impurity ions decreases the absorption length of the silicon reducing the depth of the heated layer. During the first laser shot, numerous small defects are nucleated in this thin region with the size of the defect being no larger than the width of the heated region. During subsequent laser shots the width of this heated region increases along with the size of the defects. After ten laser shots the defects are large enough to be detected in the TEM. Meanwhile, during each pulse interstitials have been making it to the surface where they recombine resulting in fewer interstitials available to form loops and contribute to TED during post-annealing.

For both the 5 keV and 1 keV implants, results show that the mobility is greater in the samples that have more loops. Logically, more defects should mean more scattering sites so the mobility should go down as the number of loops increase. What I believe is occurring is that the loops are trapping boron as well as silicon interstitials resulting in a decrease in the carrier activation. This decrease in carrier activation results in the increase in mobility measured.

CONCLUSIONS

Heavily-doped, ultra-shallow junctions in boron implanted silicon using pulsed laser annealing have been created. NLA alone results in a decrease in the sheet resistance. For the 1 keV implant this decrease in sheet resistance comes with little increase in junction depth. Also for the 1 keV implant, results show that use of the NLA prior to the RTA can result in a decrease in the defect density along with decreased junction motion. Along with choosing the energy density and pulse duration of the laser, results suggest that further optimization of the NLA can be obtained by adjusting the number of laser shots and the laser wavelength to provide the best interaction with the damage profile produced by the implanted species.

ACKNOWLEDGMENTS

The authors would like to thank the SRC and SEMATECH for providing funding for this research, Gana Rimple for assistance with the NLA, and Christina Schade for running all of the 1 keV SIMS.

REFERENCES

1. SI Association 1997.
2. Cowern N, Janssen T, Jos H, J. Appl. Phys. 68, 6191 (1990).
3. Hofker W, Werner H, Oosthoek D, Koeman N, Appl. Phys. 4, 125 (1974).
4. Solmi S, Landi E, Baruffaldi F, J. Appl. Phys. 68, 3250 (1990).
5. Tsukamoto H, Sol. St. Electron. 43, 487 (1999).
6. Zhang LH, Jones KS, Chi P, Simons DS, Appl. Phys. Lett. 67, 2025 (1995).
7. Chong Y, Pey K, Lu Y, Wee A, Osipowicz T, Seng H, See A, Dai J, Appl. Phys. Lett. 7, 2994 (2000).
8. Privitera V, Spinella C, Fortunato G, Mariucci L, Appl. Phys. Lett. 77, 552 (2000).
9. Agarwal A, Materials Research Society Spring 2000.
10. Agarwal A, Gossmann H-J, Fiory A, J. Electron. Mater. 28, 1333 (1999).
11. Downey D, Falk S, Bertuch A, Marcus S, J. Electron. Mater. 28, 1340 (1999).
12. Cowern N, Huizing H, Stolk P, Visser C, de Kruif R, Kyllesbech Larsen K, Privitera B, Nanver L, Crans W, Nucl. Instrum. Methods Phys. Res. B, 120, 14 (1996).
13. Eaglesham DJ, Stolk PA, Gossmann H-J, Poate JM, Appl. Phys. Lett. 65, 2305 (1994).

14. Earles SK, Law ME, Brindos RE, Jones KS, Materials Research Society Spring 2000.

Mat. Res. Soc. Symp. Proc. Vol. 669 © 2001 Materials Research Society

Vacancy and oxygen behavior in carbon highly doped silicon

Pierre Lavéant, Peter Werner, Norbert Engler, and Ulrich Goesele,
Max Planck Institute of Microstructure Physics
Weinberg 2, D-06120 Halle, Germany

ABSTRACT

Carbon doping of silicon has gained interest since in high concentrations, carbon can reduce or even suppress undesirable diffusion of the base dopant boron in silicon-based bipolar transistors. This behavior can only be understood in taking into account the silicon point defects i.e. vacancies and self-interstitials. In this work, we observe the oversaturation of vacancies produced by a high carbon concentration during annealing. Experiments with a vacancy diffusing dopant, Antimony, are shown and prove this effect : in a carbon rich sample, the antimony diffusion is enhanced about 8 times compared to samples with a much lower carbon concentration. We also investigate the carbon co-precipitation with oxygen. The carbon precipitation, as SiC, is facilitated with a high oxygen concentration. We explain this affinity by an exchange of point defects and a volume compensation. Finally, we show the precipitation of oxygen in relation to the vacancy oversaturation at 900°C.

INTRODUCTION

Considered as well understood and as a benign impurity, carbon gained some interest when incorporated in high concentration in the silicon lattice. Since carbon contracts the silicon lattice considerably it can be used for gap engineering or for lattice engineering, i.e. to elaborate new substrates for better epitaxy. But the main potential of high carbon concentrations may lie in the control of dopant diffusion. The suppression by carbon incorporation of the so-called "transient enhanced diffusion" of boron opened up the possibility to fabricate Si-Ge based heterobipolar transistors with a very narrow base region which is of enormous interest for ultra-high frequency applications [1]. This effect can only be fully understood and modeled in taking into account the silicon point defects i.e. vacancies (V) and self-interstitials (I) [2].
In the present study, we simulate the carbon diffusion and discuss its influence on the self-interstitials and vacancies. MBE-grown samples with a high carbon concentration and doped with a vacancy diffusing element are annealed. We modeled the carbon out-diffusion by kick-out and Frank-Turnbull mechanism. In addition, the co-precipitation of carbon and oxygen is also investigated and we try to explain the effect of oxygen on the SiC precipitation and then the behavior of oxygen in a vacancy oversaturated layer. SIMS depth profile and cross-sectional TEM are also used.

CARBON DIFFUSION

Carbon is mostly substitutionally dissolved in the Si lattice, C_S. The solubility or thermal equilibrium concentration of this electrically neutral specie is given by [3]

$$C_s^{eq} = 4 \times 10^{24} \exp(-2.3 / k_B T) \quad \text{cm}^{-3} \tag{1}$$

Nevertheless, to understand the carbon diffusion one has to take into account the carbon interstitial, denoted here by Ci, which is orders of magnitudes more mobile than Cs [4]:

$$D_i = 0.44 \exp(-0.88 \ eV \ / \ k_b T) \quad cm^2/s \tag{2}$$

The in-diffusion of substitutional carbon is fairly normal and can be described by a concentration-independent diffusivity leading to an erfc-type concentration profile [4]. This directly leads to the so-called *kick-out mechanism* [5]

$$C_S + I \Leftrightarrow C_i \tag{5}$$

where a carbon substitutional is pushed out by a self-interstitial and leads to a fast diffusing carbon interstitial. Under equilibrium conditions like in-diffusion, the transport capacities, i.e. the product of the diffusion coefficient by the concentration, of Ci and Cs are much smaller than the one of self-interstitials.

$$D_{C_i} c_{C_i}^{eq} = D_{C_S}^{eff} c_{C_S}^{eq} < D_I c_I^{eq} \tag{6}$$

As shown in equation (1), the solubility of carbon in silicon is very low in the order of 10^{15} at/cm³ at 900°C, a typical annealing temperature. This solubility is lower than the typical carbon concentration is CZ material and some applications, as cited in the introduction, need even higher concentration, in the 10^{20} at/cm³ range. Under such conditions, equation (6) is not valid anymore and the transport capacity of Ci becomes higher than the one of self-interstitials leading to an undersaturation expressed in equation (7) :

$$\frac{c_I}{c_I^{eq}} \approx \frac{D_I c_I^{eq}}{D_{C_i} c_{C_i}^{eq}} \cdot \frac{c_{Cs}^{eq}}{c_{Cs}} \tag{7}$$

and shown in figure 1a. However, this oversaturation may be exaggerated if the model just include the kick-out mechanism.

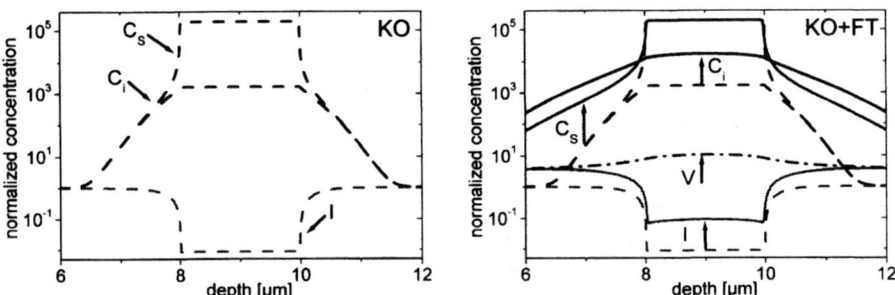

Figure 1 : *Simulation of C diffusion out of a C box profile at 900°C after 3 hours. On the left hand side the simulation is performed with only the kick-out (KO) mechanism, on the right hand side with kick-out and Frank-Turnbull (FT) mechanism. With the Frank-Turnbull and the kick-out mechanisms, the C out-diffusion is faster and a V oversaturation is created. The Cs is about 1×10^{20} at/cm³ and Ci 1×10^{8} at/cm³*

$$D^{eff} = D_I \cdot \frac{c_I}{c_I^{eq}} + D_V \cdot \frac{c_V}{c_V^{eq}} \qquad (9)$$

and this leads to an supersaturation of vacancies as shown in fig 1b. The undersaturation of self-intertitials due to a high carbon concentration affects the dopants diffusing by self-interstitials as seen in the reduction of the transient enhanced diffusion of boron. One can then also expect an enhancement of the diffusion of vacancy diffusing species, like antimony or germanium. We show in figure (2) two samples grown by MBE with a antimony box profile near in one sample a highly doped carbon box. After annealing for 6 hours at 900°C, the sample with about 3×10^{20} at/cm³ carbon shows an antimony diffusion about 8 time faster than in the carbon-poor sample.

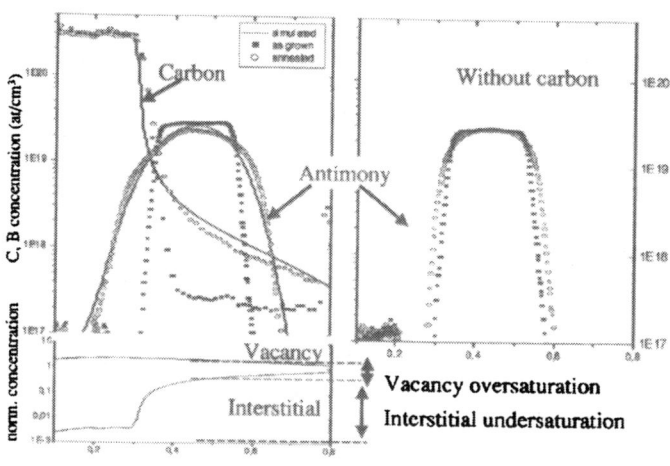

Depth (μm)

Figure 2 : SIMS depth profile of MBE grown samples before (full dots) and after(open dots) annealing at 900°C for 6 hours. The simulation show the expected oversaturation of vacancies.

PRECIPITATION

Our simulation model do not contain the carbon precipitation. The typical carbon concentration in CZ ranges around 10^{16} cm⁻³, which is above the carbon solubility at melt temperature. By CVD or MBE even higher concentrations can be achieved, up to 10^{21} cm⁻³. One can expect an immediate carbon precipitation to occur during annealing but it seems not to be the case. Two basic reasons appear to be responsible for this reluctant precipitation. The first reason is the very high interface energy σ_C associated with the SiC/Si interface which is much higher than the analogous quantity for oxygen precipitates [6]. A second one would be a volume factor. Indeed, whereas oxygen precipitation is associated with a volume increase of about a factor of two, carbon precipitation involves a volume shrinkage of close to one atomic silicon volume for each carbon incorporated in the C-agglomerate or SiC-precipitate. An undersaturation of self-interstitials will even make carbon precipitation more difficult, as expressed with the critical radius r_{crit} :

$$r_{crit} = \sigma_c \Omega / kT \ln (C_s C_i / C_s^* C_i^*) \quad , \qquad (13)$$

high interface energy σ_C associated with the SiC/Si interface which is much higher than the analogous quantity for oxygen precipitates [6]. A second one would be a volume factor. Indeed, whereas oxygen precipitation is associated with a volume increase of about a factor of two, carbon precipitation involves a volume shrinkage of close to one atomic silicon volume for each carbon incorporated in the C-agglomerate or SiC-precipitate. An undersaturation of self-interstitials will even make carbon precipitation more difficult, as expressed with the critical radius r_{crit} :

$$r_{crit} = \sigma_c \Omega / kT \ln (C_s C_i / C_s^n C_i^n) \quad , \tag{13}$$

It is immediately obvious that the co-precipitation of carbon (associated with a volume shrinkage) and oxygen (associated with a volume expansion) is favorable and will lead to a co-precipitation in the ratio of two oxygen atoms for each carbon atom precipitated, as experimentally observed [7]. It appears that oxygen and carbon also interact on an atomic level. A SiO_2 precipitate would eject a self-interstitial whereas the SiC precipitation would need one as expressed in equation 14 :

$$\begin{cases} 2Si + 2O_i \Leftrightarrow SiO_2 + I \\ I + C_S \Leftrightarrow SiC \quad or \quad Si + C_S \Leftrightarrow SiC + V \end{cases} \tag{14}$$

where O_i notes an oxygen atom in an interstitial site.

In samples with a high carbon concentration after a long annealing small precipitates, typically with a diameter smaller than 10 nm can be detected. Those SiC precipitates show typical Moiré fringes (fig 3). SIMS data show also an enhancement of the oxygen concentration, diffusing out of the substrate. In the antimony rich region, larger precipitates can be seen, with a diameter of about 20 nm as well as a large enhancement of the oxygen concentration. Since this effect is seen only in the carbon rich sample (fig. 2a), we think that the vacancy oversaturation due to the high carbon concentration would affect the oxygen behavior leading to large SiO_2 precipitates.

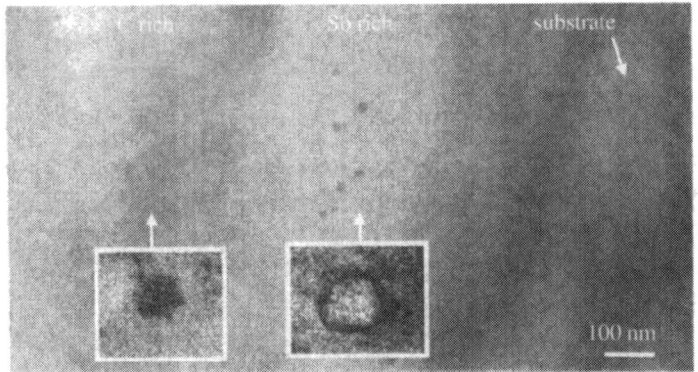

Figure 3 : *Cross-sectional TEM of a MBE grown sample with a Sb and C doping after annealing for 6h at 900°C (SIMS profile shown in fig. 2).*

CONCLUSION

We have shown the possibility to simulate accurately the carbon diffusion out of doping profiles, even for concentrations order of magnitude above the solubility. To understand fully the carbon behavior the kick-out combined to the Frank-Turnbull mechanisms are needed. It influences the silicon point defects creating an oversaturation of vacancies and an undersaturation of self-interstitials.

The retardation of the carbon precipitation can be understood as a lack of self-interstitials due to the carbon out-diffusion as well as a volume shrinkage and a high interface energy. The oxygen precipitation, leading to the emission of self-interstitials, favors the SiC formation. Finally we show the precipitation of oxygen enhanced by the vacancy oversaturation created by the carbon diffusion.

ACKNOLEDGMENTS

The authors are very grateful to A. Frommfeld, G. Gerth, S. Hopfe and B. Lausch for technical assistance.

REFERENCES

1.   S. C. Jain, H. J. Osten, B. Dietrich, and H. Rücker, Semicond. Sci. Technol. **10**, 1289 (1995).
2.   R.F. Scholz, P. Werner U. Gösele and T.Y. Tan, Appl. Phys. Lett. **74**, 392 (1999).
3.   A.R. Bean, R.C. Newman, J. Phys. Chem. Solids **32**, 1211 (1971).
4.   U. Gösele, Mat. Res. Soc. Symp. Proc. **59**, 419 (1986).
5.   U. Gösele, W. Frank, and A. Seeger, Appl. Phys. **23**, 361 (1980).
6.   W. Taylor, T. Tan and U: Gösele, Appl. Phys. **62**, 3336 (1993)
7.   Q. Sun, K. H. Yao, J. Lagowski, and H. C. Gatos, J. Appl. Phys. **67**, 4313 (1990).

Mat. Res. Soc. Symp. Proc. Vol. 669 © 2001 Materials Research Society

Modeling Threading Dislocation Loop Nucleation and Evolution in MeV Boron Implanted Silicon

Ibrahim Avci[1] and Mark E. Law
Craig Jasper[2], Hernan A. Rueda and Rainer Thoma
[1] Swamp Center, Department of Electrical Engineering,
NEB Room # 535, University of Florida, Gainesville FL 32601
[2] Motorola, Digital DNA Laboratories, Mesa, AZ 85202.

ABSTRACT

A single statistical point defect based model for the nucleation and evolution of dislocation loops during annealing in Si is developed. The model assumes that the radius and the density of dislocation loops follow a log normal distribution. The loop nucleation part of the model also assumes that all the loops come from {311} unfaulting. The model is verified with the experimental results obtained by studying the formation of dislocation loops and threading dislocation loops as a function of implant condition in boron implanted silicon by varying the dose from 1×10^{13} to 5×10^{14} cm^{-2} at an energy of 1.5 MeV. Due to the statistical nature of the model, the threading dislocation loop density is easily obtained from simulation results. The dramatic change in the threading dislocation loop density with the increasing implant dose is also predicted by the simulations.

INTRODUCTION

High energy non-amorphizing implants are commonly used in integrated circuit (IC) manufacturing. The main advantages of ion implantation are to introduce a desired impurity into a target material, accurate dose control, reproducibility of the impurity profiles, lower process temperatures and the ability to tailor the doping profile [1]. They are commonly used to form retrograde wells for CMOS latch-up immunity improvement and buried layers for bipolar transistor subcollectors [2-4]. However, heavy lattice damage can be generated near the projected range of the implanted dopant [5]. The defects near this region are typically interstitial in nature and will evolve in threading dislocation loops (TDL), {311}s [6-8] and stable dislocation loops. The threading dislocation loops are long dislocation dipoles that can grow to the surface and thread the surface [8]. If these defects are located in the active device region, they can cause leakage currents and be detrimental to the electrical properties of the device [9].

Understanding and modeling of the unique material aspects of damage/defect accumulation during very high energy implantation and subsequent annealing cycle is critical in improving the device performance. A model would also help device/process engineers to change and adjust their device structures and process conditions without having to build costly test lots. In this article we explain a model for the nucleation and evolution of dislocation loops. Using this model, threading dislocation loop density is derived from simulation results. Simulation results have been validated against experimental data [10]. The results are in good agreement with the data. The threading dislocation loop density increases with increasing implant dose to a maximum at a dose of 1×10^{14} cm^{-2} and decreases with the increasing dose afterwards. This dramatic change in TDL density at the critical dose of 1×10^{14} cm^{-2} is also captured by the simulation.

EXPERIMENT

The implants were carried out in 7μm thick lightly doped (1×10^{15} cm^{-3}) P- type epitaxial silicon grown on (100) P+ silicon. Extended defect formation was studied in boron implanted silicon at various implant doses. The dose was varied from 1×10^{13} to 5×10^{14} cm^{-2}. Boron implant energy was held constant at 1.5MeV. After implantation the samples were annealed at 800°C for 90 minutes followed by a 550°C anneal for 60 minutes and finally a 950°C anneal for 10 minutes. Etch pit samples were produced using a Schimmel etch consisting of a 2:1 49% HF: 1M CrO$_3$ solution. An etch of 20 seconds was used.

To examine the projected range defects both plan-view and cross-sectional TEM was done on the samples after annealing. The plan-view samples were pre-thinned (when necessary) using chemical mechanical touch polishing (CMP) with a silica slurry, to position the peak of the projected range damage approximately 3000-4000Å below the surface. A high resolution TEM operating at 200kV was used to study the projected range defects. Micrographs were taken in bright field using g$_{220}$ imaging conditions. These conditions allow one to observe defects as deep as 5000Å to 10,000Å deep. TEM of unimplanted control samples show that the CMP process does not introduce any observable defects [11].

MODEL

As a result of ion implantation, large amount of excess interstitials is created around the projected range of the implant. Upon annealing, defects such as {311}s or dislocation loops form where the excess interstitial concentration is high. The developed loop nucleation model assumes that these excess interstitials are the source of the defects. First, UT-Marlowe [12], a Monte Carlo simulation program, is used to calculate the excess interstitials and vacancies created in silicon substrate due to ion implantation. Second, interstitial and vacancy clusters such as di-interstitials (I$_2$) , di-vacancies (V$_2$) and sub microscopic interstitial clusters (SMICS) are created upon annealing. Third, {311} defects are nucleated from SMICS [13]. During this nucleation, large amount of excess interstitials is consumed. Some of these nucleated {311}s become thermally unstable and unfault to dislocation loops. {311}s become the source of dislocation loops [14]. Then, while the remaining {311}s start dissolving, dislocation loops start evolving. In the developed loop nucleation model, unfaulting process is called the nucleation stage of the dislocation loops and the evolution of the loops is called the Ostwald ripening stage.

The model assumes that dislocation loop density (D$_{all}$) and average radius of loops (R) follow a log normal distribution function (f$_D$(R)). The capture and emission rate of interstitials by the dislocation loops can be expressed in terms of the rates of emission and absorption of point defects at the loop layer boundaries modulated by a log normal distribution function. The rate also depends on the unfaulting rate of {311}s during the nucleation phase (N_{rate}^{Nall}) and can be expressed as:

$$\frac{dN_{all}}{dt} = N_{rate}^{Nall} + K_{IL} \int_{0^+}^{\infty} (C_I - C_{Ib}) f_D(R) dR \Big|_{\text{at loop layer boundaries}} \quad (1)$$

where N_{all} is the number of interstitials bounded by the loop. K_{IL} is the constant of a reaction between the interstitials and the dislocation loop assemble. K_{IL} is the function of the loop radius and the diffusivity of interstitials. It is used as calibration parameter during the simulations. C_i is the concentration of interstitials and vacancies. C_{Ib} is the effective equilibrium concentrations of interstitials at the loop layer boundaries. The first term in equation 1 is the nucleation term and the second and third terms are the evolution terms and explained in [15]. Assuming that all loops are circular, the density of dislocation loops can be expressed by:

$$\frac{dD_{all}(R)}{dt} = N_{rate}^{Dall} - \frac{2D_{all}(R)}{R^2} K_R \tag{2}$$

The constant, K_R, is the coarsening rate of dislocation loops and used as a fitting parameter in the simulations. The first term in equation 2 is the nucleation term and the second term is the coarsening term. It is also possible to extract a relation between σ (standard deviation of data [12]) and average loop radius R from TEM data as an analytic function of R (cm) as follows:

$$\frac{\sigma}{R} = 0.33 + 5e4R \tag{3}$$

Equation 3 is substituted in equations 1 and 2. The nucleation terms shown in equations 1 and 2 are given by

$$N_{rate}^{Nall} = K_{311}C_{311} \tag{4}$$

$$N_{rate}^{Dall} = K_{311}D_{311} \tag{5}$$

C_{311} and D_{311} are the number of interstitials in $\{311\}$'s and density of $\{311\}$'s, respectively. K_{311} is the unfaulting rate. K_{311} is the same in both equations because when a $\{311\}$ defect unfaults to dislocation loop, the number of interstitials bounded by that $\{311\}$ is transferred to the unfaulted loop.

If the distribution of loops with respect to their radius is known, it is possible to obtain TDL information from this distribution. Figure 1 shows how to obtain TDL density from a distribution function. In the process simulators, physical shape of the device is modeled using meshes. Every mesh is composed of nodes that hold the physical data as shown in figure 1. The developed loop nucleation model assumes that loop density and radius follow the log normal distribution function and each node represents a different distribution function. Dislocation loops whose radii are greater than their depth from the surface are considered as TDLs. In figure 1, R_{c1}, R_{c2}, and R_{c3} represent the critical radii to be considered as a TDL at each loop depth. The total density of threading dislocation loops can simply be calculated by integrating each distribution function from the critical radius to infinity (shaded areas in figure 1) and adding them.

SIMULATION AND EXPERIMENTAL RESULTS

In order to simulate loop nucleation, damage profiles are generated for each implant dose using UT-Marlowe. The profiles showed that all the excess interstitials are generated around the

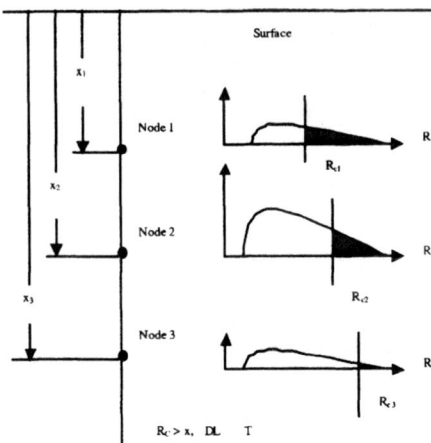

Figure 1. Schematic representation of TDLs in a distribution function

projected range ($R_p \approx 2.3$ μm for 1.5 MeV boron implant). Excess vacancy profile is also obtained in a similar way.

The excess interstitial and vacancy profiles provide the basis for the nucleation of interstitials clusters (I_2, SMICS) and vacancy clusters (V_2), eventually leading to the nucleation of {311}s and the dislocation loops. Figure 2 shows the simulation results of the changes in the defect densities with time after a 1.5 MeV boron implantation with a dose of 1×10^{14}cm^{-2} during annealing at 800°C. As seen in the figure 2, density of {311}s, D_{311}, increases very rapidly at short times then it starts decreasing. Total number of interstitials trapped by {311}, C_{311}, follow the same trends. The nucleation of dislocation loops is slower than the nucleation of {311}s since the loops nucleate from unfaulted {311}s. Total number of interstitials bounded by dislocation loops, N_{all}, increases very rapidly at short times and continues to increase while{311}s dissolve. This shows that dislocation loops capture some of the interstitials released by {311}s. Once the excess interstitials are consumed by loops and some of these excess interstitials diffuse away from the damage region, loops go into the Ostwald ripening process. No significant change can be seen in N_{all} for longer annealing times. The Ostwald ripening process can be seen in the density of dislocation loops (D_{all}) profile as well. D_{all} increases very fast at the short times when the nucleation rate is high. When the nucleation rate slows down, D_{all} stays almost constant due to the fact that the excess interstitial concentration is still high. Thus, dislocation loops follow a pure growth process during this time period. As soon as the excess interstitial concentration drops, D_{all} starts decreasing. Meanwhile, N_{all} stays constant. Thus, the bigger loops grow at the expense of small ones.

Figure 3 shows the density of all dislocation loops and the threading dislocation loops (D^{TDL}) as a function of boron implant dose along with the simulation results. As the implant dose increase, D_{all} and D^{TDL} increase with the increasing dose to a maximum at a dose of 1×10^{14} cm^{-2}. This is often referred to as the critical dose for threading dislocation loops. This rapid increase in dislocation loop growth is due to the increased dose that results in increased number of trapped interstitials in the dislocation loops. Increasing the dose of the implant will increase the excess interstitial population in the silicon substrate. Thus, this will increase the growth of the loops. At doses beyond 1×10^{14} cm^{-2},

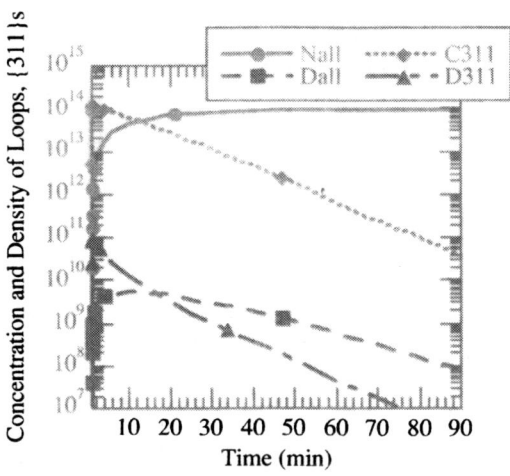

Figure 2. Changes in defect densities with time after implantation of boron with a dose of 1×10^{14} cm^{-2} and an energy of 1.5 MeV during 800°C anneal.

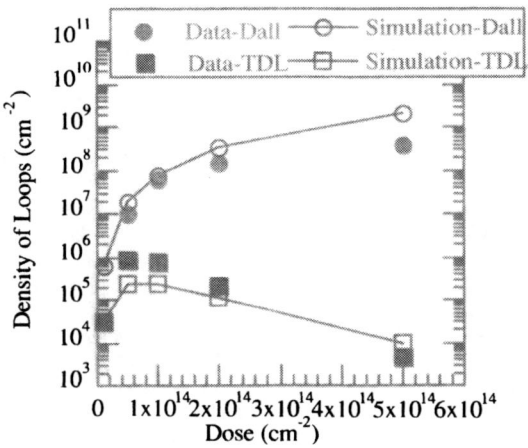

Figure 3. Density of all dislocation loops and threading dislocation loops vs. boron dose with implant energy of 1.5 MeV.

while D_{all} keeps increasing, D^{TDL} decreases back to close to the minimum detection limit

$(5\times10^3$ cm^{-2}). Similar trends can be seen in the simulation results including the dramatic change in the threading dislocation loop density at the critical dose of 1×10^{14} cm^{-2}. Simulation results show that the defects around the projected range are smaller and numerous for the high dose implant conditions. This can also be observed from the TEM images. When a sufficient number of defects nucleate, they can absorb the interstitials with forming TDLs. As a result the threading dislocation loop density decreases above the dose of 1×10^{14} cm^{-2}. The simulation predicts a higher density of dislocation loops at the high implant dose, but the results at the other implant dose values are in good agreement with the experimental results.

CONCLUSIONS

A statistical based model to predict loop nucleation and evolution is modeled. A single set of differential equations is used to characterize the loop behavior through the nucleation and Ostwald ripening stages. The model assumes that all the loops come from {311} unfaulting. The excess interstitial and vacancy populations required for the nucleation are obtained from UT-Marlowe. They are utilized to generate interstitial and vacancy clusters, eventually leading to the nucleation of {311}s and dislocation loops. Since the model keeps track of dislocation loop distribution through the substrate, the density of threading dislocation loops is easily calculated using these profiles. Simulation results are verified with the experimental data. A good agreement between the experimental data and simulation results is obtained. Simulations predict the increase in the TDL density with increasing dose up to a dose of 1×10^{14}cm^{-2} and the decrease in TDL density beyond this dose. The simulation also shows that the decrease in the TDL density at high implant doses is due to the increased number of defects since they can consume the interstitials with forming TDLs.

REFERENCES

1. E. Chason, S. T. Picraux, J. M. Poate, J. O. Borland, M. I. Current, T. Diaz de la Rubia, D. J. Eaglesham, O. W. Holland, M. E. Law, C. W. Magee, J. W. Mayer, J. Melngailis, A. F. Tasch, Journal of Applied Physics, **81**, 1997.
2. H. Y. Lin and C. H. Ching, Nucl. Instrum. Methods Phys. Res. B **37/38**, 960, 1989
3. J. O. Borland and R Koelsch, Semicond. Sci. Technol. **36**, 28, 1993.
4. H. Wang, N. W. Cheung, P. K. Chu, J. Lin, and J. W. Mayer, Applied Physics. Letters, **52**, 1023, 1988
5. K. S. Jones, S. Prussin, and E. R. Weber, Applied Physics, Vol. A 45, pp. 1-34, 1988
6. R. J. Schreutelkamp, K. S. Custer, J. R. Liefting, W. X. Lu, and F. W. Saris, Material Sci. Rep. **6**, 275, 1991
7. J. Y. Cheng, D. J. Eaglesham, D. C. Jacobson, P. A. Stolk, J. L. Benton, and J. M. Poate, Journal of Applied Physics, **80**, 2105, 1996.
8. D. J. Eaglesham, P. A. Stolk, H. J. Gossmann, and J. M. Poate, Applied Physics Letters, **65**, 2305, 1994.
9. J. Washburn, Defects Semicond. **2**, 209, 1980.
10. C. Jasper, K. S. Jones, Data will be published later.
11. C. Jasper, A. Hoover, K. S. Jones, Applied Physics Letters, **75**, 17, 1-3, 1999
12. B. Obradovic, G. Wang, Y. Chen, D. Li, C. Snell, A. F. Tasch, UT-MARLOWE 5.0 with tomcat, 1999
13. M. E. Law and K. S. Jones, IEDM 2000, pp. 511-514, 2000
14. J. Li, and K. S. Jones, Applied Physics Letters, Vol. 73, No 25, pp. 3748-3750, 1998.
15. I. Avci, H. A. Rueda, M. E. Law, SISPAD 2000, pp. 210-213, 2000

Mat. Res. Soc. Symp. Proc. Vol. 669 © 2001 Materials Research Society

Characterization of Damage Induced by Cluster Ion Implantation

Takaaki Aoki[1,2], Jiro Matsuo[2] and Gikan Takaoka[2]

[1] Osaka Science and Technology Center,
Utsubo-honmachi, Nishi-ku, Osaka 550-0004, JAPAN
[2] Ion Beam Engineering Experimental Laboratory, Kyoto University,
Sakyo, Kyoto 606-8501, JAPAN

ABSTRACT

Molecular dynamics simulations of boron monomer and small clusters (B_4 and B_{10}) impacting on Si(001) were performed in order to investigate the damage formation by monomer/cluster impact. These monomer and clusters show similar implant depth and efficiency, but different damage structures. At the impact of B monomer with 230eV of incident energy, some point-defects such as vacancy-interstitial pairs are mainly formed. On the other hand B_{10} produces several times larger number of vacancies and interstitials compared with B_1, This damage structure is different from one by B_1 implantation and due to high yield amorphization of implanted region. This characteristic damage formation process is expected to cause different annihilation process.

INTRODUCTION

As the scale of LSI device decreases, the formation of ultra shallow p-type junction becomes more important. Cluster ion implantation using small boron cluster, decaborane ($B_{10}H_{14}$), has been introduced as a candidate for ultra shallow junction formation. Both experiments [1,2] and molecular dynamics (MD) simulations [3] of small B cluster and monomer implantation has been performed. When a B cluster, with the size below 10 and the energy of several hundreds eV per atom, impacts on Si substrate, the cluster breaks up in the substrate and each B atom penetrates into the substrate independently. Therefore, the depth profile of B atoms by cluster implantation was as same as those by B monomer ion with same energy per atom.

Additionally, it is important to understand the damage formation process, because the annealing after ion implantation causes transient enhanced diffusion (TED) of dopant [4,5], which is one of the problems to fabricate high-quality shallow junction. It has been considered that, the TED has close relation to the structure of damage formed by ion implantation. As in previous work [6], when a $B_{10}H_{14}$ is implanted with 3keV, the TED is suppressed at 900°C of annealing temperature. On the other hand, the recent work about low-energy boron monomer implantation reported that the TED is also reduced by 500eV of B monomer implantation [7]. It is needed to discuss the similarity and difference between monomer and cluster implantation processes.

In this paper, MD simulations of B_1, B_4 and B_{10} clusters impacting on Si(001) substrate are performed. The defects induced by ion impact are classified into vacancies and interstitials. The distribution and structures of these defects were examined. The difference of damage formation and annihilation mechanism between monomer and cluster impact will be discussed.

SIMULATION MODEL

In order to examine the damage formation process by small boron cluster impacts, the MD simulations of B_1, B_4 and B_{10} monomer/cluster impacting on a Si(001) substrate were performed. The Stillinger-Weber potential model [8] was applied to the inter-atomic potential of Si-Si, and the Ziegler, Biersack, and Littmark (ZBL) model [9] was applied to B-B and B-Si potentials. A Si(001) substrate was prepared as a target material, which consists of 32768 atoms with a cube side of about 90Å and the periodic boundary conditions were applied on this target.

B_1, B_4 and B_{10} are radiated on the Si substrate with an incident energy of 230eV/atom. B_1 monomer is implanted with an incident angle of 7° to the surface normal and rotated 30° to the (001) direction to avoid channeling. B_4 and B_{10} have square and spherical structures, respectively, and are implanted at normal direction to the surface. In previous work [3], it is confirmed that B_4 and B_{10} clusters shows similar penetration depth and implant efficiency. In order to obtain statistical properties, 50 simulations for B_1 and 25 simulations for both B_4 and B_{10} were performed respectively at different impact points. Each simulation was performed for 8ps, which is enough time that incident atom its kinetic energy and transient displacements by ion impact recovers. Displacements produced by ion impact are investigated after the model by Perez-Martin et al. [10], where a vacancy is defined as a lattice point that has no atoms bound within 1.6Å, and an interstitial is an atom that is farer than 1.6Å from any lattice points.

RESULTS AND DISCUSSION

Figure 1 shows the distribution of distance between Si atom and its nearest lattice site of diamond structure, after the impact of B monomer and cluster. Each distribution is normalized by the number of total incident B atoms, so that each profile indicates the distribution of displaced length cased by one incident B atom the cluster. Most of Si atoms are set within 0.8Å around the lattice site, which means that most of lattice site will occupied by Si atoms later. The distribution by B_1 shows a peak at 1.6Å. This peak corresponds to split vacancies, where an atom is between to lattice points and this means that there is at least a vacancy [10]. On the other hand, at the impact of B_4 or B_{10}, larger number of displacements are produced compared with B_1 impact. It is shown that the peak of distribution at B_{10} impact shifts to shorter than one at B_1 impact. This result indicates that the structure of damage by B_{10} is different from that by B_1.

The structures around vacancy and interstitial are investigated. Figure 2 shows radial distributions of substrate Si atom around vacancy and interstitial. In the case of B_1 impact, the distribution around vacancies shows distinguished peaks and valleys. Comparing with the radial

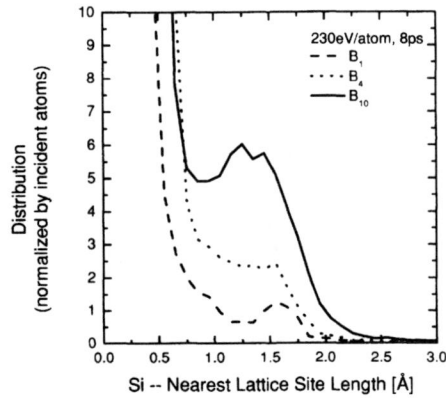

Figure 1. Distribution of Si atom between nearest lattice site after the impact of B monomer and cluster

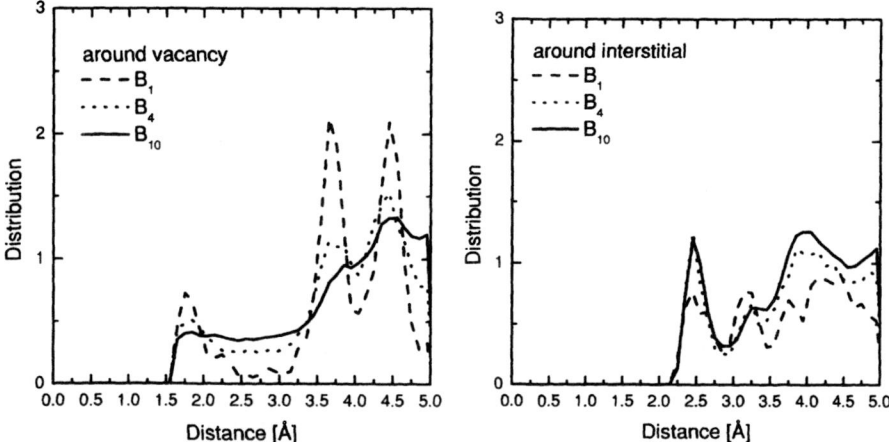

Figure 2. Radial distribution of substrate Si atoms around vacancy (left) and interstitial (right)

distribution function of diamond structure, each peak in B_1 distribution shifts to nearer side, but shows good correlation with one in diamond structure. As for the distribution around interstitial at B_1 impact, the second peak is observed 3.1Å with similar intensity to the first-nearest peak. Those two distributions indicate that the damage by B_1 impact tend to cause point-defect, where only one atom is knocked-on from its lattice site but other surrounding atoms keep lattice structure, and knocked-on atom remain as fourfold interstitial [10].

At the impact of B_4 and B_{10}, the peak-and-valley structure is not shown around vacancy within 3Å. This result means that the lattice structures before and after impact have no correlation with each other. This result supports the result of distribution around interstitial. As shown in right of figure 2, the distribution is broad as the distance increases.

Table 1 summarizes the bonding status of interstitial and surrounding lattice Si atoms. The cut-off length for bonding is 2.7Å, which is between the first and the second nearest length in Si crystal. Each value in the tables is normalized by total incident B

Table 1. Bonding status of interstitial and surrounding lattice Si atoms. Cut-off bond length is 2.7Å.

B1

Lattice atoms / interstitial atom	Interstitials					
	1	2	3	4	5	5~
0~1	0.04				0.10	
1~2	0.48	0.20	0.06			
2~3	0.80					
3~4	0.66					
4~5	0.04					
5~						
sum	2.02	0.20	0.06	0.00	0.10	0.00

B4

Lattice atoms / interstitial atom	Interstitials					
	1	2	3	4	5	5~
0~1	0.11	0.06	0.09	0.21	0.26	0.14
1~2	0.22	0.60	0.34	0.38	0.10	0.14
2~3	1.04	0.48	0.03			
3~4	2.13	0.06				
4~5	0.36					
5~						
sum	3.86	1.21	0.47	0.58	0.36	0.27

B10

Lattice atoms / interstitial atom	Interstitials					
	1	2	3	4	5	5~
0~1	0.11	0.12	0.76	0.69	0.44	1.07
1~2	0.62	1.20	1.54	0.93	0.30	0.11
2~3	2.22	1.60	0.38	0.10		
3~4	4.76	0.06				
4~5	0.90					
5~	0.03					
sum	8.64	2.98	2.68	1.71	0.74	1.18

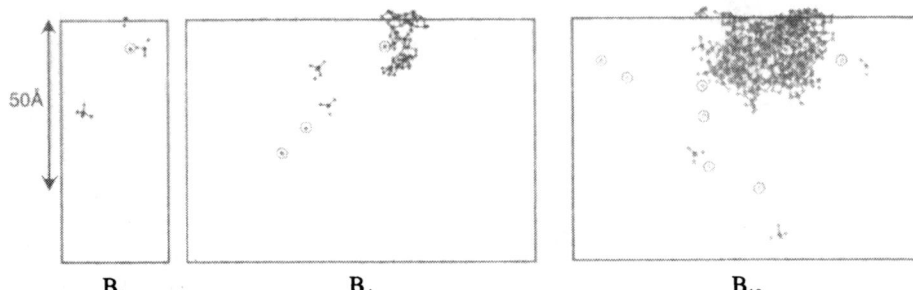

B_1 B_4 B_{10}

Figure 3. Snapshots of interstitial and bonded lattice Si implanted with B_1, B_4 and B_{10}. Incident B atoms are indicated with circles

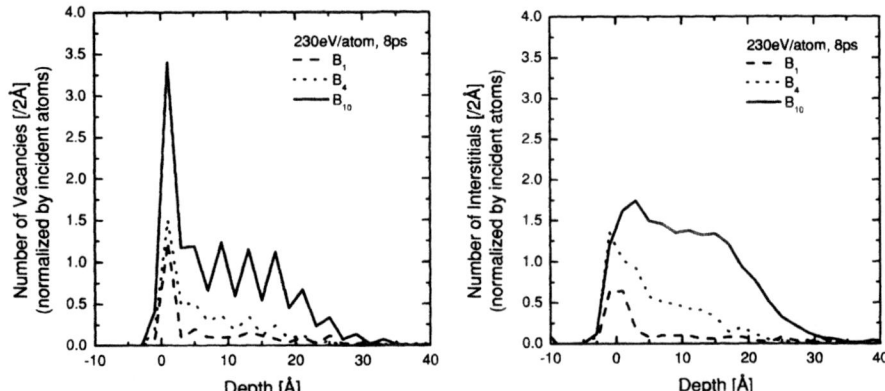

Figure 4. Depth profile of vacancies and interstitials by B monomer and cluster implantation

atoms. The legend in each row means the ratio of the number of surrounding Si atom to the size interstitial cluster. For example, the value at the second column and the second row indicates the formation probability of I_2Si_3 and I_2Si_4 (where I and Si means an interstitial and a lattice Si, respectively).

As shown in table 1, B_1 impact mainly causes individual interstitial, which does not form interstitial clusters. The interstitials are stable as three- or four-fold and this result supports one shown in figure 2. When B_4 and B_{10} are implanted into Si target, larger number interstitial clusters and individual interstitials are formed compared to B_1. As for large interstitial cluster, the ratio of surrounding Si atoms to interstitial cluster atoms is about 1, which means mixture of interstitial and lattice Si are formed.

Figure 3 shows the snapshots of incident B atoms, interstitial and bonded Si lattice atoms at the impact of B_1, B_4 and B_{10}, 8ps after impact. The incident B atoms are indicated with circles. At the impact of B_1, each interstitial atom is separated from others and mainly shows tetrahedral structure with surround lattice Si atoms. At the impact of B_{10}, however, most of interstitial atoms are found in near surface region of the impact point, and separated interstitials, which are mainly shown for B_1 impact, are formed at far from the surface, in other words, 'end-of-range.' The incident B atoms are distributed in wider range compared with the range where the high-density interstitials reside.

Figure 4 shows depth distribution of vacancies and interstitials by B monomer and cluster implantation. These profiles are similar to the depth profiles of damage modeled by M. J. Catula et al. [11], where a displacement as a Si atom which have a potential energy of 0.4eV above the bulk state [3].

As shown in figures 3 and 4, B_{10} cluster impact creates a high density of vacancies and interstitial at the impact point. This high-density damaged region ranges from the surface to the depth of 20Å, which is comparable to the mean implant depth of B monomer and cluster at 230eV/atom. Considering the density of interstitials shown in figure 4, B_{10} impact causes about 15 atoms per 2Å thickness. On the other hand, most of interstitials reside within 20Å from the impact point as shown in figure 3. It can be concluded that the implant region is fully amorphised with only one B_{10} impact with 230eV/atom. As reported in the previous works [4,5], it is known that TED can be reduced by pre-amorphization of the surface region enough deeply compared to the boron implant range. Thus characteristic damage formation by B_{10} is expected to result in advantages in shallow junction formation. However, the experimental result reports [7], when the incident energy is as low as 500eV/atom. Both boron monomer and cluster implantation show similar diffusion length, and enhanced diffusion is less probable compared to that for several keV B ion implantation. It suggested that, reduction of TED by low energy ion implantation is due to the proximity to the surface. As the incident energy and the projection range decrease, the surface can act as the sink of the vacancy, which suppress the generation of interstitial Si atom to cause TED. In the viewpoint of LSI fabrication process, cluster ion implantation technique is available as an alternative to low-energy boron implantation.

SUMMARY AND CONCLUSIONS

Molecular dynamics simulations of boron monomer and small clusters were performed and the damage caused by these ion irradiations were investigated.

For B_1 implantation, the damage is mainly due to vacancy-interstitial pair, where one lattice Si atom is knocked-on and surrounding Si atom keeps lattice structure. The knocked-on atom tends to remain as tetrafold interstitial. These point-defect structures are considered to cause the transient enhanced diffusion by annealing at low temperature.

On the other hand, B_{10} cluster creates a high density of displacements on the surface at the impact point because of the high-density energy irradiation effect. This damaged region is considered to be amorphized and appears as a box-like shape from the surface to a depth of 20Å, which is comparable with the mean implant depth of the B atoms. When an incident energy is as low as sub-keV, this characteristic damage formation process gives no difference from monomer implantation in annihilation process because the proximity to the surface is more dominant to characterize diffusion of dopant compared to the damage structure in the target. These results supports the availability of decaborane ion implantation as low-energy boron implantation technique for LSI fabrication.

ACKNOWLEDGEMENTS

This work has been supported by NEDO (New Energy and Industrial Technology Development Organization, Japan).

REFERENCES

1. K. Goto, J. Matsuo, T. Sugii, H. Minakata, I. Yamada and T. Hisatsugu, IEDM Tech. Digst., 435 (1996).
2. K. Goto, J. Matsuo, Y. Tada, Y. Momiyama, T. Sugii and I. Yamada, IEDM Tech. Digst., 471 (1997).
3. T. Aoki, J. Matsuo, Z. Insepov and I. Yamada, *1998 International Conference on Ion Implantation Technology Proceedings,* 1254 (1999).
4. D. J. Eaglesham, P. A. Stolk, H. J. Gossman and J. M. Poate, Appl. Phys. Lett., **65**, 2305 (1994).
5. K. S. Jones, P. G. Elliman, M. M. Petravic, and P. Kringhoj, Appl. Phys. Lett., **68**, 3111 (1996).
6. T. Kusaba, N. Shimada, J. Matsuo and I. Yamada, *1998 International Conference on Ion Implantation Technology Proceedings*, 1258 (1999).
7. A. Agarwal, H. –J. gossmann, D. C. Jacobson, D. J. Eaglesham, M. Sosnowski, J. M.Poate, I. Yamada, J. Matsuo and T. E. Haynes, Appl. Phys. Lett., **73**, 2015.
8. F. H. Stillinger and T. A. Weber, Phys. Rev., **B31**, 5632 (1985).
9. J. P. Ziegler, J. P. Biersack and U. Littmark, *The stopping and range of ions in solids;* New York: Pergamon Press (1985).
10. A. M. C. Perez-Martin, J. Dominguez-Vazquez, J. J. Jimenez-Rodoriguez, Nucl. Instr. and Meth., **B164-165**, 431 (2000).
11. M. J. Caturla, T. D. de la Rubia and G. H. Gilmer, Nucl. Instr. and Meth. **B106**, 1 (1995).

Mat. Res. Soc. Symp. Proc. Vol. 669 © 2001 Materials Research Society

AB-INITIO MODELING OF C-B INTERACTIONS IN SI

Chun-Li Liu, Wolfgang Windl, Len Borucki, Shifeng Lu*, Xiang-Yang Liu**
Advanced Process Development and External Research Lab., Motorola
*Process and Materials Characterization Laboratory, Motorola
**Physical Sciences Research Laboratory, Motorola

ABSTRACT

We present the results of ab-initio calculations for the structure and energetics of small boron-carbon (BCI) as well as carbon-carbon (C_2I) clusters in Si , a continuum model for the nucleation, growth, and dissolution of the clusters, and experimental investigation by SIMS. The modeling results suggest that these clusters may play a role in controlling B diffusion in Si and SiGe systems and the experimental results seem to support the modeling findings.

I. INTRODUCTION

SiGeC has several beneficial materials properties over SiGe which include the compensation of film strains by adjustment of the Ge/C ratio, enhanced thermal stability, increased critical film thickness, suppressed transient enhanced diffusion (TED) of boron [1], and preservation of the narrowed band gap of strained SiGe. One particular device application of SiGeC films is the construction of heterojunction bipolar transistors (HBT) using Si/SiGeC/Si heterostructures. SiGeC HBTs have increased performance (higher frequency) due to the smaller band gap [2] and stability of the base profile.

The currently prevalent explanation of suppression of diffusion of boron by carbon is that carbon reduces the free silicon interstitial (I) concentration by forming a C-Si complex, resulting in fewer B-Si complexes, which are believed to be responsible for B TED in Si [1,3]. Reference [1] further showed that the diffusion of carbon incorporated in silicon well above its solid solubility causes an undersaturation of silicon self-interstitials, which further retards boron diffusion.

In this work, we further explore the atomic mechanisms for the effect of carbon on B diffusion through ab-initio investigation of C-B interactions by focusing on C-B split interstitial pairs (CBI). For the first time the results of our ab initio calculations indicate that carbon and boron can interact directly by forming CBI pairs. For the modeled case, CBI is the most important cluster containing B, while the predominant cluster capturing and storing Si interstitials is the C2I cluster. Our modeling predicts that the CBI concentration may not be sufficient to surpass the effect of carbon in reduction of Si interstitials by forming C2I clusters, and thus that the direct interactions between C and B by forming C-B pairs may play a secondary role in suppressing B diffusion. However, a comparison to experiment indicates that BCI complexes might play an even more central role than C2I clusters (our ab initio calculations have error bars of ~0.3 eV).

II. COMPUTATIONAL METHODS AND ASSUMPTIONS

VASP (the Vienna Ab-initio Simulation Package) [4] was used for the calculations in this work. The ultra-soft pseudopotentials with a plane wave basis supplied with VASP and the generalized gradient approximation (GGA) were used. A supercell of 64 Si atoms and a cutoff energy of 21 Ry corresponding to the cutoff energy of the C pseudopotential were chosen for all calculations. Diffusion barriers were determined using the nudged elastic band method (NEBM) [5] implemented in VASP, which determines a (local-) minimum energy diffusion path once the initial and final geometries are chosen. We have used this method in the past, e.g., to determine dopant diffusion at Pt grain boundaries [6], the B diffusion mechanism [3], or C incorporation mechanisms by surface diffusion processes [7], and have found it to be accurate and efficient. Total energies for charged states were calculated similarly to Ref. [3].

III. RESULTS AND DISCUSSION

3.1 Ab-Initio Modeling Results
Charge State of Substitutional Carbon and Boron in Si
Carbon is a group-IV element and, as such, not an immediate candidate for a Fermi-level dependence of its energy in substitutional sites. However, substitutional C in Si has been shown to give rise to an antibonding C s-derived state which has been proposed to be slightly higher or lower than the conduction band edge of Si [8,9]. In the latter case, a charge-state dependence would exist. Therefore, we calculated the total energies of substitutional carbon in Si at different charge states. As expected, neutral substitutional carbon is the most stable charge state. In comparison, the most stable charge state for B in Si is −1 [3].

Migration Energy of C in Si
In order to benchmark our calculations against the previous work by Capaz *et al.* [10], we recalculated the migration energy of interstitial carbon using the nudged elastic band method. We obtained a value of the migration energy of 0.56 eV, consistent with the value of 0.50 eV from Capaz *et al.* and comparable to the experimental values of 0.73-0.88 eV [11]. The fact that the migration energy is smaller than experiment is consistent with calculations for other interstitial defects in Si and has been suspected to be due to the known gap problem in density-functional calculations, which predict a theoretical band gap of about 0.6 eV, half as big as the experimental value. This should have as a consequence that all configurations with electronic states above the valence band edge might be too low in energy, with error bars of ~0.3 eV.

Configuration of the C-B Split Interstitial in Si (CBI)
Finding the minimum configuration for clusters involving different atomic species and native point defects can be an extremely tedious and time-consuming process. Luckily, extensive investigations have been performed in the past for B-B [12] and C-C interactions [9,10,13]. For both elements it has been found that substitutional pairs on neighboring sites have a repulsive interaction energy, ~0.9 eV in the case of B [9], and ~1.8 eV in the case of C [13]. Therefore, we can safely exclude substitutional B-C pairs as the dominant configuration. Since both impurities are known to form pairs with self-interstitials rather than with vacancies and have an interstitial-assisted diffusion mechanism, we thus conclude that the dominant B-C pair must involve a Si self-interstitial. Since the lowest-energy structure for B2I ([12]; for this structure, extensive minimizations have been performed and many competing configurations have been examined) and C2I pairs (for the latter, see below) has been found to be a ⟨100⟩ split dumbbell, we assumed that the same structure will have at least a competitively low energy for the mixed CBI pair (Fig. 1). However, we examined different possible

orientations of the CB dumbbell to have at least a sanity check that our dumbbell assumption is reasonable.

The total energies of the C-B split interstitials in $\langle 100 \rangle$ and $\langle 111 \rangle$ orientation were calculated to be 0.55 eV and 1.59 eV higher than that of the $\langle 100 \rangle$ dumbbell, respectively. The results indicate that the $\langle 100 \rangle$ oriented CBI pair has the lowest energy and thus is the most stable dumbbell orientation. We find the neutral charge state to be most stable at midgap, followed by the negative (0.05 eV higher) and positive (0.67 eV higher) ones.

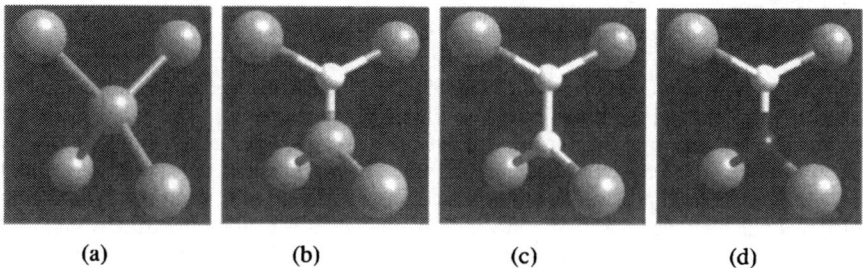

| (a) | (b) | (c) | (d) |

Figure 1. (a) Perfect Si with four nearest neighbors. (b) CI cluster. (c) C_2I cluster. (d) BCI cluster.

Binding Energies

A: C-I split interstitial pair

We find for CI pairs in the $\langle 100 \rangle$ split configuration the neutral charge state to have the lowest energy, 0.19 (0.56) eV lower than the -1 ($+1$) state at mid-gap. Taking the Si self-interstitial pair (two Si atoms occupying a single Si lattice site) and a neutral substitutional carbon as references, we calculated the binding energies for a neutral C-I split interstitial to be 1.75 eV. In comparison, the theoretical binding energy for a B-I split interstitial is 0.80 eV [3].

B: C-B split interstitial pair (CBI)

Kinetically, there are two ways to form C-B split interstitials, B-I + Cs = CBI and C-I + Bs = CBI. Therefore, there are two sets of references to calculate the binding energies of the CBI pair. For the first process, we find the binding energy for the lowest-barrier decay process (the superscripts indicate charge states) at mid-gap

$$Eb\ (CBI^0 \rightarrow C^0 + BI^0) = 2.36\ eV.$$

For the second process, we find (h is a hole)

$$Eb\ (CBI^0 \rightarrow CI^0 + B^- + h^+) = 1.11\ eV.$$

C: Small Clusters

In order to simulate the other small clusters that compete with the BCI in a continuum model, we also considered the smallest stable clusters larger than the impurity-interstitial pairs, which are B2I and C2I. Previously, Capaz *et al.* have identified two types of C2I clusters in Si [14]. In the "type A" configuration, a C-I split interstitial is slightly perturbed by a nearby substitutional carbon. The "type B" configuration consists of a Si bond interstitial between two substitutional carbon atoms. However, since we found that for BCI and B2I clusters a $\langle 100 \rangle$ dumbbell configuration has the lowest energy, we also tried this structure for the C2I cluster and found its energy to be 0.3 eV lower than the two Capaz structures. The binding energies calculated for the C2I $\langle 100 \rangle$ dumbbell as well as for the B2I cluster [12] at midgap are listed below.

$$CI + C = C2I \qquad\qquad Eb=1.28\ eV$$
$$BI + B = B2I \qquad\qquad Eb=1.49\ eV$$

3.2 Results of Continuum Modeling and Experiment

Despite the efforts to understand the mechanisms for the effect of the presence of carbon on reduction of boron diffusion [1,2,4], modeling has not been completely successful to explain the effect based on the mechanisms proposed so far, suggesting a more complex nature of the matter. For example, when Rücker *et al.* simulated B and C diffusion profiles in Si after annealing at 900 C for 45 minutes, they were only able to match the measured profiles by assuming a constant supply of C from the C-rich region [15], even though the undersaturation of Si interstitials was predicted correctly.

Our ab-initio results suggest the possibility that CBI pairs form in addition to the one-species clusters and thus influence the annealing behavior, e.g., by reducing the population of mobile B in Si and therefore directly reducing B diffusion. We want to examine this influence in the following in comparison to experiment.

The SIMS analysis of the C and B concentration profiles in SiGe seems to indicate that C and B profiles peak at the same location after annealing. The structures were grown between 500 and 650 ºC by a reduced-pressure chemical vapor deposition (RPCVD) epitaxial reactor. As seen in Fig.2 (a), the C profile is flat as deposited and the B profile is confined in a very narrow region (Note the C and B profiles were obtained separately in SIMS analysis for as-deposited films. We chose to use Fig.2 (a) as the starting profiles for modeling). After rapid thermal annealing at 1042 ºC for 20 s, the peaks in the C and B concentration profiles are found at the same location, although C has diffused significantly as indicated by the broadening of the C profile [Fig. 2 (b)]. The strong broadening of the C concentration profile should be due to the significantly higher binding energies of the C-I pairs over BI. In order to understand this behavior and to test our above calculations, we combined our results into a continuum model, starting from a well-tested model for B diffusion and clustering [12], and applied it to a model structure similar to the one in Fig. 2(a). Our diffused distribution for B and C [Fig. 2(b), 1042 ºC, 60 s (we use an extended time since our model without Ge and without further calibration of the ab-initio numbers and initial conditions underestimates C diffusion)] looks similar to the experimental findings, and also indicates that the major reason for the peaking of the C is an undersaturation of self-interstitials in the area of the original C maximum due to the formation of mainly C2I clusters, to a lesser degree of BCI clusters. Nevertheless, our model predicts BCI clusters to be the second-most important cluster in this set-up, with a post-anneal peak concentration ~1/5 as high as the C2I peak.

The ratio C2I / BCI depends sensitively on the exact values for their respective formation energies, which are not exactly known due to the previously discussed error bars in defect ab initio calculations of energies and which need to be calibrated. Indeed, there is one important difference between experiment and simulation that indicates a still incomplete modeling picture: Whereas B and C peak at exactly the same location in

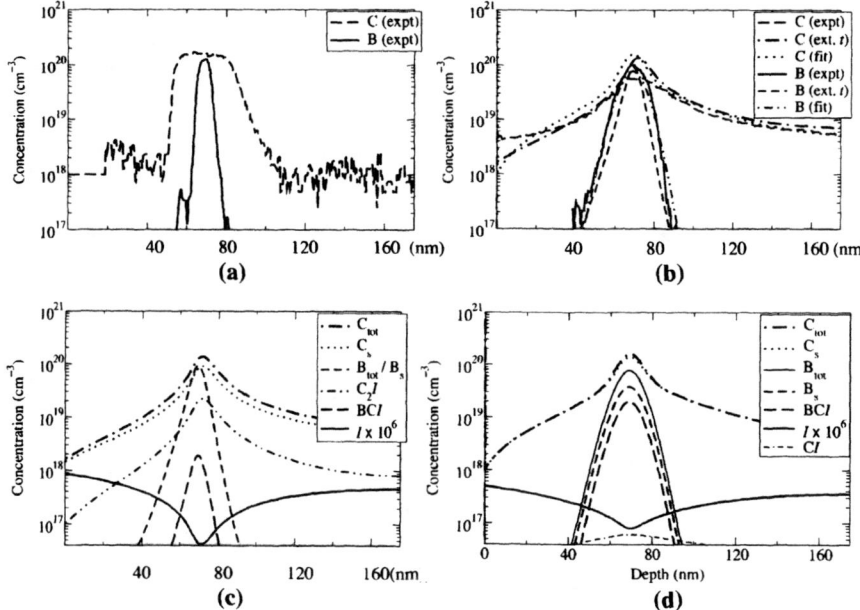

Fig. 2 (a) As-deposited experimental concentration profiles for boron and carbon used as initial profiles for the continuum modeling. (b) Experimental and simulated profiles for the total boron and carbon concentrations after annealing at 1042 C for 20 seconds. (c), (d) Different contributions to the simulated total boron and carbon profiles shown in (b).

experiment [Fig. 2(b)], the simulated peaks are slightly shifted due to an initial shift in the centers of the as-deposited distributions [Fig. 2(c) and Fig. 2(d)]. This can be a clear indication of the dominance of BCI clusters (in contrast to our uncalibrated model), or maybe a stress-compensation effect (B and C cause compressive, Ge tensile stress) that is not captured in our model. However, if our interpretation is right, it clearly rules out the previous suggestion of the C-I interaction as the only reason for suppression of TED. A simple experiment without Ge can clarify this point further.

Finally, Kimerling et al. [16, 17] have conducted a series of deep-level transient spectroscopy studies involving samples of controlled boron, oxygen, and carbon content in Si. They concluded that CBI pairs appeared at 150 °C and increased in amount up to 400 °C as the annealing temperature was raised, while other types of interstitial pairs (C-C, C-I, B-I, C-O, B-O) decreased in amount. They suggested that C-B complex clusters exist when both C and B are in Si, in agreement with our results.

CONCLUSIONS

In conclusion, we have performed initial atomistic calculations for C-B split interstitials and obtained the fundamental parameters. We have also formulated a continuum model based on these calculations and applied it to the annealing of a deposited B-C structure. Lacking Ge and the final calibration, our model qualitatively predicts experimental findings for a similar SiGeC structure reasonably well, which suggest strong C diffusion, nearly immobile B, and formation of a peak in the C concentration. Our model finds the BCI cluster to be the second-most important cluster after C2I clusters for the examined concentrations and conditions. However, we interpret the different peak positions in simulation and experiment to indicate a dominance of BCI clusters, or, to a lesser degree, a stress-compensation effect, but to rule out the previous suggestion of the C-I interaction [1] as the sole responsible reason for suppression of B TED. An experiment comparable to ours, but without Ge, could clarify this point further. Our result of a significant role of BCI clusters is in agreement with previous deep-level spectroscopy findings of these clusters [16,17].

REFERENCES
1. R. Scholz, U. Gosele, J.-Y. Huh, T. Y. Tan, Appl. Phys. Lett. **72** (2), p. 200 (1998).
2. P. A. Stolk, D. J. Eaglesham, H.-J. Gossmann, and J. M. Poate, Appl. Phys. Lett. **66** (11), p. 1370 (1995).
3. W. Windl, M. M. Bunea, R. Stumpf, S. T. Dunham, and M. P. Masquelier, Phys. Rev. Lett. **83**, 4345 (1999).
4. G. Kresse and J. Hafner, Phys. Rev. B **47**, 558 (1993).
5. H. Jonsson, G. Mills, K. W. Jacobsen, *Nudged Elastic Band Method for Finding Minimum Energy Paths of Transitions*, ed. B. J. Berne, G. Ciccotti, and D. F. Coker, World Scientific, Singapore, 1998.
6. R. Stumpf, C.-L. Liu, and C. Tracy, Appl. Phys. Lett. **75**, 1389 (1999).
7. C.- L. Liu, L. J. Borucki, T. Merchant, M. Stoker, and A. Korkin, Appl. Phys. Lett. **76**, 885 (2000).
8. H. P. Hjalmarson, P. Vogl, D. J. Wolford, and J. D. Dow, Phys. Rev. Lett. **44**, 810 (1980); J. D. Lorentzen, G. H. Loechelt, M. Meléndez-Lira, J. Menéndez, S. Sego, R. J. Culbertson, W. Windl, O. F. Sankey, A. E. Bair, and T. L. Alford, Appl. Phys. Lett. **70**, 2353 (1997).
9. W. Windl, O. F. Sankey, and J. Menéndez, Phys. Rev. B **57**, 2431 (1998).
10. R. B. Capaz, A. Dal Pino, Jr., and J. D. Joannopoulos, Phys. Rev. B **50**, 7439 (1994).
11. T. Y. Tan and U. Gösele, in *Handbook of Semiconductor Technology-Electronic Structure and Properties of Semiconductors*, ed. by K. A. Jackson and W. Schroeter, vol. 1 (Wiley & Sons, New York, 2000), p. 231.
12. X. -Y. Liu, W. Windl, and M. Masquelier, Appl. Phys. Lett. **77**, 2018 (2000).
14. R. B. Capaz, A. Dal Pino, Jr., and J. D. Joannopulous, Phys. Reb. B. **58**, 9845 (1998).
13. W. Windl, J. D. Kress, A. F. Voter, J. Menéndez, and O. F. Sankey, in *Defects and Diffusion in Silicon Processing*, ed. by T. Diaz de la Rubia, S. Coffa, P. A. Stolk, and C. S. Rafferty, Mat. Res. Soc. Proc. **469** (Pittsburgh, PA, 1997), p. 443.
15. H. Rücker, B. Heinemann, W. Ropke, R. Kurps, D. Kruger, G. Lippert, and H. J. Osten, Appl. Phys. Lett. **73**, 1682 (1998).
16. L. C. Kimerling, M. T. Asom, and J. L. Benton, P. J. Drevinsky, and C. E. Caefer, Materials Science Forum, **38-41**, 141 (1989).
17. P. J. Drevinsky, C. E. Caeler, S. P. Tobin, J. C. Mikkelsen, and L. C. Kimerling, Proc. Mater. Res. Soc. **104**, 167 (1988).

Mat. Res. Soc. Symp. Proc. Vol. 669 © 2001 Materials Research Society

Computer Simulation of Decaborane Implantation into Silicon, Annealing and Re-crystallization of Silicon

Zinetulla Insepov and Isao Yamada
Laboratory of Advanced Science &Technology for Industry, Himeji Institute of Technology, 3-1-2 Kouto, Kamigori, Ako, Hyogo 678-1205 Japan

ABSTRACT

Molecular Dynamics (MD) and Activation-Relaxation Technique (ART) models of decaborane ion implantation into Si and following rapid thermal annealing (RTA) processes have been developed. The B and Si atomic positions for implantation of accelerated decaborane ions, with total energy 3.5- 15 KeV, into Si substrate were obtained by MD simulation. The main difference between monomer and decaborane ion implantation with the same doses is the formation of a large amorphized area in a subsurface region for the decaborane case. The number of displaced Si atoms shows non-linear energy dependence at low impact energies. At higher energies of the investigated range of the decaborane energy range, however, a linear dependence is observed in accordance with the prediction of the Kinchin-Pease formula. A new method that incorporates Activation-Relaxation Technique (ART) with MD has been developed and used to study re-crystallization of Si amorphized in the implantation process.

INTRODUCTION

Implantation of a decaborane cluster ion, $(B_{10}H_{14})^+$, has first successfully been realized in experiments by a Kyoto University group, in cooperation with Fujitsu company [1].

Implanted B dopant diffusion in Si has recently attracted much attention because of the anomalously high diffusion rate, the transient enhanced diffusion (TED) effect that is typical for conventional implantation method that uses single ion implantation. This effect has immediately been revealed after the Si samples heavily doped with decaborane ions were rapidly thermally annealed at a temperature of about 800 - 1000 °C for 10s. [2]. A physical model of TED states that every implanted B^+ ion creates at least one Si-interstitial (the so called +1 model), and that the highly migrating B interstitials are due to the kick-out mechanism caused by the Si self-interstitials [3].

Dopant diffusion in Si is a well studied area of semiconductor physics [4-10]. The need for modeling of low energy decaborane implantation has emerged only recently, following experimental confirmation of this process as a potential replacement for low energy single B^+ ion implantation. To the authors' knowledge, there has been only one simulation of 1.5 and 4 keV decaborane implantation into a Si substrate with Molecular Dynamics (MD) [11].

Development of new modeling methods of re-crystallization of amorphized Silicon substrate area is an important current topic in fundamental materials science [13-17]. The main deficiency of MD method in studying the kinetics of re-crystallization of amorphous materials is that the simulation time in MD, usually less than 20 ps, is too short compared with the times of crystal ordering that are many orders of magnitude longer.

The aim of this paper is to report on the development of new computation models of decaborane implantation into Si and of dopant diffusion in irradiated Si samples. We also combined MD and Activation-Relaxation Technique (ART), and using the best features of both methods studied the thermal annealing process.

MODELS

A. Implantation model

Implantation of energetic cluster or heavy ions makes an extreme demand for the development of new MD models because periodic boundary conditions (PBC) that are usually used in MD simulations may severely restrict the total computation time to $t < 2L/c$, where L is a typical system size, and c is the speed of sound. For a typical system $L \sim 100$Å, and $c \sim 5$ km/s for Si, that gives $t < 4$ ps. If computation is continued longer that this time interval, the elastic waves generated due to cluster impact are reflected back from the system borders and cause a negative feedback to the simulation area.

In this paper, the boundary conditions were used as in our previous paper [12], where a new Hybrid Molecular Dynamics (HMD) method was proposed. This new method combines conventional atomistic MD, for the central cluster collision zone, with a continuum mechanics representation, for the rest of a system. It significantly reduces the system size and can keep the accuracy of the energy flow through the system boundaries. According to this technique, the response of the continuum part to the atomistic MD-part is represented by two components. One is determined by forces calculated from a stress tensor and it depends on the magnitude of deformation of the boundary layers. The second one controls the energy balance and is introduced by energy absorbing walls, which were simulated by thermal diffusion equations.

The decaborane molecule ion bombarding a Si surface was modeled as a B_{10} cluster. The atomic positions at B10 cluster implantation were obtained at room and higher temperatures. The basic cell size for the MD simulation was determined from the cluster energy between up to 15 keV and experimental dose of 10^{13-16} ion/cm^2 [1, 22]. A cubic slab of (80x40x40) Å^3 was cut about the z-axis and a cylindrical sample was used together with our thermal boundary conditions (TBC). The MD system has consisted of about 200,000 Si atoms and 10 B atoms, with a total kinetic energy of 3.5-15 KeV. Atoms in the four bottom atomic layers were fixed.

The ZBL potential at short distances combined with the Stillinger-Weber potential at equilibrium distances was used to evaluate interactions between two and three Si atoms, as usual [11,12]. Interaction between B and Si atoms was modeled via the ZBL screened Coulomb potential at short distances, $r < r_1 = 0.52$ Å, and with a Morse-type potential at long distances, $r > r_2 = 0.86$ Å [11]. The binding energy of B in the Si lattice was used as an adjustable parameter with values between 0 to 1.5 eV. Interaction between two B dopants were modeled with an (exp-6)-type potential, with the depth of 0.2 eV, $r_{eq} = 1.5$ Å.

In MD simulation, all B and Si atomic positions and velocities are calculated as output data. A distribution of displaced Si atoms, with potential energies > 0.2 eV than average potential energy, as well as other defects can be obtained from the MD result.

In addition, the structure of the irradiated Si substrate was studied by the radial distribution function, rdf, for the heavily disordered area of decaborane impact.

B. Diffusion model

As the B atomic mass is only 1/3 of the Si atomic mass, the implanted B atoms undergo rare but violent collisions with neighboring Si atoms. Therefore, the motion of B atoms could not be considered as a Brownian one. This makes finding the B diffusion constant a challenging problem for theory. Molecular Dynamics can in principle find the diffusion coefficient, but it is incapable of treating a realistic system for a long computation time. This is usually limited to tens of ps, which is not long enough to simulate diffusion of B during RTA processing. Nevertheless, MD could have been successfully used to study the details of dopant diffusion in different directions, e.g. in xy-plane and in z-direction.

To calculate the Boron and Silicon diffusion coefficients, we have used a Green-Kubo formula, by calculating velocity autocorrelation functions (vacf), $f(t)$, for Boron and Silicon atoms. A frequency dependent transport coefficient has been obtained by a Fourier transform of the autocorrelation function [20].

$$f(t) = \left(\frac{m}{3k_B T} \right) \sum_{i=1}^{N} \langle \vec{v}_i(0) \vec{v}_i(t) \rangle_0,$$
$$D(\omega) = \left(\frac{k_B T}{m} \right) \int_0^\infty f(t) \cos(\omega t) dt,$$

$$(1)$$

here, m is the atomic mass for B or Si atom, k_B is the Boltzmann constant, T is the absolute temperature, $\langle ... \rangle_0$ means averaging a scalar product of the atomic velocities over an equilibrium ensemble (or time). As usual, the computing time that should not exceed $t \sim 4$ ps because of the restricted system size.

In this paper, we have calculated the diffusion constants, $D_i(\omega=0)$, (i means B, Si), for Boron and Silicon atoms separately. We have also studied the diffusion motion in xy-plane and along the z-direction, by computing $f(t)$ for velocity components in different directions, namely in xy-plane and along the z-direction. This is justified because the surface could influence the diffusion of atoms.

C. Re-crystallization model

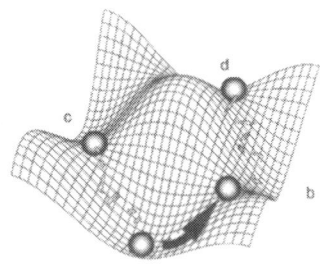

Fig. 1. A potential energy map illustrating the principle of our ART-MD model: an atom at a potential minimum (a) may walk to a nearby saddle point position (b) by an activation procedure (big arrow). Atoms at (c), and (d) may relax down the potential by MD method (small zigzag arrows).

The recent development of new computation models for studying diffusion and re-crystallization processes in amorphous materials [13-17] guides further examination of diffusion and ordering processes by combination of conventional MD and MC methods that together could, at least in principle, treat such slow processes as crystal ordering. Activation–Relaxation Technique (ART) [15-17] has successfully been applied for crystallization of a small amorphous system of 216 Si atoms in a cube with PBC in [15]. Energy stabilization was not needed because a static energy method was used. In a recent paper [17], a procedure very similar to ART has been incorporated into an MC method and used for structural optimization of an amorphous network and for hydrogen diffusion in diamond.

ART has been used for an activation step, to deliver a displaced Si atom and it's nearest neighbors to a nearby saddle point (see Fig. 1). After reaching saddle point, the atoms were allowed to evolve according to Newton's equations of motion. The advantage of our approach is that it allows studying very large systems that are typical for ion implantation problems. We have successfully applied this method to a Si system with Stillinger - Weber potential containing several thousand atoms within the amorphous region and with the rest of system containing of about 10^5 Si atoms. As we discuss further, the proposed method could simulate re-crystallization of the amorphized zone generated by implantation. To the authors' knowledge, the ART has not been used before in combination with MD for the simulation of re-crystallization of amorphous areas.

In ART, the conjugate gradient (CG) forces G_i are applied to all atoms in an amorphous state:

$$\vec{G}_i = \vec{F}_i - (1 + \alpha)(\vec{F}_i \cdot d\vec{r}_i) \cdot d\vec{r}_i, \quad i = 1, N \tag{2}$$

where dr_i are unit vectors drawn in random direction. These CG forces are equal to the appropriate MD forces F_i acting on an atom i, in the direction perpendicular to the unit vector dr_i, and they are equal to a small portion α of F_i, in the direction opposite to dr_i. They are equal to zero at a saddle point of the potential map (positions (b), (c) and (d) in Fig. 1). The ART procedure applied iteratively to all displaced atoms and to all their nearest neighbors (movers) would lead them to the nearest saddle points and therefore this technique is suitable to our problem.

The most difficult part in combing ART with MD is the problem of absorption of excess energy that has to be released by a mover after it has forcibly been moved to saddle point. To stabilize the total energy of the system, we have put to zero all the velocities of the movers, after each successful ART move.

If necessary, this method could easily find all the nearest saddle points for a displaced atom as it was shown in [17], and then this knowledge could be used to develop a very powerful technique for studying ordering and thin film deposition and formation processes.

RESULTS

A. Implantation and amorphization

Several Boron atoms were reflected back into vacuum at a lower energy of 3.5 keV and most of them were implanted energies higher than 4 keV. This result agrees with the previous paper [11] where a lower decaborane energy region of 1.5 and 4 keV has been studied.

A large amorphized pocket has been obtained directly under the surface. Fig. 2 shows energy dependence of the total number of displaced Si atoms, 7 ps after of B10 cluster implantation. The displaced Si atoms that were obtained from their potential energy: $U_i > U_{av} + 0.2$ eV, where U_{av} is the

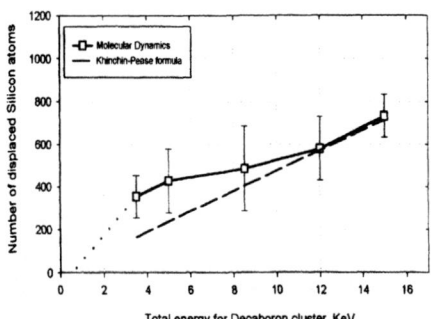

Figure 2. The number of displaced Si atoms vs total energy of B_{10} cluster. Squares – these MD results; dashed line represents Kinchin-Pease formula, with a E_{th} as an adjusting parameter [9].

average potential energy in the system. The simulation of B^+ implantation has shown that there are almost no amorphized areas in the Si substrate. This figure shows that in the low B10 energy region, between 3.5 and 10 KeV, this relation has non-linear behavior. For energies higher than 10KeV, this dependence approaches a linear prediction of the Kinchin-Pease formula [21].

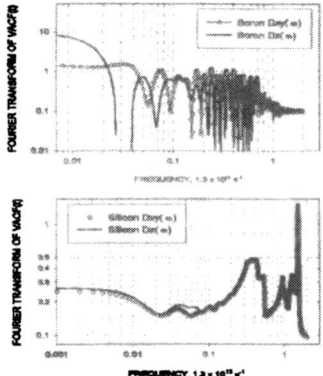

Fig. 3. Fourier transforms of the velocity autocorrelation functions for Boron (a) and Si (b) atoms at 700K and after 15KeV B10 implantation.

B. Velocity auto-correlation functions and diffusion constants

We have calculated *vacf* of Boron and Si atoms separately, both for velocity components in *xy*-plane, $f_{xy}(t)$, and in *z*-direction, $f_z(t)$, at temperatures of 300, 500, 700, and 1300K, for samples implanted with B10 clusters with energies 3.5-15 KeV, for revealing any asymmetry in the atomic movement before the re-crystallization procedure has been applied. The calculated vacf for Boron atoms show high value for the *vacfz*, compared with that of in *xy*-plane. At a low energy of 3.5 KeV, in *xy*-plane diffusivity for implanted Boron was higher than that in *z*-direction. Almost all implanted Boron atoms resided inside an amorphous pocket. At 10-15 KeV of implantation energy, all Boron atoms deeply penetrated inside the substrate and none of them were left inside the amorphous zone. We assume that the mechanism for diffusion might be quite different at a high energy than at energies lower than 10 KeV. *Vacf* for Silicon atoms shows a usual oscillating behavior that means that there is almost no diffusion motion at such low temperature and during such short time interval.

Surprisingly high diffusion coefficient was obtained for the Boron diffusion when only *z*-component (upper plot in Fig. 3) of the velocity has been used and this is shown in this figure.

C. ART+MD results

Figure 4 shows a radial distribution function for all Si atoms inside the amorphous region, including all displaced atoms and all their nearest neighbors that have also been moved according to formula (2) derived from ART-MD technique. As is seen in this figure, crystallinity of the region has been improved after of about 6000 activation-relaxation moves. We have also checked the total potential energy for the region and found that the energy become much lower but it is still slightly higher than in perfect SW-crystal. We have traced all trajectories for all the movers in time. The data show that almost all

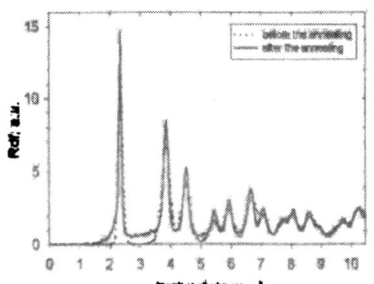

Figure 4. The radial distribution function for amorphous region before the annealing procedure (dotted line) and after ART procedure (solid line).

movers are close to their initial positions.

CONCLUSIONS

Molecular Dynamics and Activation-Relaxation Technique models of B cluster implantation into Si substrate were developed which allow simulation of implantation and subsequent rapid thermal annealing. A large amorphized area directly beneath the Si surface was obtained when this substrate was implanted with a B_{10} cluster. Several Boron atoms are reflected back into vacuum at lower energy and most of them were implanted at high kinetic energy. The number of disordered Si atoms shows non-linear dependence on implantation energy at low energies of 3.5-10 KeV but approaches a linear prediction of the Kinchin-Pease formula above 10 keV, with the threshold displacement energy used as an adjustable parameter. Transport coefficients for B and Si atoms obtained as Fourier transforms of the velocity autocorrelation functions show a high diffusivity for B atoms in z-direction. A new re-crystallization method combining ART and MD has been developed that allows to simulate annealing of amorphized regions in Si after implantation of energetic B_{10} clusters.

REFERENCES

1. I. Yamada, J. Matsuo, E.S. Jones, D. Takeuchi, T. Aoki, T. Goto, T. Sugii, in *"Materials Modification and Synthesis by Ion Beam Processing"*, edited by D.E. Alexander, N.W. Cheung, B. Park, and W. Skopura (Mat. Res. Soc. Symp. Proc. **438**, Pittsburg PA 1997), pp. 363-374.

2. A. Agarwal, H.J. Gossmann, D.C. Jacobson, D.J. Eaglesham, M. Sosnowski, J.M. Poate, I. Yamada, J. Matsuo, and T.E. Haynes, *Appl. Phys. Lett.* **75**, 2015 (1998).

3. E.Chason, S.T.Picraux, J.M.Poate et al., *Appl. Phys. Rev.* **81**, 6513 (1997).

4. C.S.Nichols, C.G.Van de Walle, and S.T.Pantelides, *Phys. Rev.* **B40**, 5484 (1989).

5. P.M.Fahey, P.B.Griffin, and J.D.Plummer, *Rev. Mod. Phys.* **61**, 289 (1989).

6. G.H.Gilmer, T.D. de la Rubia, D.M.Stock, M.Jaraiz, *Nucl. Instr. Meth. in Phys. Res.* **B102**, 247 (1995).

7. M.Jaraiz, G.H.Gilmer, J.M.Poate, T.Diaz de la Rubia, *Appl. Phys. Lett.* **68**, 409 (1996).

8. S. Tian, M. Moris, S.J. Morris, B. Obradovic, A.F. Tasch, in *"Materials Modification and Synthesis by Ion Beam Processing"*, edited by D.E.Alexander, N.W.Cheung, B.Park, and W.Skopura (Mat. Res. Soc. Symp. Proc. **438**, Pittsburg PA 1997), pp. 83-88.

9. M.Tang, L.Colombo, J.Zhu, T.Diaz de la Rubia, *Phys.Rev.* **B55**, 14279 (1997).

10. P.J.Bedrossian, M.-J.Caturla, and T.Diaz de la Rubia in *"Materials Modification and Synthesis by Ion BeamProcessing"*, edited by D.E. Alexander, N.W. Cheung, B. Park,and W. Skopura (Mat. Res. Soc. Symp. Proc. **438**, Pittsburgh PA 1997), pp. 715-720.

11. R.Smith, M.Shaw, R.P.Webb, M.A.Foad, *J. Appl. Phys.* **83**, 3148 (1998)

12. Z. Insepov, M. Sosnowski, I. Yamada, *Nucl. Instr. Methods Phys. Res.* **B127/128**, 269 (1997).

13. A.Voter, *J. Chem. Phys.* **106**, 4665 (1997)

14. F. Wooten, K. Winer, D. Weaire, *Phys. Rev. Lett.* **54**, 1392 (1985)

15. G.T. Barkema, N. Mousseau, ibid, **77**, 4358 (1996)

16. N. Mousseau, G.T. Barkema, *Phys. Rev.* **E57**, 2419 (1998)

17. M. Kaukonen, J. Perajoki, R.M. Nieminen, G. Jungnickel, and Th. Frauenheim, *Phys. Rev.* **B61**, 980 (2000)

18. H.J.C.Berendsen, W.F.Van Gunsteren, In *"Molecular liquids, dynamics and interaction"*, edited by A.J.Barnes, W.J. Orville-Thomas, and J. Yarwood (NATO ASI series C135, Reidel, NY 1984) pp. 475-500

19. M.P. Allen and D.J. Tildesley, *"Computer simulation of liquids"*, Clarendon, Oxford, 1987.

20. A. Rahman, *Phys.Rev.* **136**, A405 (1964); D. Levesque and L. Verlet, *Phys.Rev.* **A2**, 2514 (1970).

21. M.J. Norget, M.T. Robinson and I.M. Torrens, *Nucl. Eng. And Des.* **33**, 50 (1975).

Mat. Res. Soc. Symp. Proc. Vol. 669 © 2001 Materials Research Society

Effect of the Ge preamorphisation dose on the thermal evolution of End of Range defects

B. Colombeau[1], F. Cristiano[2], J-C. Marrot[2], G. Ben Assayag[1] and A. Claverie[1]
[1] CEMES/CNRS, 29 rue J.Marvig, 31055 Toulouse Cedex, France
[2] LAAS/CNRS, 7 av. colonel Roche, 31077 Toulouse Cedex, France

ABSTRACT

In this paper, we study the effect of the Ge^+ preamorphisation dose on the thermal evolution of End of Range (EOR) defects upon annealing. Amorphisations were carried out by implanting Ge^+ at 150 keV to doses ranging from 1×10^{15} ions/cm^2 to 8×10^{15} ions/cm^2. Rapid Thermal Annealing (RTA) was performed for various time/temperature combinations in nitrogen ambient. Plan view transmission electron microscopy under specific imaging conditions was used to measure the size distributions and densities of the EOR defects. We found that for a fixed thermal budget, the increase in the Ge ion dose results in an increase in the defect density but has no effect on the defect size distribution. This invariance of the mean size of defects with respect to the initial supersaturation introduced in the matrix is an expected characteristic of a conservative Ostwald ripening mechanism. Moreover, the total number (N_b) of Si interstitial atoms bound to the EOR defects is a monotonically increasing function of the Ge ion dose. Furthermore, we found that N_b is directly proportional to the number of Si atoms in excess of the vacancies found below the a/c interface as calculated by Monte Carlo simulations. This is consistent with the "excess interstitial" model which explains the origin of the EOR defects.

INTRODUCTION

In order to realise ultra-shallow junctions compatible with advanced CMOS technology, it is now admitted that preamorphisation of the wafer prior to low energy dopant implantation has several advantages [1]. After annealing of such preamorphised implants, End of Range (EOR) defects are formed below the a/c interface. As these defects strongly influence dopant diffusion [2], it is necessary to know how process parameters such as the preamorphisation ion dose affect their thermal evolution.

The goal of this paper is to study the effect of the Ge^+ preamorphisation dose on the thermal evolution of End of Range (EOR) defects upon annealing. In other words, we investigate the influence of the amplitude of the initial supersaturation of Si interstitial atoms (Si(int)s) located below the a/c interface on the kinetics of EOR defects. Indeed, previous studies [3,4] have investigated the effect of the preamorphisation dose on the defect density of EOR defects, but no conclusions have been drawn on the separate evolution of the mean size and defect density upon annealing. Moreover, until now, no publications have clearly shown the so-called "initial supersaturation effect" [5]. To this purpose, we have studied by TEM samples implanted with various Ge amorphisation doses. The thermal evolution of the EOR defects from these analyses has been compared to the Ostwald ripening theory. Finally, these results have been explained following the "excess interstitial" model according to which the EOR defects are formed by the precipitation of the interstitial excess induced by the implantation process.

EXPERIMENTAL DETAILS

Czochralski-grown n-type Si(100) wafers of 2-4 ohm.cm resistivity were used in our experiments. Amorphisations were carried out by implanting germanium ions at 150 keV to doses ranging from 1×10^{15} ions/cm^2 to 8×10^{15} ions/cm^2. These implantations result in the formation of amorphous layers of various thickness (from 173 to 200 nm). Samples were cut and annealed in a Rapid Thermal Annealing (RTA) system at 1000°C for 30 and 100 sec, under nitrogen ambient. During this annealing, solid phase regrowth of the amorphous layer occurs while a band of End of Range (EOR) defects is formed below the a/c interfaces. Plan view and cross sectional samples for Transmission Electron Microscopy (TEM) analysis were prepared by mechanical thinning followed by ion beam milling. Both Weak Beam Dark Field (WBDF) and Symmetrical Bright Field (SBF) imaging conditions were used to measure the size distribution and density of the EOR defects. This allows the number of Si atoms bound to the defects, N_b, to be calculated.

RESULTS AND DISCUSSIONS

Thermal evolution of EOR defects

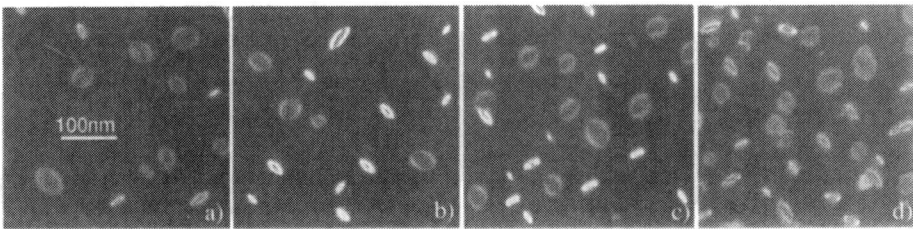

Figure 1: Weak Beam Dark field micrographs showing samples implanted with Ge 150 keV to a dose of a) 1×10^{15} ions/cm^2 b) 2×10^{15} ions/cm^2 c) 4×10^{15} ions/cm^2 d) 8×10^{15} ions/cm^2, after RTA annealing at 1000°C for 100 sec.

Figure 1 shows plan view TEM micrographs taken from samples implanted with 150 keV Ge$^+$ ions to various ion doses after annealing at 1000°C for 100 sec. For doses below 8×10^{15} ions/cm^2, only faulted dislocation loops can be observed. The size distributions of the dislocation loops for the different Ge ion doses are presented in figure 2. It is striking to note that the increase in the Ge ion dose only results in a vertical shift of the size distributions. As a consequence, the shape of the size distributions and the defect mean size are unaffected, while the defect density increases, as shown in figure 3a. Therefore, the total number of Si(int)s bound to the loops, N_b, also increases, which is not surprising since the number of available Si(int)s (the initial supersaturation) increases as a function of the Ge ion dose (figure 3b).

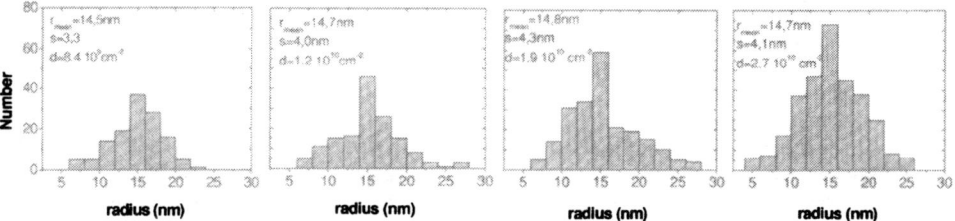

Figure 2: *TEM measurements of size distributions of samples shown in Figure 2.*

Similar results have been obtained after a shorter anneal (1000°C, 30 sec). Furthermore, we have found that, for a given dose, the total number of Si(int)s stored in the loops remains constant upon annealing at 1000°C, indicating that the EOR defects evolve following a conservative Ostwald ripening process [6].

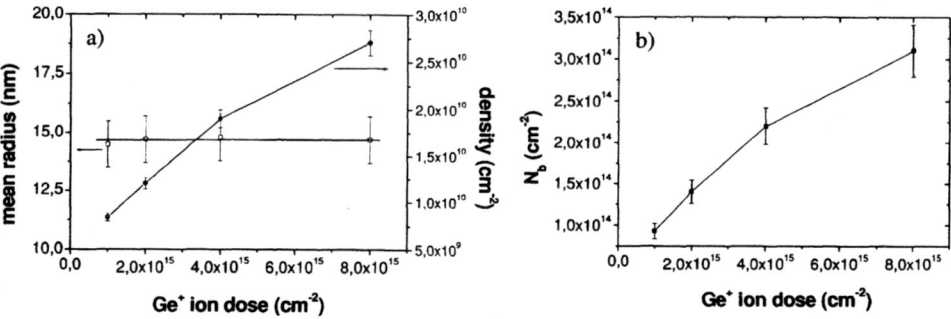

Figure 3: *Evolution of (a) the mean loop radii, loop densities and (b) the total number (N_b) of interstitials bound to the loops, after RTA annealing at 1000°C for 100 sec as a function of the Ge ion dose.*

The invariance of the mean size of "precipitates" with respect to the initial supersaturation introduced in the matrix is an expected characteristic of a conservative Ostwald ripening mechanism. Few theoretical studies have predicted this effect [5], but until now, no experimental results have clearly confirmed these predictions. Recently, Monte Carlo simulations by Jaraiz et al. [7] and atomistic simulations by Bonafos et al. [8], have indirectly confirmed the so-called "supersaturation invariant" effect. They have found that during a conservative Ostwald ripening process of precipitates diluted in the matrix and thus not interacting directly between themselves but through a "mean field", the sizes of the particles do not depend on the implantation dose i.e., on the initial supersaturation.

Our experiments, which have been carried out under almost "ideal" conditions (same type of precipitates, located far from an external sink) clearly show that the Ostwald theory perfectly describes the defect evolution upon annealing.

From a technological point of view, these results mean that it is possible to modify the density of particles but not their sizes by tuning a single process parameter, that is the implant dose. On the other hand, in order to control the defect size it is sufficient to modify the thermal budget only (mostly the annealing temperature).

Excess interstitial model

We first consider samples implanted with 150 keV Ge$^+$ ions to a dose of 1×10^{15} ions/cm^2. The number of excess interstitials as a function of depth, as calculated using the TRIM code [9] is shown in figure 4a. As in our previous work [4], we assume that total recombination of interstitials and vacancies occurs at the very early stages of the anneal, and thus the excess interstitials concentration is obtained by subtracting the total number of vacancies from that of interstitials in the same region. For the particular implantation conditions considered in figure 4a, the total concentration of excess interstitials below the a/c interface and available for the precipitation process, is about 4×10^{14} int/cm^2.

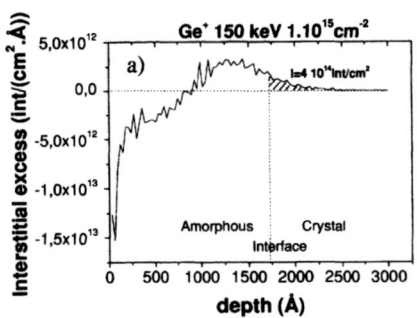
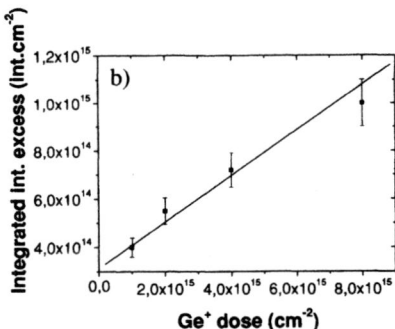

Figure 4: a) Depth profile of the interstitial excess after Ge$^+$ 150 keV 1×10^{15} ions/cm^2 using Monte carlo simulation. b) Integrated interstitial excess below a/c interfaces for the various Ge$^+$ ion doses studied in this work.

The calculated values of the excess interstitial concentrations found below the amorphous/crystalline (a/c) interfaces are reported in table I and plotted in figure 4b as a function of the implanted dose. The number of excess Si(int)s generated at a given depth by the Ge

	Ge$^+$ 150 keV → Si			
	Dose (ions/cm^2)			
	10^{15}	2×10^{15}	4×10^{15}	8×10^{15}
Amorphous layer (nm) (TEM measurements)	173	183	192	200
Interstitial excess below a/c interface (int/cm^2)	4×10^{14}	5.5×10^{14}	7.2×10^{14}	1×10^{15}

Table I: Measured thickness of the amorphous layer (TEM) and corresponding calculated interstitial excess for the various Ge$^+$ ion doses studied in this work.

implant is proportional to the dose. Since, in the mean time, the a/c interface shifts towards larger depths, the total number of Si(int)s found below the a/c interface is a sublinear function of the preamorphisation dose. Indeed, the increase in the dose by a factor 8 only results in the increase in the initial supersaturation by less than a factor 3.

We have plotted in figure 5 the variation of the total number of Si(int)s stored in the loops (from TEM analyses) after annealing as a function of the number of excess interstitials calculated (from Monte Carlo simulations) before annealing.

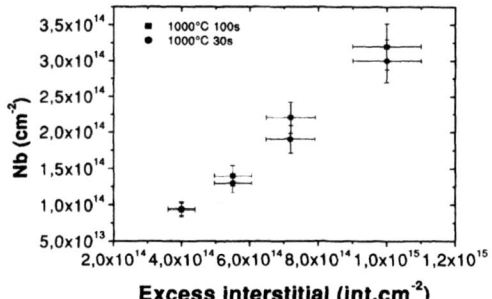

Figure 5: *Number of atoms in the loops (extracted from TEM images) as a function of the number of excess interstitials (calculated by TRIM).*

This figure clearly shows that the total number of atoms bound to the loops is directly proportional to the number of Si(int)s in excess of the vacancies found below the a/c interface as calculated by Monte Carlo simulations before annealing. This is consistent with the "excess interstitial" model [10,11] according to which the EOR defects are formed by precipitation of the interstitial excess induced by the implantation process.

It is to be noted that the "excess interstitial" model is formally similar to the "+1" model [12] but that, in the amorphizing case, the net difference between vacancies and interstitials is relevant only on the crystalline side of the layer.

CONCLUSIONS

In this paper we have studied by TEM the effect of the Ge^+ preamorphisation dose on the thermal evolution of End of Range (EOR) defects upon annealing. We found that, for a given thermal budget, the size distribution is not affected by the Ge ion dose, only the defect density increases, as expected by the Ostwald ripening theory. That means that the number (N_b) of Si(int)s bound to the EOR defects is a monotonically increasing function of the Ge ion dose. Furthermore, N_b is directly proportional to the number of Si atoms in excess of the vacancies found below the a/c interface before annealing, as calculated by Monte Carlo simulations. The "excess interstitial" model correctly explains the formation of the EOR defects.

ACKNOWLEDGMENT

This work was finalised shortly after the completion of the RAPID Project (ESPRIT 23481) and just before the beginning of the IST/FRENDTECH Project. However, we are glad to thank the almost continuous support from the European Commission.

REFERENCES

1. S. Thomson, P. Packan, and M. Bohr, *Intel Technology Journal Q3*, (1998).
2. A. Claverie, B. Colombeau, G. Ben Assayag, C. Bonafos, F. Cristiano, M. Omri and B. de Mauduit, *Mat. Sci. in Semic. Proc.*, **3**, 269 (2000).
3. K.S. Jones and D. Venables, , *J. Appl. Phys.*, **69,** 2931 (1991).
4. L. Laanab, C. Bergaud, M.M Faye, J. Faure, A. Martinez and A. Claverie, *Mat. Res. Soc. Symp. Proc*, **279**, 381 (1993).
5. H. Schroeder, P.F.P. Fichner, H. Trinkaus, *Fundamental aspects of inert gases in solids*, ed. S.E. Donnely and J.H. Evans, Plenum Press, New-York, **279**, 289 (1991).
6. C. Bonafos, D. Mathiot and A. Claverie, *J. Appl. Phys.*, **83,** 3008 (1998).
7. M. Jaraiz, L. Pelaz, E. Rubio, J. Barbolla, G.H. Gilmer, D.J. Eaglesham, H.J. Gossmann and J.M. Poate, *Mat. Res. Soc. Symp. Proc*, **532**, 3 (1998).
8. C. Bonafos, B. Colombeau, M. Carrada, A. Altibelli, G. Ben Assayag, B. Garrido, M. Lopez, A. Perez-Rodriguez and A. Claverie, *Nucl. Instr. Meth. in Phys. Res. B*, (2001) in print.
9. TRIM-95, after J.F. Ziegler, J.P. Biersack, and D. Littmark. *The Stopping and Ranges of Ions in Solids*, **vol.1,** ed. J.F. Ziegler (Pergamon, New York) (1985).
10. A. Claverie, C. Bonafos, M. Omri, B. de Mauduit, G. Ben Assayag, A. Martinez, D. Alquier and D. Mathiot, *Mat. Res. Soc. Symp. Proc*, **438**, 3 (1997).
11. E.G. Roth, O.W. Holland and D.K. Thomas, *Appl. Phys. Lett.*, **74**, 679 (1999).
12. M.D. Giles, J. Electrrochem. Soc., **138**, 1160 (1991).

Mat. Res. Soc. Symp. Proc. Vol. 669 © 2001 Materials Research Society

Evolution Of Defects Induced By High Energy He Implantation In Gold-Diffused Silicon

R. El Bouayadi, G. Regula, B. Pichaud, M. Lancin, J. J. Simon, E. Ntsoenzok [1]
Laboratoire TECSEN, Aix-Marseille III, Service 151, Marseille, F-13397
[1] CERI-CNRS, 3A, rue de la Férollerie, Orléans cedex, F-45071

ABSTRACT

Silicon samples were gold-diffused at different temperatures, implanted with He ions at 1.6 MeV and then annealed at 1050°C for 2 hours. The implantation induced-defect structure and their distribution in the depth of the sample, studied by conventional and high resolution cross section electron microscopy (HRXTEM) depend on the gold level introduced in the wafer prior to the gettering process. A high concentration of gold in silicon seems to influence the defect configuration in the cavity zone. Indeed, gold chemisorbed at cavities can homogenize the surface energy of their planes in different orientations, and can increase the cavity critical diameter beyond they become facetted. Secondary ion mass spectroscopy (SIMS) profiles exhibit a shouldered shape and a width closely related to the presence of the defects (observed by XTEM) which are very efficient sinks both for gold and copper atoms. Unfortunately, the electrical improvement of the material (checked by minority carriers diffusion length measurements MCDL) is not achieved by this gettering process, probably due to the high metal impurity concentrations remaining out of the gettering zone, to the presence of AuCu complexes and η-Cu₃Si precipitates identified by deep level transient spectroscopy (DLTS) measurements and HRXTEM observations respectively.

INTRODUCTION

Metal impurities are known to degrade dramatically the performances of silicon based devices even in concentration as low as 10^{12} at.cm^{-3} [1]. Dissolved in the silicon bulk, these impurities can strongly reduce the minority carrier diffusion length. With the size reduction of the device structure, it is more and more interesting to create very localised gettering zones at, or just below, the active zone. This so-called proximity gettering have been carried out in Si [3,4] by implantation of different kind of ions (C^+, Si^+ etc...) but there is also a great deal of interest in He or H implantation [5-7]. In silicon, He and H segregate into small gas-vacancy complexes which favor bubbles or cavity formation depending on the implantation and on the subsequent heat treatment conditions [8]. The gettering ability of such cavities have been demonstrated for transition metals at keV energies [6-9] but only few work has been carried out at MeV energies [10]. In this study, in the latter energy range of He implantation, we have investigated the cavities (and implantation related defects) and we have checked their gettering efficiency for Au and Cu atoms by two means: i) chemically by SIMS measurements and ii) physically by showing a nanostructure change of the cavities in highly gold-doped samples. Moreover, this combination of metal impurities is interesting because each metal diffuses in a different way in silicon using either an interstitial (Cu) or hybrid (Au) mechanism [11]. They have different precipitation behaviors as well: Au does not form stable silicides while Cu does, giving mismatched precipitates in silicon [2,12].

EXPERIMENTAL DETAILS

Two kinds of (111) 500 μm thick materials were studied : epitaxially-grown (E) and float zone (F) silicon. E samples consist of 75 μm n-type ([P]=1.4 x 10^{14} atoms cm^{-3}) epitaxially grown on a n-type Czochralski substrate. F samples are p type ([B]=2 x 10^{15} atoms cm^{-3}). A 100 nm thick Au layer was plated and diffused at temperatures ranging from 870°C to 1050°C during four hours (see table 1). These diffusion conditions giving high gold concentrations were chosen to improve the detection of this impurity. The samples were then mechanically polished to obtain a constant level of gold in the bulk by removing the near-surface zones in which the Au concentration falls down rapidly (U curve [13]). Then, they were implanted with 1.6 MeV helium ions provided by a Van de Graaff accelerator. The corresponding projection range R_p predicted by transport range of ions in matter (TRIM) is R_p=5.6μm. The implantation was performed at room temperature using a dose of 5 10^{16} cm^{-2} with a constant flux of about 1.2 10^{13} He $cm^{-2} s^{-1}$ (2 μA cm^{-2}). The samples were annealed at 1050°C for 2h under argon gas flow in a quartz tube using a conventional furnace, and cooled down to room temperature. About 5 minutes are needed to cool the sample from 1050°C to 700°C. To check the gettering efficiency of the implantation induced defects, SIMS profiles were carried out with a Cs^+ primary source. Moreover, minority carriers diffusion length measurements (MCDL) were also performed using metal-insulator-semiconductor (MIS) diodes on F wafers before and after the gettering process. DLTS on Schottky diodes were carried out to identify the centers responsible for the decrease of the minority carrier diffusion length. In addition, HRXTEM observations coupled with energy dispersive spectroscopy (EDS) were performed at the end of the process, to characterize the cavities and related defects.

Table 1. Process steps and analysis undergone by the different samples and their Cu contamination.

specimen	diffusion conditions	Cu contamination	2nd annealing	analysis
E0	870°C, 4h	yes	1050°C, 2h	TEM (EDS), SIMS
E1	Au : 870°C, 4h	yes	1050°C, 2h	SIMS
E2	Au : 915°C, 4h	yes	1050°C, 2h	SIMS
E3	Au : 950°C, 4h	yes	1050°C, 2h	TEM (EDS), SIMS
F1	Au : 870°C, 4h	yes	1050°C, 2h	SIMS, DLTS, MCDL
F2	Au : 1050 °C, 4h	yes	1050°C, 2h	TEM (EDS), SIMS, DLTS, MCDL

RESULTS AND DISCUSSION

   i) **Gold trapping :** the ability of the cavities to trap gold is demonstrated by SIMS analysis. Indeed, in all the SIMS profiles, a great amount of gold atoms is seen in the vicinity of R_p. Nevertheless, the shape and width of the profiles depend on gold level introduced in the sample.

Low starting level of gold atoms (10^{14} at cm^{-3}<[Au]< 910^{14} at cm^{-3})
Figure 1 shows typical SIMS measurements performed on the Au-diffused silicon samples at low temperatures for 4 hours. All gold profiles show a 1μm wide asymmetric shouldered peak with a maximum at 5.6μm, which is the exact location of R_p. Low gold diffusion temperature gives strictly identical SIMS gold profiles on both kinds of silicon E and F which indicates that there is

no influence of the doping type (p or n) and doping level on the defect distribution. An estimation of the gold trapping efficiency at cavities was carried out in sample E1:

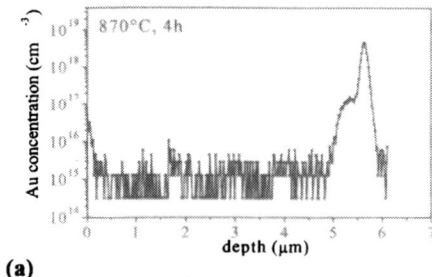
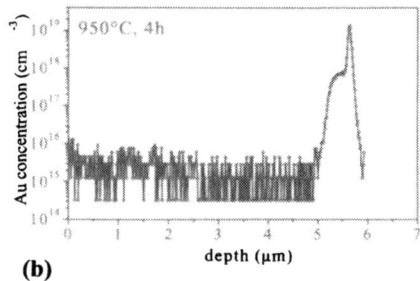

(a)
(b)

Figure 1. SIMS depth profiles for Au in samples : a) E1 and b) E3.

after having removed the near surface regions of sample E1, a constant gold level introduced by a 870°C-4h diffusion annealing can be assumed to be about 1×10^{14} at cm^{-3} as was calculated and measured [14] in similar silicon samples (same substrate As doped). From this value, a surface concentration of $n_i = 4.4 \times 10^{12}$ at cm^{-2} gold atoms can be derived. By integrating below the whole peak in figure 1 (curve b), a number of gold trapped atoms $n_t = 2.7 \, 10^{12}$ at cm^{-2} was obtained, which gave a gettering efficiency $\varepsilon = n_t/ n_i$ close to 0.6. If the in-diffused gold level is increased, the amount of gold trapped at cavities becomes larger which means that the cavity sites are not saturated in sample E1 (compare curve a and b in figure 1).

High starting level of gold atoms ([Au] = 910^{15} at cm^{-3})
In the case of gold diffusion at 1050°C (figure 2a), both gold and copper SIMS profiles show a double peak with a maximum at R_p. The whole peak width is at least two times broader than in the previous case. Two kinds of SIMS profiles, obtained for high and low gold diffusion temperature, imply two different gettering mechanisms and/or two defect types and/or microstructures. What are they? To get a better insight into the process, both DLTS and HRXTEM experiments were carried out. AuCu complex at 0.31eV above the valence band was found in both F1 and F2 samples. Gold precipitates were not observed while they were detected by Wong-Leung et al.[15] but in a rather different experimental configuration. Up to now, gold was not detected by nano EDS, probably because of the low amount of gold present at cavity surfaces added to the low thickness of the foil observed by HRXTEM. In spite of the good gettering efficiency of implantation-related defects as shown above, the remaining gold level is still high and the initial MCDL can not be recovered.

ii) Copper contamination : the samples were copper contaminated during He implantation in a concentration below the solubility limit at 1050°C of about [Cu] $\approx 1 \, 10^{16}$ at.cm^{-3} deduced from SIMS measurements (figure 2b) and Evans' reference samples. Copper was introduced during implantation from the back surface via the sample holder, and /or during the polishing step. The shape of the Cu SIMS profile follows the Au one, in accordance with the AuCu complexes detected by DLTS. Moreover, in some samples, η-Cu$_3$Si precipitates were observed at cavities by HRXTEM and were analysed by nanoEDS. Since the accidental Cu contamination reaches the same level for all samples, copper precipitates cannot be responsible for the defect configuration inducing a large gold SIMS profile like in figure 2a.

iii) correlation between the defect configuration observed by XTEM and the pre-diffused gold level : the defect microstructure at R_p strongly depends on the gold concentration introduced in the sample prior to He implantation.

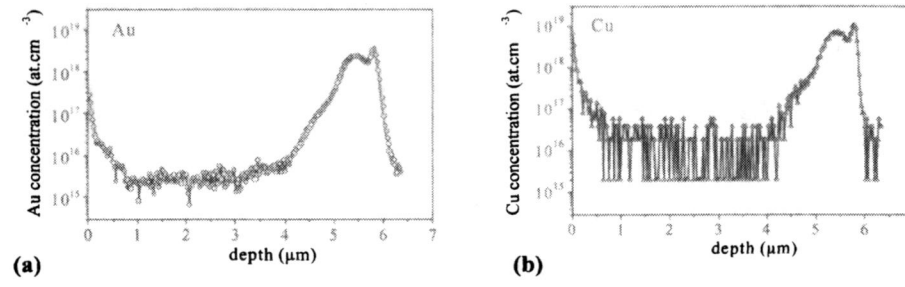

Figure 2. SIMS depth profiles of F2 samples for a) Au and b) Cu

In all the samples studied by XTEM, cavities were found at 5.6µm. The width ΔW of the whole defect zone of gold free and gold-diffused samples are presented in the micrographs of figure 3. The width ΔW_c of the cavity zone is in accordance with the straggling $\Delta R_p=0.5$µm predicted by TRIM ($0.2<\Delta W_c<0.4$). The statistics given below concern at least 300 cavities in each sample.

Gold free samples : in E0 samples, ΔW_c is about 0.4 µm. Most of cavities are spherical 10 to 50 nm in size, with a mean diameter of 35 nm, with a few large faceted cavities. It is worth mentioning that no dislocations were observed in this kind of sample except few stacking faults connecting large cavities (figure 3a).

10^{14} *at* $cm^{-3}<[Au]< 910^{14}$ *at* cm^{-3} : ΔW was found to be around 1µm, as shown in figure 3b. Close examination of cavities (figure 3c) shows that these latter present a faceted geometry mainly elongated along {111} planes. Their diameter is ranging from 10 up to 100 nm with a mean diameter of 50 nm and ΔW_c is about 0.2µm. Some smaller cavities (~2-5nm, see circle in fig.3c) are also observed up to distances of 0.5µm from the cavity chain. Gold SIMS profiles of the corresponding samples exhibit a shape in accordance with the defects (dislocations, cavities) distribution. The maximum of the peak is located on the cavity chain and thus is due to the cavity trapping ability, while the dislocations are responsible for the shouldered shape. An artificial broadening due to the dynamic SIMS method itself is unlikely.

$[Au]= 910^{15}$ *at* cm^{-3} : the damaged zone consists of two kinds of defect population (figure 3e), having a rather similar gold trapping efficiency, as described in the previous sections. First, large spherical cavities (with a diameters reaching often 110 nm) and SF are located at 5.6 µm with a ΔW_c of 0.4 µm. Second, high density of loop dislocations are mixed with bundles of small spherical cavities (figure 3d) very heterogeneously distributed. The heterogeneity of the implantation step is not totally excluded, but is unlikely, since only these samples have shown such a defect distribution. ΔW is about 3μm (figure3f) which is consistent with the SIMS width profiles in figure 2a.

The increase in the dislocation density, the scattering in the cavity diameter at R_p and the cavity shape evolution with gold concentration can be attributed to three mechanisms:
i) A recombination of self-interstitials (I), injected in the matrix during the gold diffusion, with vacancies (V) during the second annealing which could affect the cavity formation process.

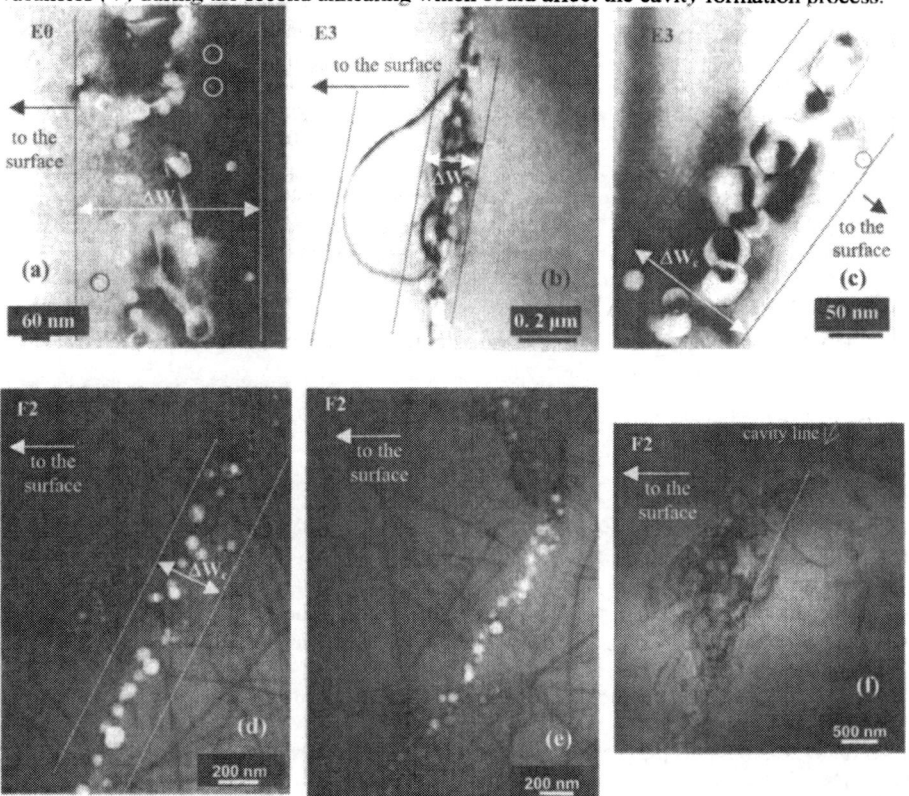

Figure 3. Defect population at Rp observed in the <110> direction for samples : a) gold free ; gold pre-diffused at a constant level : b) and c) [Au] = $9 10^{14}$ at cm^{-3} ; d), e) and f) [Au]= $9 10^{15}$ at cm^{-3}.

ii) A I-condensation nucleating SF and dislocation loops. These dislocations can be developed further by punching in high stressed regions near bundles of tiny cavities. Dislocations can also climb, consuming I atoms.

iii) Gold atoms chemisorbed at cavities can decrease the mobility of cavities, preventing or reducing their coalescence and can also change their surface energy, increasing the critical size to become facetted.

We showed that gold concentration influences the defect configuration. This effect is probably limited in the usual gettering experiments since the impurity concentrations are overly below 10^{16} at.cm^{-3}.

CONCLUSION

Au and Cu gettering efficiency of cavities due to high energy He implantation in silicon has been investigated. Au is not only trapped by the cavity chain located at the projection range, but also by the region in front of this cavity chain which contains smaller cavities and dislocations. Thus both cavities and dislocations are getter sinks for impurities with a good efficiency. The created potential gettering sites are far from being saturated, despite of Cu contamination, probably because Cu was partly trapped in a precipitate and/or in a Au-complex state. The implantation-related-defects (dislocation density, cavity shape and their diameter distribution) depend strongly on the pre-diffused gold concentration, probably because it changes the interstitial silicon supersaturation level and/or because chemisorbed gold atoms can decrease the mobility of cavities and modify their surface energy.

ACKNOWLEDGMENTS
The authors are very grateful to C. Dubois and W. Saikaly for technical support and fruitful discussions.

REFERENCES

[1] T.E Seidel, MRS 1986 Symp. Proc. 3, (1986)
[2] W. Schröter, M. Seibt and D. Gilles , Materials Science and Technology Eds. R.W. Cahn, P. Haasen and E.J. Kramer, VHC Weinheim 1991 Vol. 4, (pp 539-587)
[3] W. Skorupa, N. Hatzopoulos, R.A. Yankov and A. B. Danilin, Appl. Phys. Lett. **67**, 20 (1995)
[4] A. Cacciato, C.M. Camalleri, G. Franco, V. Raineri and S. Coffa, J. Appl. Phys. **80**, 8 (1996)
[5] G.A. Petersen, S.M. Myers and D.M. Follstaedt, Nucl. Instr. and Meth. in Phys. Res. B **127/128**, 301 (1997)
[6] J. Wong-leung, C.E. Ascheron, M. Petravic, R.G. Elliman and J.S. Williams, Appl. Phys. Lett. **66**, 1231 (1995)
[7] V. Raineri, P.G. Fallica, G. Percolla, A. Battaglia, M. Barbagallo and S.U. Campisano, J. Appl. Phys. **78**, 3727 (1995)
[8] P.F.P. Fichner, J.R. Kaschny, R.A. Yankov, A. Mücklich, U. Kreissig and W. Skorupa, Appl. Phys. Lett . **70**, 732 (1997)
[9] S.M. Myers and D.M. Follstaedt, J. Appl. Phys. **79**, 1337 (1996)
[10] G. Mariani-regula, B. Pichaud, S. Godey, E. Ntsoenzok, O. Perner and R. El bouayadi, Mat. Sci. and Engineer. B, **71**, 203 (2000)
[11] N.A. Stolvijk, H. Bracht, 'diffusion in silicon, germanium and their alloys' ed. , landot-Börnstein New Series III/33A p 196.
[12] K. Graff, Metal Impurities in Silicon-Device Fabrication, Springer Series in Materials Science, Eds. H. J. QUEISSER , Springer-Verlag Berlin Heidelberg 1995
[13] J. Hauber, N.A. Stolwijk, L. Tapfer, H. Mehrer and W. Frank, J. Phys. C, **19**, 5817 (1986)
[14] O. Boström, B. Pichaud, M. Regula, J. C. Bajard, G. Blondiaux, O. A. Soltanovich, E. B. Yakimov, A. Lhorte, and J. B. Quoirin, Mat. Sci. and Engineer. B, **71**, 166 (2000)
[15] J. Wong-leung, E. Nygren and J.S.Williams, Appl. Phys. Lett. **67**, 416 (1995)

Mat. Res. Soc. Symp. Proc. Vol. 669 © 2001 Materials Research Society

Accurate Modeling of Residual recoil-mixing during SIMS Measurements

Ming Hong Yang and Robert Odom
Charles Evans and Associates, Sunnyvale,
CA 94086, USA

Abstract

Secondary ion mass spectrometry (SIMS) is an effective and powerful analytical technique, widely used in accurately determining dopant distributions (depth profiles). However, primary ion beam induced mass transport (ion mixing), especially the residual effect during SIMS profile measurements, greatly limits the accuracy at nanometer depth resolutions by displacing and broadening the measured depth profile. In this paper, we present a simple deconvolution algorithm based on the general characteristics of the experimentally observed SIMS response function to reduce this broadening effect, thereby providing more accurate depth profiles. The results for several specific applications of this approach are presented and its strengths and limitations are discussed.

INTRODUCTION

Secondary ion mass spectrometry (SIMS) is an effective and powerful analytical technique, widely used in accurately determining dopant distributions. However, primary ion beam induced mass transport (recoil ion mixing) during SIMS measurements greatly limits the accuracy at nanometer depth resolutions by displacing and broadening the measured depth profile. A common yet dramatic illustration of this ion mixing effect is shown in Figure 1a for the case of arsenic and boron implants into silicon. Recoil ion mixing effects in SIMS profiles are traditionally described [1] by an experimentally determined response function measured on a delta doped layer as illustrated in Figure 1b. The response function is typically anisotropic with an exponential tail extending deep into the sample. The peak is also shifted slightly toward beam direction by a distance δ. Littmark and Hofer [2] showed that the SIMS response function can be described basically by cascade recoil ion mixing. Wilson et al.[3] further pointed out that the symmetric portion of the response function (Region I) corresponds to the direct removal of over-layer species from the surface, either by sputtering or by ion mixing to the subsurface layers and often is approximated by Gaussian distributions. The exponential decay (Region II) corresponds to the residual effect of continual recoil mixing and erosion of sample surface during SIMS depth profiling. This residual effect is shown more clearly in Figure 1c. For the simplicity we use 'direct mixing' and 'residual recoil mixing' to represent these two unique characteristics of the response function. Theoretically the true depth profiles can be calculated by deconvolution if the response function is known. Many deconvolution methods have been proposed in the SIMS literature [4-6]. However they are not used widely, partly because SIMS response functions from delta doped standards are often not available. Instead, most efforts are focused on lowering the primary beam energy to reduce the recoil ion mixing [7]. We recognize that though the detailed response for direct ion mixing is often not known, the response due to the residual recoil mixing can often be measured easily. This is especially important for the p-n junction depth profiling measurements and ion implant characterizations because dopant concentrations decrease over 3-4 decades within several nanometers. As illustrated in Figure 1c, the error in the measured junction depth due to the 'residual ion mixing' (Δ_{II}) is often several times larger than the 'direct mixing' (Δ_I) in these circumstances because of the slow decaying tail. Keeping this in mind, we present in this paper a simple empirical deconvolution algorithm developed specifically to remove the 'residual ion mixing' contributions to SIMS depth profiles, thereby providing more accurate measurement of junction depths.

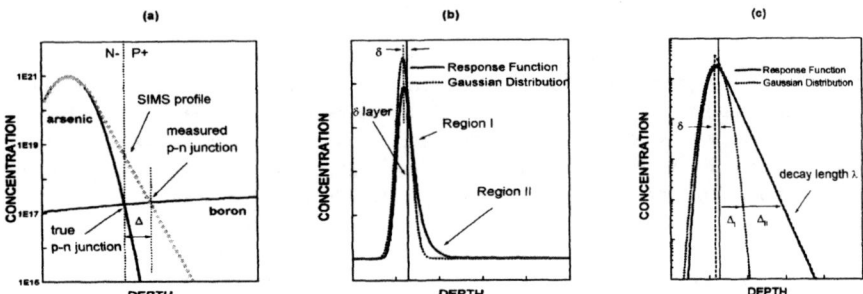

Figure 1. (a) Schematic of a p-n junction formed by ion implantation. Solid lines are actual boron and arsenic concentration distributions. The junction is where the arsenic and boron concentrations are equal. Since the measured arsenic profile (dotted line) is broadened by recoil mixing, the measured junction depths are deeper than true values. (b) SIMS response function is shown on a linear scale. A normalized Gaussian distribution is also shown to approximate the direct mixing when the residual mixing is absent. (c) The residual mixing effect is clearly shown on a semi-log scale. The distance over which the signal decreases by a factor e is defined as decay length. Δ_I and Δ_{II} are the junction depth displacements due to "direct" and "residual mixing".

EMPIRICAL RESIDUAL ION MIXING MODEL

The SIMS response function, $G(x-x')$, is defined as the measured concentration at any given depth, x, in response to a delta doped tracer at depth, x'. For a uniform material system, the measured concentration depth profile, $N(x)$, can be expressed as a convolution of the true profile, $N_0(x)$, with its response function, $G(x-x')$:

$$N(x) = \int N_0(x')G(x-x')dx' \qquad (1)$$

In our model, the 'residual ion mixing' response function given in equation (2) is used to approximate the unknown response function $G(x-x')$ and is shown in Figure 1c as a dashed line. The onset of the exponential decay function occurs at $x'-\delta$, and its decay length is λ, matching the experimentally determined peak position and decay length.

$$f(x-x') = 0 \qquad\qquad x-x' < -\delta$$

$$(2)$$

$$f(x-x') = \frac{1}{\lambda}\exp(-\frac{x-x'+\delta}{\lambda}) \qquad x-x' > -\delta$$

Solving equation (1) after substituting $G(x-x')$ with $f(x-x')$, we find,

$$N_0(x) \approx N(x+\delta) + \lambda \frac{\partial}{\partial x} N(x+\delta) \tag{3}$$

We should note that the broadening (Δ_1) due to the 'direct ion mixing', as illustrated in Fig. 1, is still present in $N_0(x)$. Equation (3) only approximates the true depth profile. The absolute and relative uncertainties for $N_0(x)$ due to uncertainties in λ are easily calculated from equation (3) and are given by

$$\Delta N_0(x) = \Delta \lambda \frac{\partial}{\partial x} N(x+\delta) \tag{4a}$$

$$\frac{\Delta N_0(x)}{N_0(x)} = \Delta \lambda \frac{\partial}{\partial x} \log(N(x+\delta)) \tag{4b}$$

From equation (3) we note that the deconvoluted profile depends only on the measured profile and its derivative. Therefore, measurement or calculation error in one part of a profile will not introduce deconvolution errors in other parts unless the separation distance δx is small (i.e. on the order of λ).

RESULTS AND DISCUSSION

To evaluate the proposed algorithm, we have tested the following four aspects of the algorithm: (1) self-consistency, (2) applicability, (3) sample transferability and (4) errors and limitations.

Self-consistency:

As shown in Figure 2a, a test depth profile (curve a) is created by super-imposing a Gaussian function $[=1 \times 10^{20} \exp(-(x-150)^2/400)]$ with random noise. Convolution profile (curve b) simulating the measured SIMS profile was calculated using equations (1) and (2) with $\lambda = 40$, $\delta = 40$. The deconvoluted profile (curve c) was then calculated using equation (3) with the same set of λ and δ values. The results are quite satisfactory and show that deconvolution method is indeed self-consistent. Since the residual mixing response to a delta layer is an exponential decay, it is expected that the convoluted profile will decay almost exponentially for depths at distance separations greater than twice the full width half maximum (FWHM) of the Gaussian peak. The decaying profile is fitted to a straight line on a semi-logarithmic plot with its slope equal to λ. This example shows that (1) the method is self-consistent, and (2) within the approximation of this approach, λ can be accurately determined from the decaying tail of a reasonably sharp SIMS profile.

Applicability:

The second experiment we did was to test if the algorithm works with a real profile measured from a real sample. An arsenic doped poly-silicon epilayer grown on silicon sample was chosen after its profile, measured with a PHI 6650 quadrupole instrument using a low energy (1keV) Cs^+ primary beam positioned at a 60° incident angle, exhibited a sharp interface as shown in Figure 2b. This is an ideal sample because we know the arsenic depth profile acquired with high-energy

Cs primary beam will produce a clean exponential decaying tail. Measurement with a CAMECA IMS-4f instrument using a 14.5keV Cs^+ primary beam in its standard configuration clearly confirmed our expectation and the 14.5keV profile is also shown in Figure 2b. This is a classic example showing how a true depth profile is distorted by residual ion mixing effects. Since the ion mixing is significantly less for the 1keV Cs primary beam, we use the 1keV profile as a reference standard to evaluate the deconvolution of 14.5keV profile. We are fully aware that this is not a perfect standard, and some ion mixing is still present for the 1keV profile. The estimated decay length for 1keV beam at 60 degree is ~10 Angstroms per decade. By contrast the λ value for a 14.5keV Cs beam is 94 Angstroms, determined by fitting the decaying tail of the arsenic profile. Using equation (3) and adjusting δ to match the reference profile, the deconvoluted profile (the open circle dotted line) for 14.5keV Cs beams is calculated, and is shown in Figure 2b with δ = 48 Angstroms. The deconvolved arsenic profile is significantly closer to the reference profile than the as-measured profile and therefore provides more accurate junction depth measurement. We should also point out that since the algorithm does not address 'direct ion mixing' effects, the disagreement near the arsenic peak between the 14.5keV and the reference depth profiles remain unchanged. This example clearly shows that (1) the deconvolution successfully removed the profile broadening caused by the 'residual ion mixing', (2) decay lengths can be determined from the exponential decaying tails of sharp interfaces, and (3) δ can be determined by comparing the deconvolved profile with the reference profile.

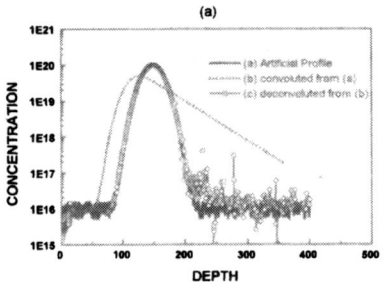
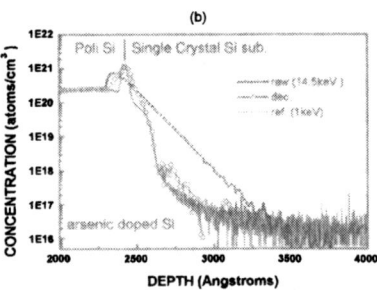

Figure 2 (a) Illustration of self-consistency of the proposed deconvolution algorithm using an artificial depth profile. (b) An arsenic depth profile (raw) measured by 14.5keV Cs beam and its deconvolution (dec.) are shown along with a reference profile by a 1keV Cs beam.

Sample to sample transferability:

The next question we asked ourselves was 'can λ and δ determined from one sample be applied to other similar samples?' when these samples are measured under the same experimental conditions. To answer this question, we tested the algorithm on two Si samples implanted with low energy phosphorus at 2 and 5keV, respectively. We compare the deconvolved profile from the 5keV phosphorus sample using parameters (λ and δ) determined from 2keV sample with the true phosphorus distribution. Profiles measured with a 3keV O_2 primary beam on a CAMECA IMS 4f instrument are used to test the deconvolution and profiles measured with a PHI 6650 quadrupole instrument using 1keV Cs primary beam are used to approximate the true distributions in these samples. 2keV phosphorus implants can be approximately treated as an delta doped sample when measured with a 3keV O_2 primary beam on a CAMECA IMS 4f instrument with oxygen flooding and these data were used to obtain the λ and δ values for phosphorus in silicon under these conditions. These are clearly demonstrated by the measured phosphorus profiles shown in Figure 3a. As in the previous example for the arsenic doped sample, λ is determined from the exponential decay of the phosphorus profile and δ is determined by aligning the deconvolved profile for the 2keV phosphorous sample to the reference profile. We found $\lambda = 31$ Å and $\delta = 16$ Å for phosphorus measurements using 3keV O_2 primary beam on a CAMECA IMS 4f instrument with oxygen flooding. Using these parameters deconvolution of the 5keV phosphorus profile measured with 3keV O_2 primary beam is calculated using equation (3) and the results are shown in Figure 3b. Good agreement was obtained for the 5keV phosphorus sample at the decaying tail portion of the profiles between the deconvolved profile and the reference profile. This example clearly shows that parameters determined from one can be used over a range implants.

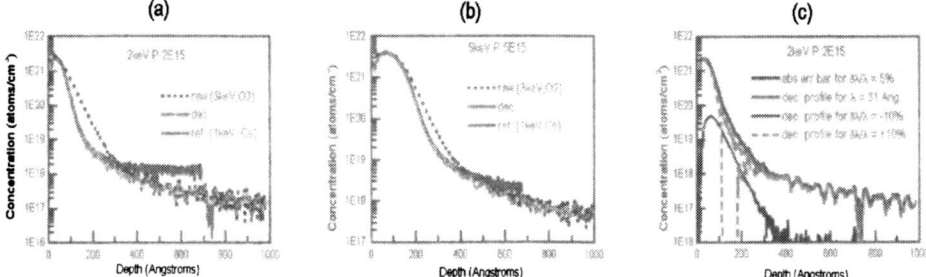

Figure 3. (a) 2keV Phosphorus implant is used to determined the parameters λ and δ. (b) Deconvolution is calculated for 5keV Phosphorus implant depth profile using parameters determined from (a). Phosphorus profiles measured with 1keV Cs primary beam are used as references. The oscillations observed in the deconvoluted profiles are calculation artifacts due to low count levels. (c) Evaluation of systematic error in λ.

Error and limitation of the algorithm:

We like to point out again that the proposed algorithm only removes the broadening caused by the 'residual ion mixing'. The proposed method should be used only for this purpose. The broadening caused by the 'direct ion mixing' is still present in the deconvolved profiles. The error we discuss hereafter will focus only on deconvolution error caused by uncertainties in λ and δ, since there are experimentally determined parameters. We first evaluated systematic errors due to uncertainties in λ. As an example, the absolute and relative error bars for the deconvoluted 2keV phosphorus implant were calculated using equation (4) for $\Delta\lambda/\lambda = 5\%$. The overall error in phosphorus concentration due to these uncertainties in λ is <10%. The absolute error due to uncertainty in λ and the deconvoluted phosphorus profile are shown in Figure 3c. The deconvoluted profiles with $\Delta\lambda/\lambda = \pm10\%$ are also shown in Figure 3c to exhibit variation and sensitivity of λ. The results show that, as expected, the deconvolution is indeed sensitive to λ and one must use caution in interpreting features when the uncertainty is large. We also note that negative concentrations occurred for a portion of the data (dashed line) when a decay length 10% larger than the experimentally determined value was used. When negative concentrations occur, one should set an upper bound to the λ value so that the deconvoluted concentrations are always equal to or greater than the measurement detection limits. In cases where λ is accurately determined, the deconvoluted profile is a great improvement to the measured profile, especially under analytical conditions requiring high primary beam ion energies. We should also point out that the wiggles observed at low concentrations in all the deconvoluted profiles are due to low counting statistics. The error in δ will simply cause a horizontal shift of the profile and will not change the relative distance. One can not determine δ from any measured profile without to a standard. Fortunately, in many SIMS applications, only relative distance is needed. We should always use caution when trying to determine the absolute shift in δ.

Conclusion

These results show that the residual ion recoil mixing is a very significant error factor in the p-n junction measurements using SIMS depth profiling. This error can be easily removed from SIMS profiles by using the suggested deconvolution method. This approach also improves the depth resolution of SIMS profiles by removing the residual ion recoil effect. Therefore, this method will provide improved accuracy in junction depth measurements where the residual recoiling mixing significantly alters the true profile shape. Parameters λ and δ of the response function for any experimental conditions should be derived experimentally from depth profiles obtained from delta-doped or step doped samples.

This method is particularly useful for measurements requiring high depth resolution within small areas in which high-energy primary beams are necessary for achieving good spatial resolution. Simple deconvolution procedures will also be invaluable to low primary beam energy SIMS of materials in which segregation is unavoidable. It should also be noted that this deconvolution method only depends on the local depth information (equation 3) and does not require knowledge of the entire profile. Unlike many deconvolution methods developed for SIMS, errors in one part of a profile do not introduce errors in other parts unless the separation distance δx is on the order of (λ).

ACKNOWLEDGEMENT

We thank Mr. Chuck Hitzman for acquiring 1keV arsenic data on a Phi 6650 quadrupole SIMS instrument and thank Dr. Yupu Li for acquiring 14.5keV arsenic data on a CAMECA ims 4f instrument.

References

1) M.G. Dowsett, in *Secondary Mass Spectrometry SIMS X*, eds. A. Benninghoven et al., (John Wiley & Sons, New York, NY, 1995) p. 355.
2) U. Littmark and W. Hofer, Nucl. Instr. Methods, *168* 329 (1980).
3) R. Willson, F. Stevie, and C. Magee, *Second Ion Mass Spectrometry*, (John Wiley & Sons, New York, NY, 1989) p.2.1-1.
4) King and Tsong, Nucl. Instr. Methods, **B7/8**, 793 (1985).
5) D. P. Chu and M.G. Dowsett, Phys. Rev. **B56**, 15167 (1997).
6) Gautier B. et al., Surf. Interface Anal. **24**, 733 (1996).
7) *Fifth International Workshop on Measurement, Characterization and Modeling of Ultra-Shallow Doping Profiles in Semiconductors*, eds. R. B. Fair et. al., (American Vacuum Society, Research Triangle Park, NC, 1999) p 117-136.

Mat. Res. Soc. Symp. Proc. Vol. 669 © 2001 Materials Research Society

Modeling Of Boron Implantation Into Si With Decaborane Ions

Zinetulla Insepov[1]), Marek Sosnowski[2]), and Isao Yamada[1])
[1]) Himeji Institute of Technology, 3-1-2 Koto, Kamigori, Ako, Hyogo 678-1205, Japan
[2]) New Jersey Institute of Technology, USA

ABSTRACT

A molecular Dynamics (MD) model of B implantation into Si and of sputtering the Si surface with energetic B_{10} clusters has been developed. The goal was to simulate the implantation of ions of decaborane ($B_{10}H_{14}$), which may become an important process for the formation of ultra shallow junctions in future MOS devices. The simulations, carried out for the cluster ion impact energy from 3.5 keV to 15 keV, have revealed the formation of a large amorphized region in a subsurface region. At low cluster impact energies in this range some of the B atoms were recoiled back from the surface, but at the energy of 12 keV and above almost all of the Boron atoms were successfully implanted into Si. The sputtering yield of Si has been also computed and found to increase with energy, reaching the value of 6 Si atoms per B_{10} cluster at 15 KeV. The number of displaced surface atoms correlates well with the sputtering yield, and between 3.5 and 10 keV it has a non-linear dependence on energy. At higher energies the number of displaced atoms increases linearly with energy, in agreement with the Khinchin-Pease formula [9]. The sputtering yield at 12 keV was also measured by the amount of Si removed by a decaborane beam from a thin Si film deposited on a carbon substrate. The predicted sputtering yield agrees well with this experiment.

INTRODUCTION

Implantation of decaborane ($B_{10}H_{14}$) cluster ions has recently attracted much attention because this new method of introducing B into Si has important advantages over the conventional technique of implanting single atomic ions to obtain very shallow doping. This method has been first successfully realized by a group at Kyoto University in collaboration with Fujitsu Corporation [1]. In these studies, low-energy decaborane ions have been implanted into Si and thermally annealed at a temperature of about 800 - 1000 °C by RTA. It has been shown that very shallow junctions have been obtained nd that this method could be a promising new technique for future small-scale PMOS devices. After the first experiments with decaborane have been published, it has been realized that the TED effect, the anomalously high boron diffusion rate that has been inherent in conventional implantation, still may persist [2]. One of the feasible explanations of the TED is based of the fact that every implanted B^+ ion creates at least one Si-interstitial (the so called +1 model), and that the highly mobile B interstitials are created by the kick-out mechanism caused by the Si self-interstitials [3].

Ion implantation and dopant diffusion in Si is a well-studied area of semiconductor physics (see e.g. [4]). In contrast, there have been very little published papers to date on modeling of low energy decaborane implantation [5]. Unfortunately, MD modeling of cluster impacts can only be carried out for a limited time interval, typically less than 20 ps, which is not long enough to simulate diffusion of B during RTA processing. MD, however, can be effectively used to study the impact processes and elucidate the differences between single atom ions and cluster ions collision mechanisms. To our knowledge, there were no computer simulations for the sputtering yield from a Si surface bombarded with decaborane cluster ions. The aim of this paper is to present recent MD simulation of decaborane implantation into Si and of the sputtering process in a wide range of ion energies of interest to shallow B implantation and to compare the results of computation with experiment.

MOLECULAR DYNAMICS MODEL

A model of decaborane ion impacts, with ion velocity normal to the surface was developed to study implantation of B and sputtering of Si target. The decaborane molecule was modeled as a B_{10} cluster. We have simulated the implantation in the energy range of 3.5 - 15 keV because of the interest in modeling of very shallow pn-junction formation. The basic cell size for the MD simulation was chosen based on the cluster energy and ion fluence, 10^{13} - 10^{16} ion/cm^2. A tall cylinder, consisting of about 200,000 Si atoms, represented the target. The ZBL potential at short distances combined with the Stillinger-Weber potential at equilibrium distances was used to evaluate interactions between two and three Si atoms, as usual [6]. Interaction between B and Si atoms was modeled via the ZBL screened Coulomb potential at short distances, $r < r1 = 0.52$ A, and with a Morse-type potential at long distances, $r > r2 = 0.86$ A, as recommended in [5]. The binding energy of B in the Si lattice was used as an adjustable parameter with values between 0 to 1.5 eV. Interaction between two B dopants were modeled with an (exp-6)-type potential, with the depth of 0.2 eV, req = 1.5 A.

The MD model assumes a constant temperature and a constant pressure and utilizes a linked list procedure combined with the Verlet-list method [7]. The boundary conditions were used as in our previous paper [8], where a new Hybrid Molecular Dynamics (HMD) method was proposed. This new method combines conventional atomistic MD, for the central collision zone, with a continuum mechanics representation, for the rest of a system. The method significantly reduces the required system size and can account accurately for the energy flow through the system boundaries. According to this technique, the response of the continuum part to the atomistic MD-part could be represented by two components. One is determined by the displacements calculated from strain tensor and it depends on the magnitude of deformation of the boundary layers caused by cluster impact. The second controls the energy balance and is related to energy absorbing walls, which were simulated by thermal diffusion equations. Atoms in the two bottom atomic layers of the cell are fixed.

Molecular Dynamics simulation was carried out up to 7 ps. All B and Si atomic positions and velocities were calculated. A distribution of displaced Si atoms, with potential energies 0.2 eV above the average potential energy, was obtained. The emitted flux was characterized in terms of mass distribution over their kinetic energies, dN/dE, for different time instants. In addition, the structure of the irradiated Si target as well as its defects can be obtained from the radial distribution function, rdf, for the heavily disordered region near the point of the cluster impact.

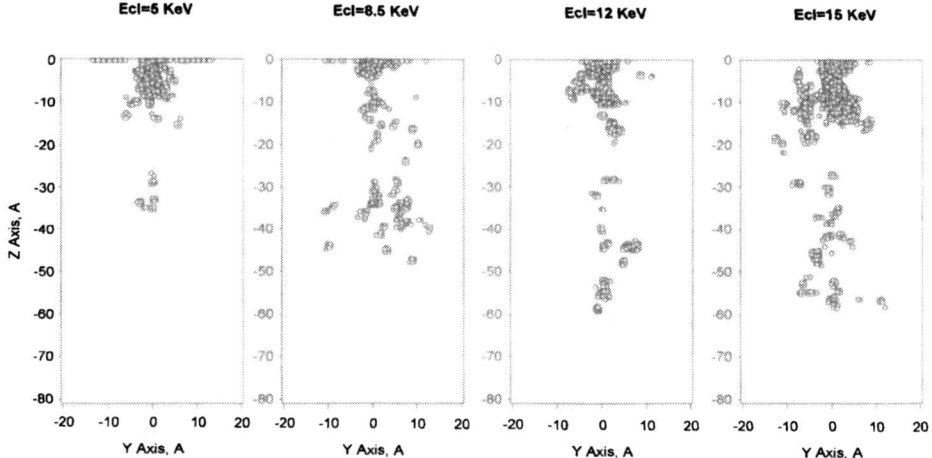

Fig. 1 Side-views of the displaced Silicon atoms at four time instants of 0.9, 1.85, 2.8 and 3.7 ps after the implantation of B_{10} cluster with 15 KeV of total energy, at room temperature.

AMORPHIZATION

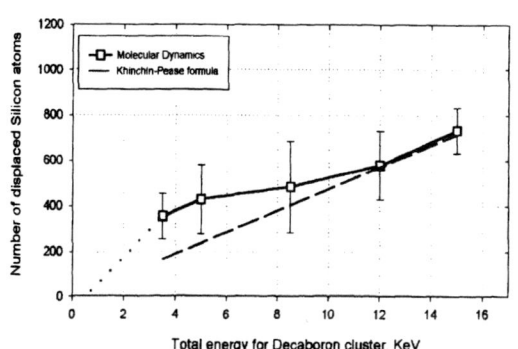

Fig. 2 The number of displaced Si atoms vs total energy of B_{10} cluster. Squares – these MD results; dotted line represents Khinchin-Pease formula, with a E_{th} as an adjusting parameter [9].

MD simulation shows that in a low impact energy range, approximately 3.5-10 keV, almost 30% of boron atoms are recoiled back from the surface but above 10 keV all B atoms are implanted. A large amorphized pocket is created directly under the surface. Figure 1 represents a side-view of 4 time instants 0.9, 1.85, 2.8 and 3.7 ps after the impact for a 15 keV B10 implantation. The volume of the target near the point of impact is heavily disordered. The total number of disordered Si atoms was estimated from the potential energy by counting atoms with energies > 0.2 eV above the average energy for the system. Figure 2 shows the energy dependence of the total number of disordered silicon atoms after the simulation time of 7 ps for the system of 210,338 Si atoms. This figure shows that at higher energies, the total number of displaced Si atoms increases linearly, in accordance with the Khinchin-Pease formula [4]. This means that at energies above 10 KeV, collision of B10 with Silicon could be considered as a collision of separate Boron atoms with the surface.

SPUTTERING YIELD

The experimental sputtering yield was obtained by measuring the amount of Si removed by a 12 KeV decaborane beam from a thin Si film deposited on a carbon substrate. Irradiation of the film with mass analyzed decaborane ions was done at NJIT and the Si film thickness before and after irradiation was measured by RBS. Decaborane fluence was measured by integration of ion current and verified by nuclear reaction analysis, using (p, α) reaction with [11]B isotope implanted in the sample [10].

Figure 3a) and 3b) presents the main result of this simulation: the energy dependence of the total sputtering yield of Si bombarded with B_{10} cluster ions at different energies in 3.5-15 keV range. The experimental value of the sputtering yield, 4 sputtered Si atoms at 12 keV per decaborane cluster ion implantation, for total decaborane cluster energy of 12 keV, agrees quite well with the simulation. Unfortunately, the threshold energy for sputtering Si substrate with decaborane ions has not been found in this simulation because. This task may take too much computation efforts and should be modeled separately.

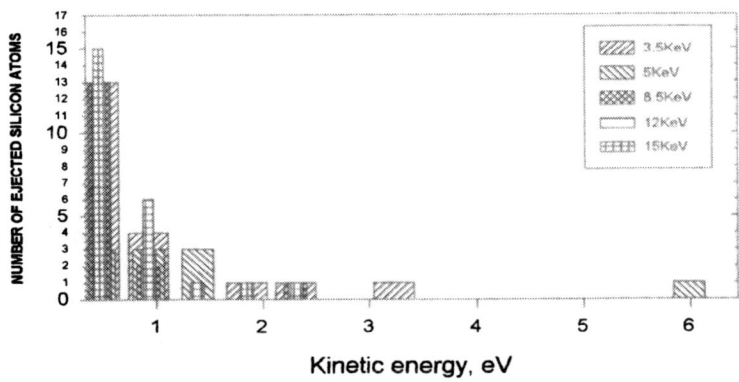

Fig. 3a) The kinetic energy dependence of the number of sputtered Si atoms 7 ps after the impact of a B10 cluster, with energies of 3.5-15 keV. The ejected Si atoms were considered as sputtered if their distance from the surface was more than the cutting radius of the potential (4.3 Å), and the z-component of their velocity corresponded to kinetic energy higher than 1.5 eV.

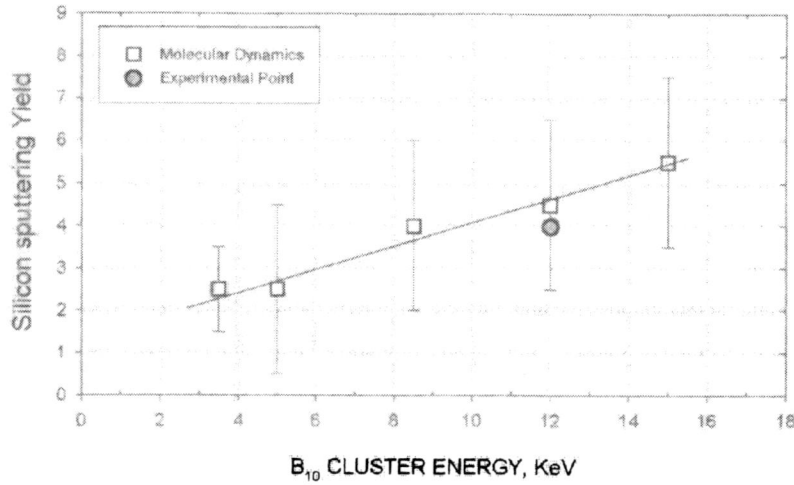

Figure 3b) Si sputtering yield caused by B10 cluster implantations, with energies of 3.5-15 KeV, calculated in this paper. The error bars were obtained for the simulations with different B10 cluster orientations, relative to the Si (100) surface. Experimental data point was obtained for a 12KeV implantation of a decaborane (B10H14) cluster ion into a thin Si film deposited on a carbon substrate.

CONCLUSIONS

A Molecular Dynamics models of decaborane cluster ion implantation into Si was developed which allow simulation of implantation, recoiling of the projectile atoms and sputtering of surface atoms. A large amorphized area directly beneath the Si surface was obtained after a B_{10} cluster impact. Some of the B atoms were recoiled back into vacuum but most were implanted deep into the substrate. None of the Boron atoms were trapped inside the amorphized area as they are for a low energy implantation. Very good agreement was obtained between the calculated sputtering yield and an experimental result at 12 keV. The number of disordered Si atoms shows non-linear increase with energy for energies between 3.5-10 keV and a linear dependence for higher impact energies. This behavior fits the linear prediction of the Khinchin-Pease formula [9]. This suggests that at energies above 10 keV the impact of a B_{10} cluster can be treated as simultaneous impacts of 10 B atoms.

REFERENCES

1. I. Yamada, J. Matsuo, E.S. Jones, D. Takeuchi, T. Aoki, T. Goto, T. Sugii, in *"Materials Modification and Synthesis by Ion Beam Processing"*, edited by D.E. Alexander, N.W. Cheung, B. Park, and W. Skopura (Mat. Res. Soc. Symp. Proc. 438, Pittsburg PA 1997), pp. 363-374.

2. A. Agarwal, H.J. Gossmann, D.C. Jacobson, D.J. Eaglesham, M. Sosnowski, J.M. Poate, I. Yamada, J. Matsuo, and T.E. Haynes, *Appl. Phys. Lett.* **75**, 2015 (1998).

3. E. Chason, S.T. Picraux, J.M. Poate et al., *Appl. Phys. Rev.* **81**, 6513 (1997).

4. P.M. Fahey, P.B. Griffin, and J.D. Plummer, *Rev. Mod. Phys.* **61**, 289 (1989).

5. R. Smith, M. Shaw, R.P. Webb, M.A. Foad, *J. Appl. Phys.* **83**, 3148 (1998)

6. H.J.C. Berendsen, W.F. Van Gunsteren, In *"Molecular liquids, dynamics and interaction"*, edited by A.J. Barnes, W.J. Orville-Thomas, and J. Yarwood (NATO ASI series C135, Reidel, NY 1984) pp. 475-500

7. M.P. Allen and D.J. Tildesley, *"Computer simulation of liquids"*, Clarendon, Oxford, 1987.

8. Z. Insepov, M. Sosnowski, I. Yamada, *Nucl. Instr. Methods Phys. Res.* **B127/128**, 269 (1997).

9. M.J. Norget, M.T. Robinson and I.M. Torrens, *Nucl. Eng. And Des.* **33** 50 (1975).

Mat. Res. Soc. Symp. Proc. Vol. 669 © 2001 Materials Research Society

Photoluminescence study of defects induced by $B_{10}H_{14}$ ions

[1,3]Noriaki Toyoda, [2]Takaaki Aoki, [2]Jiro Matsuo, [3]Isao Yamada, [1]Kazumi Wada, [1]Lionel C. Kimerling
[1]Material Processing Center, Massachusetts Institute of Technology,
Cambridge, MA 02139, U.S.A.
[2]Ion Beam Engineering Experimental Lab., Kyoto University,
Sakyo, Kyoto 606-0081, JAPAN.
[3]Laboratory of Advanced Science and Technology for Industry, Himeji Institute of Technology,
Kamigori, Ako, Hyogo, 678-1205, JAPAN

ABSTRACT

Defect formation in Si by $B_{10}H_{14}$ (decaborane) ion implantation has been investigated with photoluminescence measurement. An intense W-line was observed at photon energy of 1.018eV from as-implanted FZ-Si by 30keV $B_{10}H_{14}^+$ implantation. W-line center is considered as an interstitial aggregate and usually observed after ion implantation with subsequent low-temperature annealing in the case of atomic ion implantation. As W-line is observed from as-implanted Si, the defect formation with $B_{10}H_{14}$ is expected to be different from that of B^+ implantation with the same energy per atom. The energy dependence of W-line intensity is similar to that of diffusivity enhancement after rapid thermal annealing. Molecular dynamics simulation and Rutherford backscattering spectrometry channeling experiment suggest that one $B_{10}H_{14}$ implantation creates a larger number of dislocated Si atoms than that of B^+ implantation with the same energy per atom. This characteristic of $B_{10}H_{14}$ implantation may cause the different defect reactions in subsequent annealing process.

INTRODUCTION

With shrinking the dimension of ULSI devices, shallower junction formation is required and the energy of boron implantation has to be reduced to several hundreds eV. The $B_{10}H_{14}$ cluster ion implantation technique is a promising candidate for the ultra low energy implantation [1], because the energy per boron is reduced almost 1/10 of total implant energy. Although $B_{10}H_{14}$ implantation realizes shallow projected range, there is still transient enhanced diffusion (TED) during annealing process above implant energy of 5keV [2]. This TED is found to be associated with rod-like {311} defects, which can be observed with transmission electron microscopy (TEM) [3]. However, little is known about the small precursor of interstitial cluster that will grow into {311} defects by subsequent annealing [4]. This interstitial clustering should be dependent on the initial distribution of interstitials in the as-implanted Si. It has been reported that the dense defects are formed near the surface in the cluster ion implantation [5]. In this work, defect formation by $B_{10}H_{14}$ implantation was studied with photoluminescence (PL) measurements, molecular dynamics (MD) simulation and Rutherford backscattering spectrometry (RBS). In the PL spectra, we focused on the so-called 'W-line' (1018meV). Although there are many discussions about the origin of this W-line center, it is generally accepted that this center is (1) intrinsic defects caused by high-energy neutron or ion implantation, (2) appears after annealing between 100~300°C and disappears by higher temperature annealing, (3) No dependence on the irradiating species or bulk impurities [6]. At first, the W-line was believed to be a vacancy aggregate [7,8], however, many studies support interstitial aggregate model [9,10]. For example, Coomer et al. proposed a tri-interstitial model which is constructed by three Si interstitials between the Si-bond sites by the first principle calculations [10]. From the information of W-line in PL spectra, it will provide knowledge about the interstitial formation by $B_{10}H_{14}$ implantation and the relation between interstitial cluster and TED. Also, to support the PL measurement, the number of displaced Si atoms was measured with RBS channeling and the atomic motion in Si was simulated with MD.

EXPERIMENT

$B_{10}H_{14}$ was introduced into an ionizing chamber as vapor and ionized by electron bombardments. $B_{10}H_{14}$ ions were accelerated up to 30keV and scanned by two pairs of deflectors. An n-type FZ Si target was implanted by $B_{10}H_{14}$ at normal and the temperature of target was kept at a room temperature. The total ion dose was ranged from 1×10^{12} to 1×10^{14} ions/cm^2. After implantation, Si targets were transferred into a cryostat with liquid He circulation and photoluminescence spectra were measured at 4.2K. No subsequent annealing was performed after implantation. Si target was excited by Ar laser. The wavelength and the laser power were 488nm and 0.3W, respectively. The luminescence was passed through a spectrometer and detected by a Ge detector cooled by liquid N$_2$. RBS channeling measurement was performed with 2MeV He$^+$. The number of displaced atoms was obtained from integration of a Si surface peak in the RBS spectrum. The effect of transient enhanced diffusion was obtained from a secondary ion mass spectrometer (SIMS) depth profile before and after rapid thermal annealing (RTA). The annealing temperature and time were 900~1000°C and 10sec, respectively. In the molecular dynamics simulations, B and B_{10} were implanted into Si(100) target with 9nm long in each axis. The energy per atom was 230eV/atom. The Stillinger-Weber potential were used for Si-Si potential and ZBL potential was used for both B-B and B-Si potentials. The longest simulation time was 8ps.

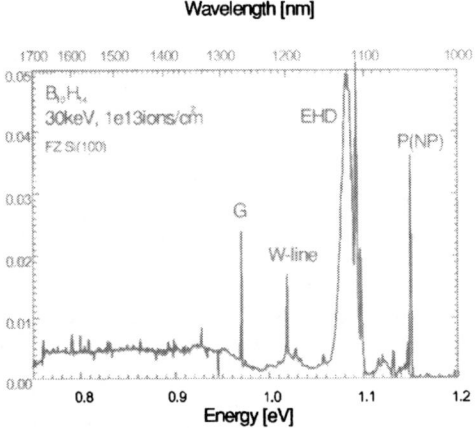

Figure 1: PL spectrum of as-implanted Si by $B_{10}H_{14}^+$ at 30keV. Ion dose was 1×10^{13} ions/cm^2.

RESULTS AND DISCUSSIONS

Figure 1 shows a PL spectrum from as implanted FZ-Si(100) target. The implantation energy of $B_{10}H_{14}$ and the atomic dose of boron were 30keV and 1×10^{14} atoms/cm^2, respectively. As the $B_{10}H_{14}$ ion current was very low (<0.5uA) at 30keV, rising of the target temperature was negligible. In the spectra, several sharp peaks and a broad one from photon energy of 0.7 to 1.0eV can be observed. This broad peak is usually observed due to ion irradiation damage. The so-called 'G-line' at photon energy of 0.97eV is luminescence from a substitutional and an interstitial carbon complex (C_sC_i) [6]. The most interesting peak in this spectrum is so-called 'W-line' at 1.018eV [6]. Usually, the W-line appears after low-temperature annealing below several hundreds of $^{\circ}$C in the case of monomer ion implantations [6,11]. But an intense W-line can be observed without annealing in the case of $B_{10}H_{14}$ implantations. Assuming the interstitial model for the origin of the W-line center, the defect formation mechanisms between atomic boron and clustered boron implantation would be different.

Figure 2 shows an implant energy dependence of W-line intensity with several $B_{10}H_{14}$ ion doses. The implant energies were 3, 10 and 30keV, and ion doses were 1×10^{12}, 1×10^{13} and 1×10^{14} ions/cm^2. There was no W-line peak at the implantation energy of 3keV in all the ion doses. However, the W-line intensity increases regardless of the ion doses with the implant energy above 10keV.

From the energy dependence of enhanced diffusion after rapid thermal annealing (RTA) [12], diffusivity enhancement $D_B/D_B{}^*$ drastically increases above the implantation energy of 5keV, where D_B is an average boron diffusivity obtained from SIMS profile and $D_B{}^*$ is an equilibrium boron diffusivity at annealing temperature. Similar to the energy dependence of W-line intensity, there was no enhancement of diffusivity in the case of 3keV implantation. It seems that there are close relations between appearance of W-line and TED.

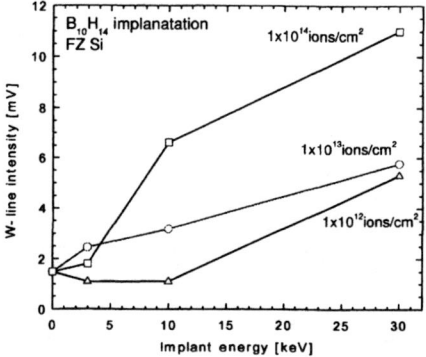

Figure 2 : Energy dependence of W-line intensity at various implant doses of $B_{10}H_{14}{}^+$.

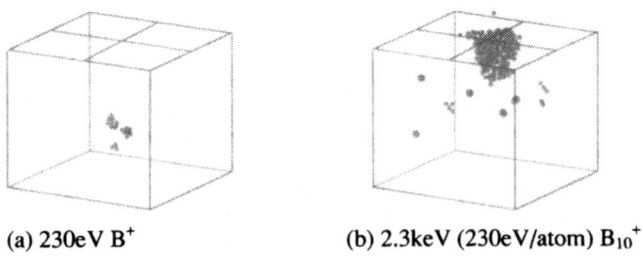

(a) 230eV B$^+$ (b) 2.3keV (230eV/atom) B$_{10}^+$

Figure 3 : Snapshot of defect distribution with B and B$_{10}$ cluster implantation into Si(100). Implantation energy was 230eV/atom and snapshot was taken at 8ps after the impacts.

To study the difference of the defect formation mechanism in Si with B$_{10}$H$_{14}$ implantation, molecular dynamics simulation was performed. Figures 3 (a) and (b) show snapshot of dislocated Si atoms after 8ps of B$^+$ or B$_{10}^+$ impacts with an energy of 230eV/atom. The total implant energy of B$_{10}$ was 2.3keV. A dark circle represents dislocated Si atom and a light one an implanted B atom. The Si target was a cubic, 9nm each side. There is no significant difference of the penetration depth between B and B$_{10}$ in MD simulations. It is already reported from experimental results that the penetration depth of boron remained the same when the energy per atom of B$^+$ and B$_{10}$H$_{14}^+$ was the same [14]. Although the penetration range is the same, the defect formation seems to be different between B and B$_{10}$ implantation. There are isolated defects in the case of B$^+$ implantation as can be seen in figure 3 (a), however, there is a dense defect region near the surface in the case of B$_{10}^+$ implantation shown in figure 3 (b).

To make this difference clear, the depth profiles of dislocated Si are shown in figure 4. The solid line shows number of the displaced Si atoms per boron atom by 230eV B$^+$ implantation and the dotted line represents those by 2.3keV B$_{10}^+$ implantation (230eV/atom). From figure 4, each boron atoms in B$_{10}$ cluster cause larger number of defects than B$^+$ implantation even though the implant energy per atom is the same. As ten boron atoms bombard a Si surface at the same time, the energy density around implanted area becomes quite high and spike-like collisions are

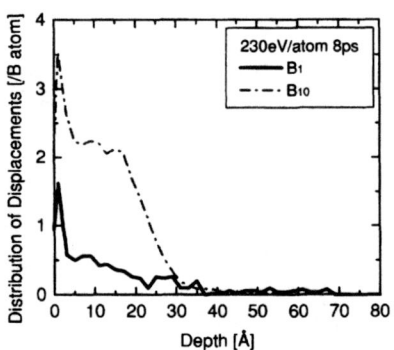

Figure 4 : Depth profiles of displaced Si atoms by B or B$_{10}$ ions with energy of 230eV/atom.

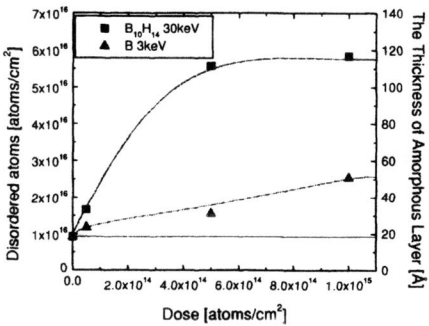

Figure 5 : Boron dose dependence of the number of disordered Si atoms by 3keV B$^+$ and 30keV B$_{10}$H$_{14}$$^+$ measured with RBS channeling.

expected in the B$_{10}$ implantation. This difference of defect formation mechanism simulated by MD will explain the intense W-line center from as-implanted Si by B$_{10}$H$_{14}$.

To evaluate the defect formation mechanism with B$_{10}$H$_{14}$ experimentally, the number of disordered Si atoms was measured with a RBS channeling technique. Figure 5 shows a dose dependence of the number of disordered Si atoms by 3keV of B$^+$ (triangle symbol) and 30keV of B$_{10}$H$_{14}$$^+$(square symbol). The energy per atom was the same (3keV/atom). The numbers of disordered atoms were obtained from an integration of Si surface peak of RBS spectrum. The horizontal axis shows an atomic dose of boron. The highest boron dose was 1x10^{15}atoms/cm^2. In figure 5, the number of disordered atoms increases almost linearly with the dose in the case of B$^+$ implantation. However, the number of disordered atoms by B$_{10}$H$_{14}$$^+$ implantation is much higher than that of B$^+$ implantation. Also it reaches a saturated value with very low ion doses. It means that one boron atom in B$_{10}$H$_{14}$ creates larger number of disordered Si than those by B$^+$ with the same energy per atom. This experimental result shows the same tendency as the MD simulations. From these experimental and simulation results, it is clear that B$_{10}$H$_{14}$ implantation forms shallow and dense defects in Si. It would be one of the reasons why intense W-line center was observed without annealing of Si implanted by B$_{10}$H$_{14}$.

CONCLUSION

Defect formation by B$_{10}$H$_{14}$ implantation was studied with photoluminescence, RBS channeling measurements and MD simulations. Intense W-line was observed from an as-implanted FZ-Si with B$_{10}$H$_{14}$ implantation. W-line intensity increases above the energy of 10keV, which is similar to the energy dependence of the diffusivity enhancement after RTA. This W-line appearance indicates interstitial Si formation and would have relation to the transient enhanced diffusions. W-line is usually observed after low-temperature annealing, however, it is observed without annealing in the case of B$_{10}$H$_{14}$ implantation. From the MD simulation and RBS channeling measurements, one B$_{10}$H$_{14}$ ion creates larger number of dislocated Si near the surface than that of B$^+$ implantation. There are many dislocated Si atoms in the implanted area by B$_{10}$H$_{14}$$^+$, which would cause different defect reactions from B$^+$ implantation. In the future, the direct comparison with the same energy per atom of B$^+$ and B$_{10}$H$_{14}$$^+$ should be performed in PL measurements.

REFERENCE

[1] K.Goto, J.Matsuo, Y.Tada, T.Tanaka, Y.Momiyama, T.Sugii and I.Yamada, IEDM Tech. Dig. 471 (1997).

[2] A.Agarwal, H-J.Gossmann, D.C.Jacobson, D.J.Eaglesham, M.Sosnowski, M.Poate, I.Yamada, J.Matsuo and T.E.Haynes, Appl. Phys. Lett., **73**, 2015 (1998).

[3] D.J.Eaglesham, P.A.Stolk, H-J.Gossmann and J.M.Poate, Appl. Phys. Lett., **65**, 2305 (1994).

[4] S.Coffa, S.Libertino and C.Spinella, Appl. Phys. Lett., **76**, 321 (2000).

[5] T.Aoki, J.Matsuo, Z.Insepov and I.Yamada, Nucl. Instr. and Meth. B, **121**, 49 (1997).

[6] G.Davis, Phys. Rep., **176**, 83 (1989).

[7] C.G.Kirkpatrick, J.R.Noonan and B.G.Streetman, Raidat. Eff., **30**, 97 (1976).

[8] N.S.Minaev, Phys. Status. Solidi. B, **108**, K89 (1989).

[9] Y.H.Lee et al, Appl. Phys. Lett., **73**, 1119 (1998).

[10] B.J.Coomer, J.P.Goss, R.Jones, S.Oberg, P.R.Briddon, Physica B, **273-274**, 505 (1999).

[11] H.Feick and E.R.Weber, Physica B, **273-274**, 497 (1999).

[12] T.Kusaba, N.Shimada, T.Aoki, J.Matsuo, I.Yamada, K.Goto and T.Sugii, *Proc. of Ion Implantation Technology 98*, 1258 (1999).

Dopant Defect Clustering

Mat. Res. Soc. Symp. Proc. Vol. 669 © 2001 Materials Research Society

EFFECT OF ARSENIC ON EXTENDED DEFECT EVOLUTION IN SILICON

R. Brindos, K. S. Jones and M. E. Law
SWAMP Center, Univ. of Florida, Gainesville, FL 32611

Abstract

The effect of arsenic on {311} defect formation was determined for temperatures ranging from 700°C to 800°C. Arsenic well structures were formed at arsenic concentrations of $3x10^{17}$, $3x10^{18}$, and $3x10^{19}$ cm^{-3}. A 40 keV $1x10^{14}$ cm^{-2} silicon implant, that is known to form {311} defects, was then incorporated into the structures. Extended defect evolution and dissolution was then studied after furnace annealing at 700°C, 750°C and 800°C for various times. It was determined that arsenic has a strong affect on the nucleation of extended defects. However, once the defects were formed, the dissolution time constant was the same for all concentrations considered. The activation energy for defect dissolution was found to be 3.4eV and was also independent of arsenic concentration. Using a newly developed {311} model in the FLOOPS process simulation software, the effect of the arsenic on {311} formation and dissolution was simulated. It was found that by using a pair model with an arsenic-interstitial binding energy of 0.95eV, the experimental results were able to be simulated.

Introduction

As devices continue to be scaled to smaller and smaller dimensions, the dopant diffusion begins to control the depth of the electrical junction. Transient Enhanced Diffusion (TED) adversely affects the diffusion of dopants and becomes an important parameter to consider in the process design of future devices.[1] TED from self-implants has been a heavily studied area for many groups and a correlation has been made between TED and extended defect formation and dissolution.[1-4] Eaglesham et al.[2] has suggested that extended defects serve as storage sites for excess interstitials and that during the dissolution of the defects interstitials are released. Once released, the interstitials are free to interact with any dopant atoms present. Common dopant atoms are known to fully or partially diffuse via interstitials and therefore the release of interstitials propels the enhanced diffusion of the dopant atoms.[1]

The addition of excess interstitials to regions doped with either boron or phosphorus has been shown to have a measurable influence on the nucleation, growth and dissolution of extended defects.[5,6] To understand the influence the dopant has on the defect processes, studies were conducted which examined the interstitial trapping by impurity dopants. Haynes et al.[5] executed a study of boron interstitial trapping using boron-doped wells. In their experiment they formed boron well structures of varying concentration and added excess interstitials by way of silicon self-implants. Subsequent anneals were done to nucleate, grow and eventually dissolve {311} defects. It was determined that for boron concentrations above $1x10^{18}$ cm^{-3} the boron traps the interstitials and causes a reduction in the {311} formation. It was also found that once the defects formed the boron concentration did not affect the dissolution process. Similar results were found for phosphorus by Keys et al.[6]

Unlike boron and phosphorus, which are known to be pure interstitial diffusers, arsenic is known to diffuse by both interstitial and vacancy mechanisms.[1] While only a partial interstitial diffuser, arsenic will likely still have an effect on the {311} defect processes which should result in a reduction in the nucleated {311}'s. To gain knowledge on how arsenic affects the {311} nucleation, growth and dissolution process, similar experiments to those of Haynes et al.[5] and Keys et al.[6] were devised. To cover the a range of concentrations below the clustering limit, doped arsenic wells ranging in concentrations between 3×10^{17} and 3×10^{19} cm^{-3} were created and excess interstitials were added using a silicon self-implant known to cause {311} defects in undoped silicon. To study the effects of the arsenic, the formation and dissolution of {311} defects that formed upon low temperature annealing, was monitored as a function of arsenic concentration. This paper reports the results from such an experiment and a simulation methodology for the results.

Experimental Overview

Arsenic well structures were fabricated by the following process. Three p-type epi-silicon wafers were each amorphized with silicon at energies of 200 keV and 70 keV at doses of 2×10^{15} and 1×10^{15} cm^{-2}, respectively. Arsenic was implanted into the amorphized region at energies of 200 keV and 70 keV at varying doses. All three wafers were then annealed at 550°C for 60 min to regrow the amorphized layer. A 60 min anneal at 1100°C was used to form a constant arsenic concentration to a depth in excess of 1600Å. The total arsenic implant doses used were 7.3×10^{12}, 7.3×10^{13} and 7.3×10^{14} cm^{-2}, which formed wells with concentrations of 3.0×10^{17}, 3.0×10^{18}, and 3.0×10^{19} cm^{-3}, respectively. Following the well anneals the wafers were sectioned and implanted with silicon at an energy of 40 keV and dose of 1×10^{14} cm^{-2}. Undoped control wafers, one with no preamorphization and one preamorphized with the same silicon implant were also annealed at 1100°C and then subjected to the 40 keV 1×10^{14} cm^{-2} silicon implant. Transmission electron microscopy (TEM) specimens were cored and furnace annealed under nitrogen ambient at various temperatures and times. Plan-view TEM specimens (PTEM) were then prepared and analyzed using a JEOL 200CX. In order to increase the contrast of the defects in relation to the background, the samples were imaged using the g=220 g,3g reflection. This condition illuminates the {311} defects while still providing adequate brightness. Cross-sectional TEM images showed that the {311} defects were at a depth centered at 700Å and the amorphous/crystalline interface for the arsenic well was at 3900Å.

Experimental Results

Plan-view TEM images are presented in Figure 1 to show the {311} defect density at each arsenic concentration after annealing at 800°C for 10 min. In each of the micrographs, the bright rod-shaped segments represent the {311} defects. Each of the defects is made up of a number of interstitials and prior studies have shown that their structure leads to 26 silicon interstitials per nm of length under the specified annealing conditions.[2]

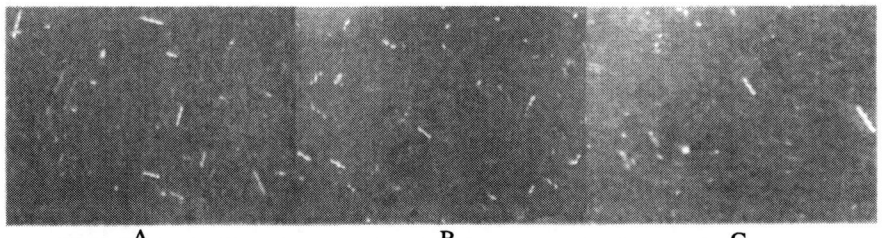

A   B   C

Figure 1: Effect of arsenic concentration on {311} defect dissolution. All samples were furnace annealed at 800°C for 10 min. A) 3x10^{17} cm^{-3} As, B) 3x10^{18} cm^{-3} As and C) 3x10^{19} cm^{-3} As.

The length of each defect was measured and by addition of each measurement the total length of defects was determined for all samples. Knowing that there are 26 interstitials per nm of length in the {311} defects, the length calculation can be converted to the total number of trapped interstitials in each sample. Presented in Figure 2 are the results from 700°C and 800°C data at each annealing time as plots of the number of trapped interstitials versus arsenic concentration. From this figure it is apparent that there is a significant effect on the number of interstitials that are trapped in {311} defects at higher arsenic concentrations. As the concentration of arsenic is raised the number of interstitials in {311} defects is decreasing. As with the other dopants studied in a similar manner, it appears that arsenic is pairing with interstitials and acting as an alternative site for the interstitials.

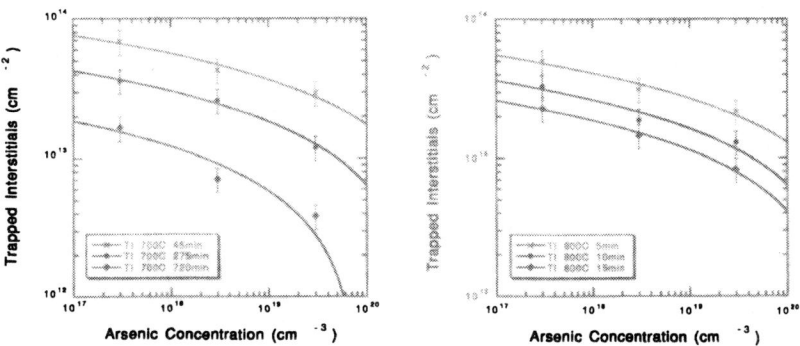

Figure 2: Net number of trapped interstitials in {311} defects as a function of arsenic concentration for various anneal times at A) 700°C and B) 800°C.

The {311} dissolution process occurs by the release of interstitials from the defects. The net loss of interstitials from the defects occurs as an exponential decay with time and can be expressed as

$$Si_i(t) = Si_0(0)\exp(-K_{311} * t) \tag{1}.$$

In Equation 1 $Si_i(t)$ is the trapped interstitial concentration per area as a function of time, $Si_0(0)$ is the concentration per area at time zero, K_{311} is the {311} dissolution rate constant and t is time. By carrying out several measurements at various temperatures and arsenic concentrations a family of decay curves may be generated. The data points were fitted with exponential least-squares fits and the dissolution rate constants were extracted. The dissolution time constant for this experimental was calculated to be 50±5 min for each concentration. The time constant obtained in this experiment is consistent with the time constants of {311} studies previously reported.[7,8] A similar dissolution time constant at all temperatures supports the idea that the {311} dissolution is not effected by the presence of arsenic. Additionally, the activation energy for dissolution was calculated to be 3.4 ± 0.2 eV and proved to be independent of arsenic concentration. It was also recognized that at each arsenic concentration the Y-intercept value is independent of annealing temperature. However, when each concentration is compared there is a large reduction in the Y-intercept value for increased arsenic concentrations.

Previous studies on {311} defects have shown that the Y-intercept or $Si_0(0)$ has related closely to the plus one model.[9] The plus one model assumes that after Frenkel pair recombination there will be a dose of interstitials that remains that is equal to the implanted dose. The plus one model appears to be independent of dose or anneal conditions. The implanted dose in this experiment was 1×10^{14} cm^{-2} and in the control samples the $Si_0(0)$ value was estimated to be 7×10^{13} cm^{-2}, which is in close agreement to the model. As the arsenic concentration was increased, a reduction in the $Si_0(0)$ value resulted. The interpretation of this data is an arsenic-interstitial complex is created during the initial stages of annealing. With less interstitials available at nucleation due to a dopant-interstitial complex formation, a decrease in the number of trapped interstitials in {311} defects is eminent.

Modeling the Arsenic Effect

Recently, the Florida Object Oriented Process Simulator (FLOOPS) was instrumented with a {311} model that predicts the formation and dissolution of {311} defects in silicon. With a few additions the arsenic-interstitial interaction can be included in the model. The simplest way to model the As-I interaction is to begin with a basic pair reaction. In this form the arsenic-interstitial reaction is treated as follows,

$$As + Si_{(I)} \Leftrightarrow AsI \tag{2}$$

This reaction can be modeled in FLOOPS using an equation of the following form,

$$C_{AsI} = C_I * C_{As} \frac{\exp\left[\dfrac{E_b}{kT}\right]}{5 \times 10^{22}} \tag{3}$$

C_{AsI} is the concentration of arsenic pairs, C_I is the concentration of interstitials added from damage of a silicon implant, C_{As} is the arsenic background concentration, E_b is the arsenic-interstitial binding energy, k is Boltzman's constant, and T is temperature in Kelvin. A silicon self-implant at 40 kev, 1×10^{14} cm^{-2} was simulated with UTMARLOWE in order to determine C_I. The damage profile is the same for all models in this study at time zero. If other energies or doses of silicon were to be simulated, a new damage profile would need to be formulated.

Simulations of the {311} model with no arsenic and with arsenic concentrations of 3×10^{17} cm^{-3} show the same dissolution assuming a temperature offset of 15°C. The dissolution rate is known to be inconsistent from experiment to experiment depending on a number of parameters such as ion implantation source, furnace type, starting material, etc.[7,8] The exact reason for the difference has not been determined at this time.

Since there is no change in the decay of the dissolution, the initial number of defects was adjusted by modifying the binding energy of the As-I pair. Figure 3A presents the simulation results when the binding energy is set to zero or above. The figure shows that increasing the binding energy decreases the number of interstitials at time zero, but has no effect on the dissolution time constant. To determine the correct binding energy for each condition, a number of simulations were run at different binding energies. Because the dissolution isn't effected by binding energy differences, the percentage decrease in time zero trapped interstitials was calculated for each case. Figure 3B shows the results at different binding energies. 0.95 eV has the best overall fit to the data obtained in each case, although only the 750°C data is presented in the figure. Presented in Figure 4 are the simulated results for the dissolution of {311} defects as a function of time. Good fits to the data at all temperatures are obtained using this model. However, at higher concentrations and at lower temperatures the simulations exhibit greater error. This may be due to one of several effects. First the {311} defects at 700°C are extremely difficult to quantify due to their small size which would lead to greater error in the statistical measurements. Another cause may be from high concentration effects. Previous studies have shown that arsenic clustering reactions begin around 8×10^{19} cm^{-3} at 700°C.[10] Therefore, it is possible that high concentration effects are starting to take shape and a simple pair model can no longer be used to model the behavior in these instances.

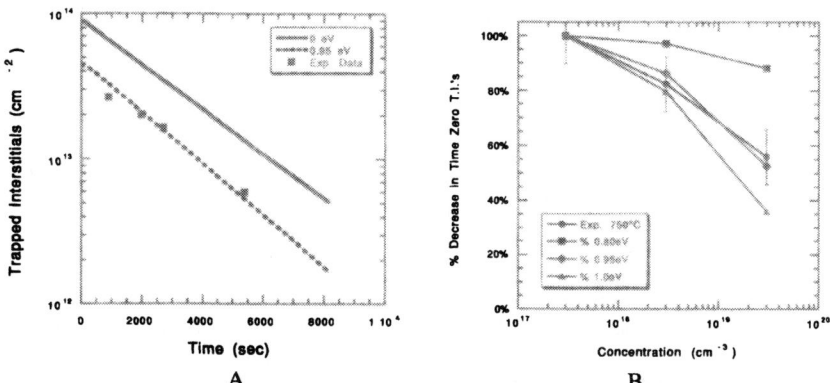

A   B

Figure 3: A) Decrease in initial starting point with increased As-I binding energy with time. A 0.95 eV binding energy best fits the experimental data for the 3×10^{19} cm^{-3} arsenic samples. B) 0.95 eV also allows for the best fits to experimental data at all concentrations considered.

Conclusions

Arsenic well structures were used to determine the effect of arsenic upon the evolution and dissolution of {311} defects. It was determined that the arsenic retarded the nucleation of defects during the initial stages of annealing. Once the defects formed and began to dissolve, arsenic no longer has an effect. This is similar to results obtained for boron and phosphorus. FLOOPS was used to model the arsenic affect. It was found that the effect could be simulated using a pair model with a binding energy of 0.95eV. This model worked well for all concentrations and temperatures studied excluding high dose/low temperature results. In the high dose/low temperature regime, it appears that solubility effects are significant and a more sophisticated model needs to be developed for those cases.

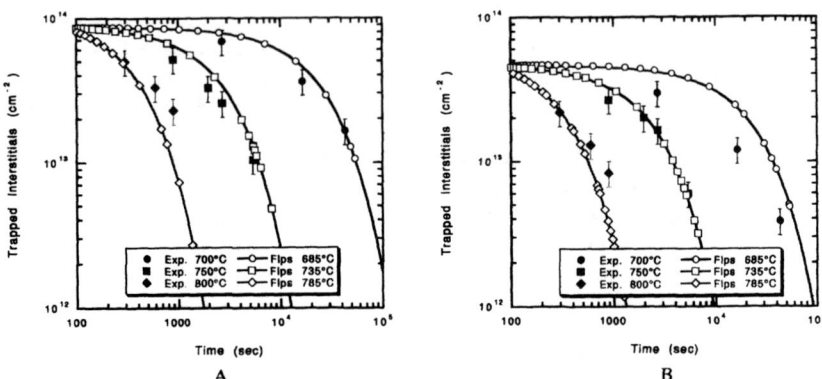

Figure 4: FLOOPS simulations as a function of temperature compared to experimental results at A) 3×10^{17} cm^{-3} As and B) 3×10^{19} cm^{-3} As..

References

[1] P. M. Fahey, P. B. Griffin, and J. D. Plummer, Reviews of Modern Physics, 61(2), p. 289(1989).

[2] D. J. Eaglesham, P. A. Stolk, J.-J. Gossmann, and J. M. Poate, Appl. Phys. Lett., 65(18),p. 2305 (1994).

[3] P. B. Griffin, R. F. Lever, P. A. Packan, and J. D. Plummer, Appl. Phys. Lett., 64(10), p.1242 (1994).

[4] J. Liu, V. Krishnamoorthy, K. S. Jones, M. E. Law, J. Shi, and J. Bennett,on Implantation Technology - 96. Proceedings of the Eleventh International Conference on Ion Implantation Technology, p. 626 (IEEE, New York, NY, 1997).

[5] T. E. Haynes, D. J. Eaglesham, P. A. Stolk, H.-J. Gossmann, D. C. Jacobson, and J. M. Poate, Appl. Phys. Lett., 69(10) , p. 1376 (1996).

[6] P. H. Keys, J. H. Li, E. Heitman, P. A. Packan, M. E. Law, and K. S. Jones, Si Front-End Processing - Physics and Technology of Dopant-Defect interactions. Materials Research Society Symposia - Proceedings, v 568, p. 199 (Materials Research Society, Warrendale, PA,1999).

[7] D. J. Eaglesham, P. A. Stolk, H.-J. Gossmann, T. E. Haynes, and J. M. Poate, Nuclear Instruments & Methods in Physical Research, Section B: Beam Interactions with Materials and Atoms, v B106, n 1-4, p. 191 (1995).

[8] K. Moller, K. S. Jones, and M. E. Law, Appl. Phys. Lett. 72, 2547 (1998).

[9] M. D. Giles, Appl. Phys. Lett., 58(21), p. 2399 (1991).

[10] S. Solmi and D. Nobili, Appl. Phys. Lett., 83(5), p. 2484 (1998).

Mat. Res. Soc. Symp. Proc. Vol. 669 © 2001 Materials Research Society

Lattice site location of ultra-shallow implanted B in Si using ion beam analysis

Hajime Kobayashi, Ichiro Nomachi, Susumu Kusanagi and Fumitaka Nishiyama[*]
Sony Corporation, Technical Support Center, Yokohama, Japan
[*]Hiroshima University, Department of Applied Physics and Chemistry, Higashi-Hiroshima, Japan

ABSTRACT

We have investigated the lattice site location of B in Si using ion channeling in combination with nuclear reaction analysis (NRA). Silicon samples implanted with Boron at an energy of 10 keV and a dose of 5×10^{14} cm^{-2} (low dose samples) or 5×10^{15} cm^{-2} (high dose samples) were annealed at 1000 °C for 10 seconds (RTA) or at 800 °C for 10 minutes (FA). The activation efficiencies of these samples were estimated from the B atomic concentration and the hole concentration obtained by secondary ion mass spectrometry (SIMS) and spreading resistance profiling (SRP), respectively. We also studied the ion implantation damage of Si crystals using ion channeling combined with Rutherford backscattering spectrometry (RBS). We found that the activation efficiency is proportional to the substitutionality, meaning that substitutional B is fully activated without any carrier compensation. We also found that B atoms go to the substitutional sites and are activated up to the solubility limit in the high dose samples. However, the ion implantation damage of the crystalline Si in the high dose samples increases somewhat after annealing.

INTRODUCTION

In scaling down the dimension of ULSI circuits, a higher concentration of B is required, and the activation efficiency becomes less than unity even below the solubility limit of B in Si [1-3]. Transient enhanced diffusion (TED) studies suggest that some of the boron becomes immobile in highly doped regions, and these B atoms are considered to form B-interstitial Si clusters (B-I clusters) [4,5]. Transmission electron microscopy (TEM) observations suggest that dissolution of {311} defects is a source of excess interstitial Si, which causes both TED and B-I cluster formation [6]. Since these clusters are assumed to be electrically inactive, they cause a decrease in the activation efficiency. The lattice site location of B is important to understand the relationship between the B clustering and the deactivation. However, little is known about the lattice site location of B. So, we have investigated the lattice site location of B using channeling-NRA. SIMS and SRP are used to estimate the activation efficiency. We also studied the ion implantation damage of Si crystals using. In this paper, we demonstrate the relationship between the lattice site location of B in Si and the activation efficiency, and suggest the relationship between the ion implant damage and the post-implant annealing.

EXPERIMENTAL PROCEDURE

B ions were implanted into n-type Si(100) substrates at an energy of 10 keV with a dose of 5×10^{14} cm^{-2} (low dose samples) or 5×10^{15} cm^{-2} (high dose samples). Post-implant anneal were performed at 1000 °C for 10 seconds (RTA) or at 800 °C for 10 minutes (FA). A 10 nm-thick layer of SiO_2 was on the Si substrates during ion implantation, and was removed in dilute HF after annealing. Depth profiles of B atoms and carriers were measured using SIMS and SRP, respectively. We used an ATOMIKA4000 for SIMS measurements using a normally incident 2.5 keV $^{16}O_2^+$ beam. $^{11}B^+$ ions were detected as secondary ions. A SSM150 manufactured by Solid State Measurements was used for SRP measurements. The surfaces of the samples were polished obliquely at approximately 1 mrad, and the depth profiling was performed by scanning a stylus in steps of 2.5 μm.

Channeling-NRA experiments were performed using a 2.5 MV Van de Graaff accelerator at Hiroshima University. We used the nuclear reaction $^{11}B(p,\alpha)^8Be$ to detect B. We identified B by detecting α particles emitted from the reaction. A 1.3 MeV H_2^+ beam (equivalent to two 0.65 MeV protons) was used. Two semiconductor detectors with an active area of 1200 mm^2 (ORTEC CU-037-1200-AS) were placed at 150° with respect to the beam direction to detect α particles. The total solid angle of these detectors was 960 msr. A 4 μm-thick aluminized Mylar film and an 8 μm-thick polypropylene film were placed in front of the detectors to eliminate the backscattered proton background. The samples were mounted on a three-axis goniometer. Channeling-NRA experiments were performed along the <100> and <110> directions. Ion implantation damage was studied using channeling-RBS using a 1 MeV He$^+$ beam. A semiconductor detector with a solid angle of 0.96 msr (ORTEC CU-011-050-300) was placed at 100° (glancing angle) to detect backscatterd He. Damage analysis was performed only along the <100> direction.

RESULTS AND DISCUSSION

Figure 1 shows the results of SIMS and SRP measurements. The activation efficiency is defined as the ratio of the total number of carriers obtained by SRP to the total number of B

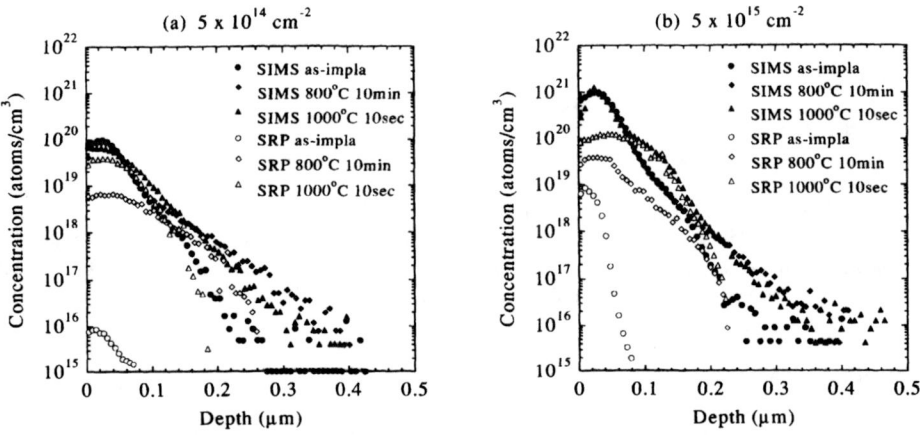

Figure 1. B atomic concentration (SIMS) and carrier concentration (SRP) of (a) the low dose samples (5 × 10^{14} cm^{-2}) and (b) the high dose samples (5 × 10^{15} cm^{-2})

Table I. Summary of electrical activation efficiencies

	5×10^{14} cm^{-2}	5×10^{15} cm^{-2}
as-impla.	0.00	0.00
FA 800°C, 10min	0.13	0.05
RTA 1000°C, 10sec	0.55	0.24

atoms obtained by SIMS. The activation efficiencies of the samples are summarized in Table I. Although the peak B atomic concentration in the low dose (5×10^{14} cm^{-2}) RTA sample is below the solubility limit of 1.3×10^{20} cm^{-3} at 1000 °C [7], the activation efficiency is smaller than unity, i.e., saturation of the carrier concentration is observed. The high dose (5×10^{15} cm^{-2}) RTA sample has a peak B atomic concentration of approximately 1×10^{21} cm^{-2}, that is far above the solubility limit, and the activation efficiency is lower (0.24) than that of the low dose RTA sample (0.55). The maximum carrier concentration in the high dose RTA sample is approximately 1.3×10^{20} cm^{-3}, almost the same as the solubility limit. This suggests that B atoms are electrically activated up to the concentration of the solubility limit by overdoping. The residual B atoms around the peak concentration are considered to form B clusters because they are neither active nor do they diffuse. Fig.1-(b) also shows that B atoms deeper than 80 nm have diffused and are electrically activated.

The channeling-NRA spectra of the low dose and the high dose samples are shown in Fig.2 and Fig.3, respectively. The minimum channeling yield of B, χ_{min}(B), is defined as the ratio of the channeling yield to the random yield integrated from 0.8 MeV to 5 MeV. Table II summarizes the χ_{min}(B). Since each sample has similar χ_{min}(B)'s in both the <100> and the <110> directions, B atoms shadowed by Si atoms, that is proportional to $1-\chi_{min}$(B), are considered to be in the substitutional sites. It is clear that the channeling yields in the high dose samples are larger than those of the low dose samples, meaning that the fractions of the substitutional B atoms are smaller in the high dose samples. Figure 4 shows the channeling-RBS spectra. The minimum channeling yield of Si, χ_{min}(Si), is defined as the ratio of the channeling yield to the random yield integrated from 570 keV to 670 keV. Table II summarizes the χ_{min}(Si). The fraction of substitutional B atoms (substitutionality : f_{sub}) is calculated from the following formula using χ_{min}(B) and χ_{min}(Si) [8].

$$f_{sub} = \frac{1 - \chi_{min}(B)}{1 - \chi_{min}(Si)} \qquad (1)$$

The f_{sub}'s of all samples are summarized in Table II. We used the values of χ_{min}(B) and χ_{min}(Si) in the <100> direction. Figure 5 shows the relationship between the substitutionality and the activation efficiency. It is clear that the activation efficiency is almost proportional to the substitutionality. It means that substitutional B atoms are fully activated without any carrier compensation by donor like defects or impurities.

The implantation damage in the low dose samples are fully recovered by annealing as shown in Fig.4-(a). In the high dose samples, the χ_{min}(Si) of the as-implanted sample is 11 % as shown in Table II. After annealing, however, χ_{min}(Si) increases rather to be 24 % (RTA) or 29 % (FA). The numbers of interstitial Si in the high dose as-implanted sample and in the high dose RTA

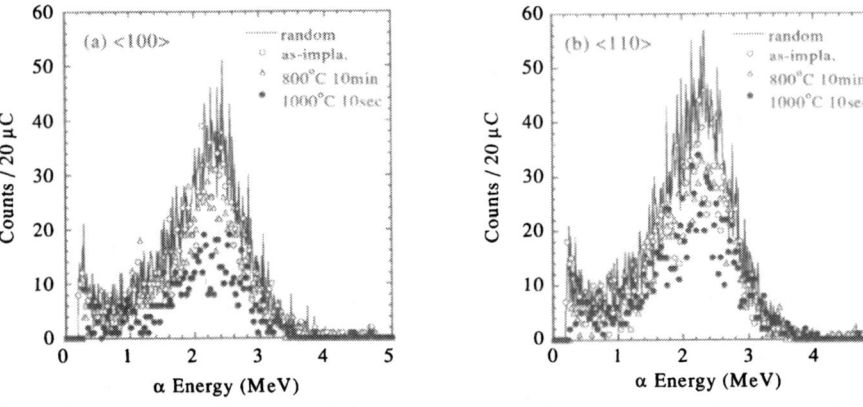

Figure 2. Energy spectra of α particles of channeling-NRA experiments for the low dose sample (5 × 10¹⁴ cm⁻²) along (a) the <100> and (b) the <110> directions.

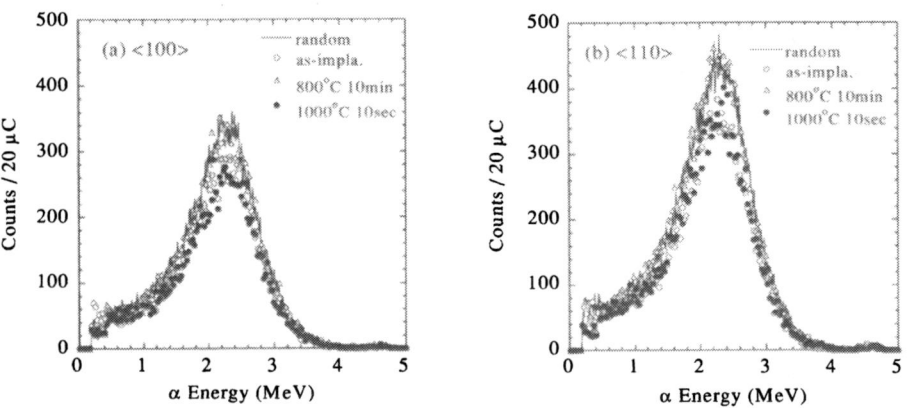

Figure 3. Energy spectra of α particles of channeling-NRA experiments for the high dose sample (5 × 10¹⁵ cm⁻²) along (a) the <100> and (b) the <110> directions.

sample are roughly estimated from the damage profiles in Fig.4-(b) to be 4.4×10^{15} cm^{-2} and 1.6×10^{16} cm^{-2}, respectively. It is found that approximately the same number of interstitial Si atoms as ion implanted B atoms are in the as-implanted sample. In the RTA sample, the number of interstitial Si atoms is approximately five times larger than that of interstitial B atoms, which is 70 % of implanted B atoms, 3.5×10^{15} cm^{-2}. Since *ab initio* calculations of the formation energies of B-I clusters suggests that B rich clusters are energetically stable [9-12], the large number of interstitial Si atoms might originated from secondary defects rather than the B-I

Table II. Summary of minimum channeling yield (χ_{min}) and substitutionality (f_{sub})

sample	5×10^{14} cm^{-2}				5×10^{15} cm^{-2}			
	$\chi_{min}(B)$		$\chi_{min}(Si)$	f_{sub}	$\chi_{min}(B)$		$\chi_{min}(Si)$	f_{sub}
	<100>	<110>	<100>		<100>	<110>	<100>	
as-impla.	0.81	0.76	0.05	0.20	0.81	0.80	0.11	0.21
FA 800 °C, 10min	0.88	0.87	0.02	0.12	0.98	1.03	0.29	0.03
RTA 1000 °C, 10sec	0.40	0.44	0.02	0.61	0.77	0.80	0.24	0.30

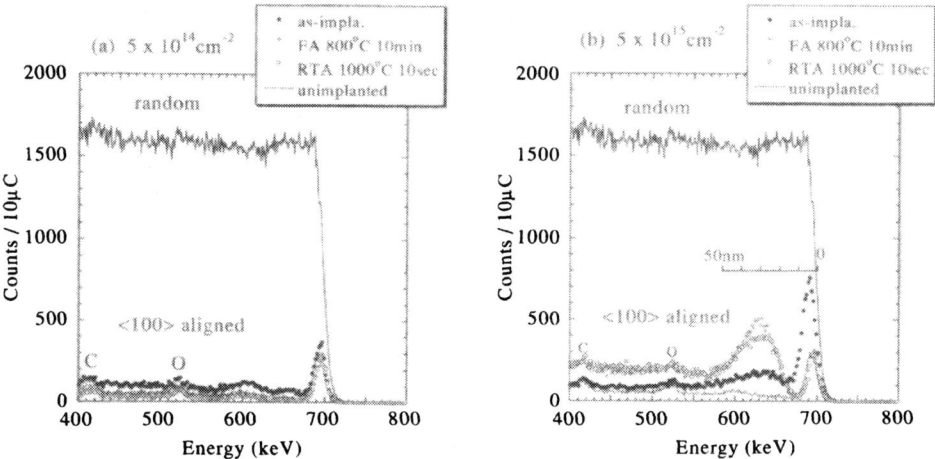

Figure 4. Channeling-RBS spectra using 1MeV He$^+$ beam for implant damage analysis of (a) the low dose samples (5×10^{14} cm^{-2}) and (b) the high dose samples (5×10^{15} cm^{-2})

clusters themselves. We consider that the B-I clusters grow with the annealing time, and generate strain in the Si lattice, then secondary defects such as dislocations are nucleated beyond a critical concentration of B atoms. In the high dose samples, the surface peak of the as-implanted sample is prominent as shown in Fig.4-(b), suggesting that a large disorder is at the surface. However, this is recovered after annealing.

CONCLUSIONS

We have investigated the lattice site location of B in Si and the implant damage using channeling-NRA, channeling-RBS, and the activation efficiencies were estimated using SIMS and SRP. We found that the activation efficiency is proportional to the substitutionality,

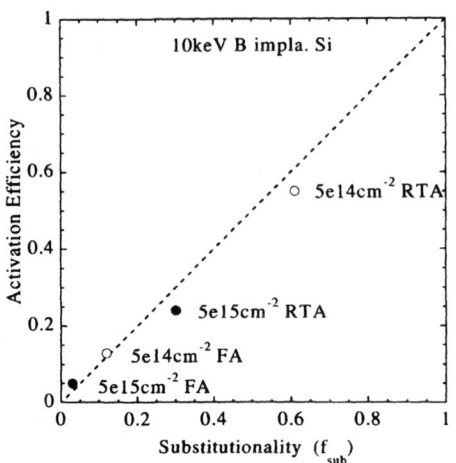

Figure 5. Relationship between the substitutionality (f_{sub}) and the activation efficiency

meaning that the substitutional B is fully activated without any carrier compensation. We also found that B atoms go to the substitutional sites and are activated up to the concentration of the solubility limit in the heavily implanted samples. The ion implantation damage of Si crystals, however, increases rather than recovers after annealing in the high dose samples.

ACKNOWLEDGEMENTS

The authors would like to thank M. Kimura for preparing samples and valuable discussions. We also thank H. Yamagata and H. Nakamura for SRP measurements, and Dr. J. F. Williams for reading manuscript.

REFERENCES

1. L. Pelaz, V. C. Venezia, H.-J. Gossmann, G. H. Gilmer, A. T. Fiory, C. S. Rafferty, M. Jaraiz and J. Barbolla, *Appl. Phys. Lett.* **75**, 662 (1999).
2. A. Nishida, E. Murakami and S. Kimura, *IEEE Trans. Electron Devices* **45**, 701 (1998).
3. P. A. Stolk H-J Gossmann, D. J. Eaglesham, D. C. Jacobson, C. S. Rafferty, G. H. Gilmer, M. Jaraiz, J. M. Poate, H. S. Luftman and T. E. Haynes, *J. Appl. Phys.* **81**, 6031 (1997).
4. E. Schroer, V. Privitera, F. Priolo, E. Napolitani and A. Carnera, *Appl. Phys. Lett.* **74**, 3996 (1999).
5. P. A. Stolk H-J Gossmann, D. J. Eaglesham, D. C. Jacobson, H. S. Luftman and J. M. Poate, *Mat. Res. Soc. Symp. Proc.* Vol. **354**, 307 (1995).
6. D. J. Eaglesham, P. A. Stolk, H.-J. Grossmann and J. M. Poate, *Appl. Phys. Lett.* **65**, 2305 (1994).
7. B. Garben, W. A. Orr-Arienzo and R. F. Lever, *J. Electrochem. Soc.* **133**, 2152 (1986).
8. L. C. Feldman, J. W. Mayer and S. T. Picraux, *Materials Analysis by Ion Channeling* (Academic Press, New York, 1982)
9. J. Zhu, T. D. Rubia, L. H. Yang, C. Mailhiot and G. H. Gilmer, *Phys. Rev.* **B54**, 4741 (1996).
10. L. Pelaz, G. H. Gilmer, H.-J. Gossmann, C. S. Rafferty, M. Jaraiz and J. Barbolla, *Appl. Phys. Lett.* **74**, 3657 (1999).
11. X.-Y. Liu, W. Windl and M. P. Masquelier, *Appl. Phys. Lett.* **77**, 2018 (2000).
12. T. J. Lenosky, B. Sadigh, S. K. Theiss, M. J. Caturla and T. D. Rubia, *Appl. Phys. Lett.* **77**, 1834 (2000).

Mat. Res. Soc. Symp. Proc. Vol. 669 © 2001 Materials Research Society

A NEW KINETIC MODEL FOR THE NUCLEATION AND GROWTH OF SELF-INTERSTITIAL CLUSTERS IN SILICON

Christophe J. Ortiz[a], and Daniel Mathiot[b]
[a] DIMES / ECTM Laboratory, TU Delft, Mekelweg 4, 2628 CD Delft, The Netherlands
[b] Laboratoire PHASE-CNRS, 23 rue du Loess, 67037 Strasbourg Cedex 2, France

ABSTRACT

A model for nucleation and growth of {311} defects is proposed on the basis of thermodynamic and kinetic considerations. Simulated results are discussed and compared to experimental results found in the literature. According to our model it is found that formation energies of self-interstitial clusters depends on the local interstitial supersaturation. Physical parameters extracted from experimental results by inverse modeling are in good agreement with recent values published in the literature.

INTRODUCTION

Nowadays, Transient Enhanced Diffusion (TED) of dopants is of importance in IC fabrication. As process simulation has become an essential part of new technology development in the silicon IC industry, it is important to have a good understanding of physical mechanisms controlling TED, in order to provide models which are as predictive and efficient as possible. To do so, a considerable effort has been devoted in the last few years to the understanding of the annealing kinetics of Dislocation Loops (DL) [1], {311} defects [2] and even of small self-interstitial clusters [3], which all maintain an interstitial supersaturation during annealing and thus play an important role in TED of dopants such as boron. In this paper we propose a kinetic model based on thermodynamic considerations for the nucleation and growth of self-interstitial clusters in silicon. As previous models [3,4,5], our model accounts for the attachment and emission of interstitials to and from clusters of different sizes and includes the interstitial recombination at the surface. It will be shown that using only five free physical parameters, our model accounts for main experimental results on TED found in the literature. Physical parameters of main importance for the understanding of dopant diffusion in silicon such as the formation energy of an isolated interstitial, the interstitial diffusion coefficient and the equilibrium interstitial concentration will be fitted separately on published experimental results and will be shown to be in good agreement with recent values found in the literature.

FORMULATION OF THE NUCLEATION MODEL

In this part we shall give a formulation of our model for the nucleation of self-interstitial clusters. Most emphasis will be placed on the rate constants for absorption and emission of interstitials, especially on the formation and dissociation energies. Thermodynamic considerations as well as the particular geometry of {311} defects will be taken into account for the calculations of these latter. For the remaining, we shall assume the surface to be a perfect sink for interstitials and so that the supersaturation is 1 at the surface, as it has been suggested by previous studies [3,6].

As it is generally assumed in the classical theory of nucleation and by many authors, clusters are expected to arise by a series of reactions of the type :

$$I_n + I_1 \underset{k_{n+1}^-}{\overset{k_n^+}{\rightleftharpoons}} I_{n+1} \tag{1}$$

where I_n represents a self-interstitial cluster of n atoms and I_1 a free interstitial (from the super-saturated solution). k_n^+ is the absorption rate of interstitials to a cluster I_n and k_{n+1}^- is the emission rate of interstitials from a cluster I_{n+1}. Expressions of these two rate constants will be derived for the particular case of {311} defects in the following.

In order to predict the evolution of the interstitial concentration, one must take into account all possible reactions between interstitials and clusters of different sizes (Eq. (1)), but also diffusion of interstitials. Then, one easily finds that the evolution of the interstitial concentration $[I_1]$ – in the region of clusters – is governed by the following equation :

$$\frac{\partial [I_1]}{\partial t} = \frac{\partial}{\partial x}\left(D_I \frac{\partial [I_1]}{\partial x} \right) - \sum_{n=1}^{\infty} \beta_n \left(k_n^+ [I_n][I_1] - k_{n+1}^- [I_{n+1}] \right) \tag{2}$$

where D_I is the interstitial diffusion coefficient. The coefficient β_n takes into account the particular case $n=1$; $\beta_n=2$ when $n=1$ and 1 otherwise.

Also, the evolution of the defects size population $[I_n]$ is expressed as :

$$\frac{\partial [I_n]}{\partial t} = k_{n-1}^+ [I_{n-1}][I_1] - k_n^- [I_n] - k_n^+ [I_n][I_1] + k_{n+1}^- [I_{n+1}] \tag{3}$$

Of course, this set of partial differential equations must be solved numerically at each time step along with Eq. (2). However, this must be done on more than one grid point. We do believe this is important for two major reasons. On one hand, the interstitial supersaturation is generally introduced by implantation. This means that the initial supersaturation is not uniform and thus, if one wants to be rigorous, the problem of nucleation must be solved everywhere, i.e., in the region including the implantation peak. On the other hand, and this is the most important reason, the surface has a strong influence on the clusters formation. Indeed, as the surface is expected to be a perfect sink for interstitials, an abrupt gradient of interstitials exists between surface and the region where clusters are located. In other words, the interstitial concentration at the surface is much lower than in the bulk. Therefore, as cluster formation is linked to the interstitial concentration (see Eqs. (2) and (3)), one expects the population of clusters beneath the surface to be very different from that in the region of maximum implantation damages. We do believe this cannot be ignored because all clusters contribute to the supersaturation we observe experimentally, the largest as well as the smallest, even those located beneath the surface.

Furthermore, we emphasize that we did not assume the steady state to solve the nucleation problem, as it can be seen in Eq. (2) and (3). Of course, this enables us to treat the transient regime, but also, as it will be shown in the following, to extract D_I separately from other parameters. Now, we shall derive expressions for the rate constants for absorption and emission of interstitials, namely, k_n^+ and k_{n+1}^-, for the particular case of {311} defects in silicon. To do so, we shall place a particular emphasis on the formation and dissociation energies.

FORMATION

The formation rate of {311} defects and more generally, of self-interstitial clusters must incorporate the diffusion of interstitial to the precipitate/silicon interface but also the probability that an interstitial has to jump from one phase (super-saturated solution) to the other (defects). This is usually done as follows :

$$k_n^+ = 4\pi R_n D_I \exp(-\Delta g_n / k_B T) \qquad (4)$$

where R_n is the effective capture radius of the defect with size n, Δg_n is the *effective* energy barrier that an interstitial must overcome and k_B the Boltzmann's constant.

When deriving expression for the formation energy one must be careful. Indeed, usually, one only considers an isolated cluster interacting with a single free interstitial for the calculation of the formation energy. We do believe this description of the system is incomplete. Actually, each cluster is surrounded by a large amount of free interstitials. Therefore, one should take into account the change in chemical potential – the driving force for precipitation – when an interstitial jumps from the supersaturated phase to a cluster. Then, when broken down into individual components, the *apparent* formation energy assumes the following form :

$$\Delta g_n = -k_B T \ln\left(\frac{[I_1]}{[I]^*}\right) + E_{fault} + E_{strain}(n) \qquad (5)$$

where $[I]^*$ is the equilibrium interstitial concentration at the absolute temperature T.

The first term is the change in free energy associated to the change in chemical potential when an interstitial jumps from the supersaturated phase to a cluster. As it can be seen, this term, which is a function of the supersaturation, is always negative and is thus the main driving force for precipitation. The second term is the energy due to the stacking fault created by the defect. It is a constant for {311} defects, as it has already been shown by several authors [3,7]. This energy has to be determined and is thus a free parameter in our model. The following term of Eq. (5) corresponds to the elastic energy per atom due to the strain induced by the defect in the crystal. The sum of the fault and strain energies represents what we usually call the formation energy $E_f(n)$, i.e. the energetic cost for adding one Si atom to a cluster with size n. In this work we assumed a rectangular shape for the {311} defects and we used the expression given by Hobler and Rafferty [5] for the calculation of the strain energy as a function of size n. The only free parameter in this expression is the amplitude of the burger vector b. As there is no trivial relation between the width W and the length L of {311} defects, we used $W = \sqrt{L.5\,\overset{\circ}{A}}$, as suggested by the same authors.

According to Eq. (5), Δg_n may be either positive or negative; this gives rise to two different situations. On one hand, if the supersaturation $[I_1]/[I]^*$ is high enough, e.g. at the early stages of the annealing, then the first term of Eq. (5) is dominant and Δg_n is negative for any size n. Thus, there is no energy barrier for the reaction (1) and the formation of defect is only diffusion limited (see Eq. (4)). On the other hand, when the interstitial supersaturation is low, i.e. after the nucleation has occurred and most of the interstitials are trapped in clusters, the first term of Eq. (5) can be counterbalanced by the formation energy $E_f(n)$ term and Δg_n becomes positive. This means that for certain sizes, especially the smallest ones for which the formation energy $E_f(n)$ is relatively high, the formation of defects becomes reaction limited when the supersaturation is low. Asymptotically, when there is no interstitial supersaturation, the energy barrier for the formation reduces to $E_f(n)$.

As it is difficult to calculate the effective capture radius R_n for a rectangular reaction volume, we considered all atoms located at the periphery of the defects as independent reaction centers.

DISSOCIATION

The rate constant k_{n+1}^- which describes the emission of I from a cluster I_{n+1} is in fact the frequency at which a silicon atom can dissociate from a cluster of $n+1$ atoms. As a first approximation we can assume this frequency to be independent of the surrounding. It can thus be written as the product of an attempt frequency and of the probability an Si atom has to dissociate from a cluster :

$$k_{n+1}^- = 6\frac{D_I}{\lambda^2}\theta_n \exp(-(E_f' - E_f(n))/k_B T) \qquad (6)$$

where λ is the average lattice distance, θ_n the number of sites from where an atom can dissociate and E_f' the formation energy of an isolated interstitial. The number of sites at the periphery were calculated according to L and W.

COMPARISON WITH EXPERIMENTAL DATA

In this section we shall compare model predictions with experimental data found in the literature. As our model has five free physical parameters (D_I, $[I]^*$, E_f', E_{fault} and b), if we want to unambiguously extract all these parameters, we need more experimental data than unknown parameters. Therefore, we used the experimental data obtained by Cowern et al [3] on the evolution of interstitial supersaturation. Briefly, Cowern et al implanted 2×10^{13} cm^{-2} Si at 40 keV and annealed the samples at 600, 700 and 800°C under N_2 ambient for times varying from 1 s up to 20 h. Using SIMS, the authors measured the TED of a boron marker layer to determine the transient interstitial supersaturation during cluster ripening. In this paper we will try to reproduce this experiment and to extract the free parameters of our model from the corresponding experimental data. To do so, the set of equations (2-3) was solved using the PDE solver PROMIS 1.5 [8] and Profile [9], a nonlinear optimizer, was used to determine the five free parameters mentioned above and to find a least-squares fit of the simulated result to the experimental data. For initial conditions we assumed that no clusters were formed at t=0, i.e., all the interstitials added by the implant (+1 model) were initially free to diffuse.

Figure 1 shows the good agreement obtained ($\chi^2<1.2$) between calculations and the experimental data at 600°C, 700°C and 800°C. As we can see, the different phases of TED, i.e. the ultrafast TED as well as the plateau region, are well predicted by our model. Table 1 summarizes values of the different parameters we extracted from experimental data. The values of D_I, $[I]^*$ and E_f' are in rather good agreement with those found by Bracht et al [10] from zinc diffusion experiments in silicon. The fault energy of {311} defects we extracted in this work, 0.83 eV, also agrees with previous measurements and predictions [3,7].

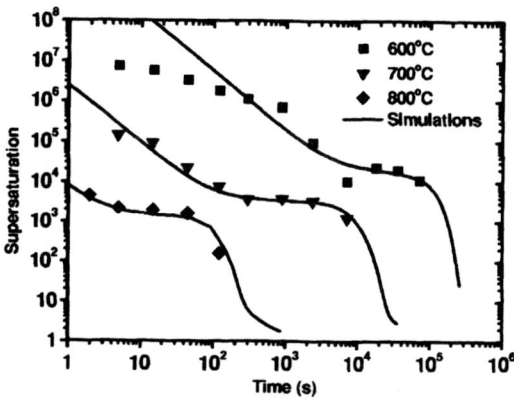

Figure 1 : Comparison between experimental data and simulations.

The value of the last free parameter of our model, i.e., the amplitude of the burger vector b, is close to that one (0.294 Å) recently published by Robertson et al [11]. The solid solubility $[I]*$ was also fitted by an Arrhenius law in order to extract the corresponding pre-exponential factor and activation energy. As expected, the activation energy we found, 3.0 eV, is consistent with E'_f, which was independently extracted.

Furthermore, we want to stress that it was not possible to obtain a good fit to experimental data with another values of D_I and $[I]^*$, even keeping a constant value of the product $D_I.[I]^*$, as it is usually done. This can be easily explained. For example, at a given temperature, decreasing D_I by a factor 10 results in a slower kinetics for the formation and dissociation of clusters and thus, in a shift to the right for the interstitial supersaturation. However, this shift in time cannot be compensated by increasing $[I]^*$ of the same factor, in order to keep a constant value of $D_I.[I]^*$. Indeed, besides the fact that this would modify the apparent formation energy (Eq. (5)), an increase of the solid solubility $[I]^*$ would mainly result in a decrease of the supersaturation $[I_1]/[I]^*$, which would not compensate the shift in time due to the decrease of D_I. Therefore, this model enables one to extract D_I and $[I]^*$ independently and unambiguously.

However, as we can see in Fig. 1, some deviation exists in the ultrafast TED region for the lowest temperature, i.e. 600°C. As it has been suggested by Cowern et al [3], formation energies of the smallest clusters can exhibit strong minima, which is not taken into account by our continuous approach. It is expected that the presence of these minima has a strong impact on the clusters formation, especially at low temperature where the flux of interstitials is lower than at 700°C or 800°C and thus leads to a sharp transition between small clusters and {311} defects.

	D_I (cm²s⁻¹)	$[I]^*$ (cm⁻³)	E'_f (eV)	E_{fault} (eV)	b (Å)
600°C	2.0×10^{-9}	4.0×10^6	2.98	0.95	0.33
700°C	8.1×10^{-9}	3.1×10^8	3.04	0.83	0.35
800°C	7.0×10^{-8}	6.7×10^9	3.02	0.83	0.35

Table 1: Optimized values of the model parameters

CONCLUSION

In this paper we proposed a kinetic model based on thermodynamic considerations for the nucleation and growth of self-interstitial clusters in silicon. We showed that the formation energy of an Si atom in a given cluster, depends not only on the cluster size, as accounted for in existing models, but also on the local free interstitial supersaturation. We demonstrated that it is possible to account for experimental observations with a limited number of parameters. Values of the extracted parameters are in good agreement with those found in the literature. However we observed some deviation between calculated results and experimental data at low temperature (600ºC). We suggested that this deviation is due to the existence of minima in the formation energies of small clusters, which cannot be taken into account by a continuous approach.

ACKNOWLEDGMENTS

C. J. Ortiz is very grateful to A. Mesli from PHASE Laboratory and N. Cowern from Philips Research for very helpful discussions. This project was partially supported by the Dutch foundation for the technology STW.

REFERENCES

[1] F. Cristiano, J. Grisolia, B. Colombeau, M. Omri, B. de Mauduit, A. Claverie, L. F. Giles and N. E. B. Cowern, J. Appl. Phys. 87 (12), 8420 (2000)

[2] D. J. Eaglesham, P. A. Stolk, H.-J. Gossmann and J. M. Poate, Appl. Phys. Let. 65 (18), 2305 (1994)

[3] N. E. B. Cowern, G. Mannino, P. A. Stolk, F. Roozeboom, H. G. A. Huizing, J. G. M van Berkum, F. Cristiano, A. Claverie and M. Jaraiz, Phys. Rev. Lett. 82 (22), 4460 (1999).

[4] A. H. Gencer and S. T. Dunham, J. Appl. Phys. 81 (2), 631 (1997)

[5] G. Hobler and C. S. Rafferty, MRS Symp.Proc. 568, 123 (1999)

[6] D. R. Lim, C.S. Rafferty and F. P. Clemens, Appl. Phys. Lett. 67, 2303 (1995)

[7] P. A. Stolk, H.-J. Gossmann, D.J. Eaglesham, D. C. Jacobson, C.S Rafferty, G. H Gilmer, M. Jaraiz, J. M. Poate, H. S. Luftman and T. E. Haynes, J. Appl. Phys. 81, 6031 (1997)

[8] P. Pichler, W. Jungling, S. Selberherr, E. Guerrero and H. W. Pötzl, IEEE Trans. Computer-Aided Design 4 (1985) 384.

[9] G. L. Ouwerling, Profile, The General Purpose Data Processor by Gertjan L. Ouwerling, Electrical Materials Laboratory, Delft University, 1988.

[10] H. Bracht, N. A. Stolwijk and H. Mehrer, Phys. Rev. B 52 (23), 16542 (1995)

[11] L. S. Robertson, K. S. Jones, L. M. Rubin and J. Jackson, J. Appl. Phys. 87 (6), 2910 (2000)

Mat. Res. Soc. Symp. Proc. Vol. 669 © 2001 Materials Research Society

Influence of Arsenic Clustering and Precipitation on the Interstitial and Vacancy Concentration in Silicon

R. Brindos[a], M. H. Clark[a], K. S. Jones[a], M. Griglione[b], Hans-J. Gossmann[b], A. Agarwal[c], B. Murto[d] and E. Andideh[e]

a) SWAMP Center, Department of Materials Science and Engineering, University of Florida, Gainesville, FL 32611-6130
b) Agere Systems, Orlando, FL 32819
c) Eaton Corporation, Beverly, MA 01915
d) International SEMATECH, Austin, TX 78741
e) Intel Corporation, Portland Technology Development, 5200 NE Elam Young Parkway, Hillsboro, OR 97124

ABSTRACT

The point defect injection from arsenic precipitation was studied using boron marker layers and antimony doped superlattices. Comparisons of arsenic and germanium amorphizing implants showed similar boron marker layer diffusion enhancements after spike annealing. The results indicate that the end of range damage caused by the implants was the source of the diffusion enhancement. Additional annealing cycles showed that there was retardation in the diffusion enhancement of the boron marker layers for precipitation range arsenic implants. Antimony marker layers showed no diffusion enhancement due to vacancy injection. The results of the experiments indicate that arsenic-interstitial complexes are the cause of the decrease flux of interstitials to the bulk.

INTRODUCTION

As the semiconductor community continues to decrease the size of the transistor to smaller and smaller dimensions, issues of dopant diffusion and dopant activation become some of the major concerns. Arsenic is generally used as an n-type dopant in silicon due to its high solubility and low diffusivity. Even though arsenic displays these desirable qualities, the increased demands are pushing the limits of its usefulness.[1] The regions with arsenic doping are generally at concentrations that exceed the solid solubility limit in silicon. It has been shown in previous studies that exceeding the solubility limit presents many problems in silicon.[2-4] Recent studies by Jones et al.[5] have shown that at low energies and moderate doses, arsenic displays dramatic transient enhanced diffusion (TED) effects after anneals at 750°C while no evidence of extended defects are found. Since arsenic is known to diffuse by a dual mechanism involving interstitials and vacancies,[6] the TED effects were suggested to be due to the formation of mobile arsenic-interstitial complexes or a vacancy release upon monoclinic SiAs precipitate formation.[7-9] These explanations are possible due to the different stages of arsenic-defect interactions available as the arsenic concentration is increased.

At low arsenic concentrations ($< 1 \times 10^{20} \mathrm{cm}^{-3}$) arsenic has been shown to affect interstitial populations by forming arsenic-interstitial complexes.[10,11] As the arsenic concentration is increased, arsenic-vacancy clusters begin to form and eventually a concentration is reached where SiAs precipitates form.[12-19] It has been suggested that the formation of SiAs precipitates occurs through the combining of arsenic-vacancy clusters.[20] When the clusters combine they would inject a vacancy into the bulk. This could cause the reduction of interstitial rich extended defects or in the motion of the arsenic profile. This study uses both boron and antimony marker layers to determine the point defect injection from SiAs precipitate formation.

EXPERIMENTAL OVERVIEW

Boron Marker Layer Samples:

A silicon epilayer with a boron spike at a depth of ~4500 Å from the surface was grown on a Si (100) substrate by chemical vapor deposition at a temperature of 800 °C and pressure of 20 torr. The silicon and boron source gases were $SiCl_2H_2$ and BH_3, respectively. An as-grown wafer with no preamorphization or implant was used to determine the initial depth versus concentration profile of the boron marker layer. A second wafer was preamorphized with a 4 keV $1x10^{15}$ cm^{-2} germanium implant to determine the effects of preamorphization only on the boron marker layer diffusion. A third set of wafers was preamorphized with a 4 keV $1x10^{15}$ cm^{-2} germanium implant followed by an 2 keV arsenic implant at doses $2x10^{14}$ to $4x10^{15}$ cm^{-2}. A fourth and fifth set of wafers were directly implanted with either arsenic or germanium at 2 keV and doses $2x10^{14}$ to $4x10^{15}$ cm^{-2} to decipher the difference between EOR damage and clustering/precipitation effects. Samples were then RTA spike annealed in a STEAG AST 3000 system at 1050 °C in 1000 ppm O_2 with a 250 °C/s ramp up followed by a 75 °C/s cool down. Secondary ion mass spectroscopy (SIMS) analysis was performed on all samples using a Cameca IMS 3f system to determine the B concentration versus depth profiles before and after annealing. Boron ions were monitored under O^+ bombardment at an energy of 10 keV and an angle of 30°. The germanium pre-amorphized samples were then furnace annealed at 850 °C for 1 hr in N_2 and the depth versus concentration profile of the B marker layer was once again determined by SIMS.

Antimony Doped Superlattices:

Doping superlattices (DSLs) with six spikes of $1x10^{14}$ cm^{-2} antimony were grown on a Si(100) substrate by a custom-made MBE system, with growth details described previously.[21] The antimony spikes were grown 10nm in width with the peak centers spaced 100nm apart. The shallowest spike was capped with 50nm of silicon. The silicon regions were grown at 450 °C, while the regions containing antimony were grown at 230 °C followed by a 650 °C 120 s rapid thermal anneal after the completion of each spike layer. Three samples were implanted with arsenic at 3 keV to doses of $5x10^{14}$, $1x10^{15}$, and $5x10^{15}$ cm^{-2}, respectively. The implanted samples were furnace annealed at 750 °C for 4 hr, followed by a second furnace annealed at 800 °C for 1 hr. An unimplanted sample underwent identical processing steps. Secondary ion mass spectroscopy (SIMS) analysis was performed on all samples using a Physical Electronics ADEPT-1010 system to determine the antimony and arsenic concentration versus depth profiles before and after annealing. Arsenic and antimony were monitored as negative molecular ions, AsSi⁻ and 121SbSi⁻, under Cs^+ bombardment at an energy of 1keV and an angle of 60°. Secondary ions were collected from the center 3% of a 450 x 450 μm rastered area. Stylus profilometry was used to determine the depth of the sputtered craters. The atomic concentrations of arsenic and antimony were calculated from relative sensitivity factors determined from standard samples. The depth scale of the antimony profiles after annealing was laterally shifted no more than 20 nm (within one standard deviation, estimated at 0.05, in relative depth scale error of SIMS[22]) so that the peaks aligned with those of the as grown profile. The antimony concentration scale of the annealed profiles was standardized by equalizing the total dosage in each peak to that of the as grown profile.

RESULTS

Boron Marker Layers: Interstitial Injection

Because boron is known to be mainly diffuse by an interstitial mechanism, boron marker layers can be used to measure the interstitial flux from an implanted layer. If arsenic were injecting interstitials from SiAs precipitate formation as suggested earlier, then an increase in the diffusion enhancement should be realized upon annealing. In this experiment the injection of interstitials from arsenic and germanium implants were compared. By comparing the diffusion enhancement of the arsenic implanted layers with that of germanium implanted layers, the effects from end of range (EOR) damage and clustering/precipitation effects may be separated.

Presented in Figure 1A are the SIMS results of the boron marker layers after the highest dose implants and spike annealing. The results show that after a 1050°C spike anneal there is no additional enhancement from arsenic precipitation when it should have the greatest affect. Additionally, SIMS results presented in Figure 1B show that there is no effect of increasing dose on the enhancement either. This shows that there is no increased EOR damage from higher dose implants and that there is no significant injection of interstitials from clustering or precipitation at these conditions.

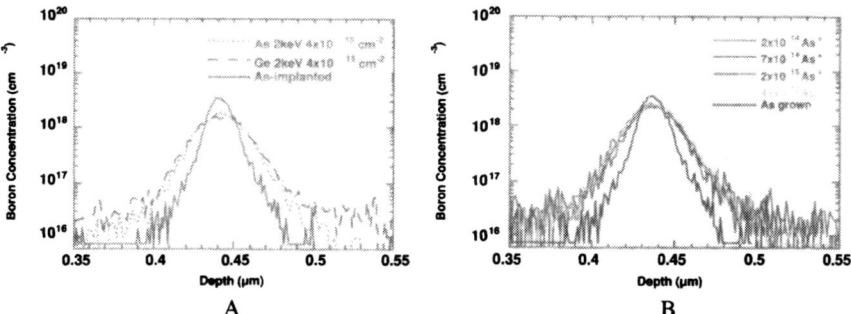

A   B

Figure 1: Diffusion enhancement in boron marker layers after spike annealing at 1050°C A) Comparison of germanium and arsenic enhancements at the some implant conditions and B) Comparison of effects from varying arsenic doses. Results show there is no additional interstitial injection from the clustering/precipitation process.

A second set of experiments was designed such that the damage form implantation was equalized across the arsenic dose range. In this case germanium implants were used to pre-amorphize the sample and subsequent arsenic implants were done. The arsenic implants that were used span the range of clustering and precipitation. Figure 2A shows the diffusion enhancements of the boron marker layers after spike annealing at 1050°C. Again the same result is seen as in the previous experiment, where the enhancement in the boron marker layers is due to the EOR damage of the amorphizing implant.

However, when the samples are annealed again at 850°C for 1 hr an interesting trend is noticed. Presented in Figure 2B are the SIMS results of the boron marker layers after the additional anneal. In this case, the lower dose samples have no effect on the diffusion enhancement compared to the germanium alone case. In comparison to the lower dose profiles, the higher doses actually show a retarded diffusion enhancement.

This result shows that the interstitial flux from the implanted region is being slowed in the higher dose arsenic cases that fall in the precipitation range. This may be due to a number of effects. Two more likely effects are vacancy injection or interstitial trapping. If during precipitation vacancies were injected then they could recombine with the interstitials and cause an effective retardation in the diffusion enhancement. Another possibility is the possible complex formation of an arsenic atom and a silicon interstitial. This process would have the same effect on the diffusion enhancement and could also explain the lack of EOR defects seen in lower energy arsenic implants in previous experiments. This experiment has no means of separating these two possibilities.

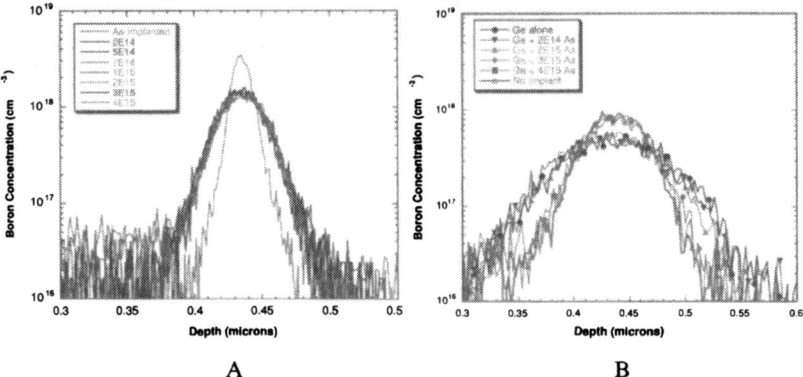

A   B

Figure 2: SIMS results of boron marker layer diffusion for A) diffusion enhancement due to germanium pre-amorphization end of range damage after spike annealing at 1050°C and B) after an additional furnace anneal at 850°C for 1hr. The results show a decrease in the interstitial flux for higher arsenic concentrations.

Antimony Doped Superlattices: Vacancy Injection

The goal of this experiment was to examine any vacancy injection upon the formation of SiAs precipitates. Antimony is known to diffuse mainly by a vacancy mechanism.[6] Therefore, if antimony marker layers are present below the surface, the enhanced motion of the marker layer would serve to monitor vacancy injection into the bulk. The low energy implants were performed to ensure the absence of end of range (EOR) defects, while using doses known to be located in the clustering and precipitation regimes. With no extended defect formation and the use of clustering/precipitation type doses, the enhancement of the antimony marker layers should be solely an effect of the clustering/precipitation process.

After the 750°C 1hr anneal, no enhanced motion was seen in the antimony spikes. The lack of motion was attributed to the low diffusion coefficient of antimony in silicon. Even though no enhanced diffusion was observed, the initial anneal did ensure the annihilation of any EOR interstitials during regrowth of the arsenic implanted region. A subsequent 800°C 1hr anneal was performed and the arsenic and antimony spikes were once again monitored. Figure 3 shows the first of the antimony spikes before and after arsenic implantation and annealing. As was seen in the first annealing cycle at 750°C there is no measurable diffusion enhancement of the antimony spikes after annealing. All the spikes in each sample moved the same amount, therefore the initial spike in each sample is only shown to enhance the effect or lack there of. The motion that was recorded was modeled using Florida's Object Oriented Process Simulator (FLOOPS). It was found that even though there was no enhancement due to the arsenic implants, there was still an overall diffusion enhancement of about 150X. The exact nature of this enhancement is not known at the time of this publication, but may be due to grown in defects or impurity elements. Whatever the case, Figure 4 shows that no additional enhancement was observed after arsenic implantation and annealing. This indicates that there is no measurable injection of vacancies due to the clustering or precipitation processes at low implant energies. This is consistent with higher energy and dose arsenic implants studied by Venables *et al.*[23]

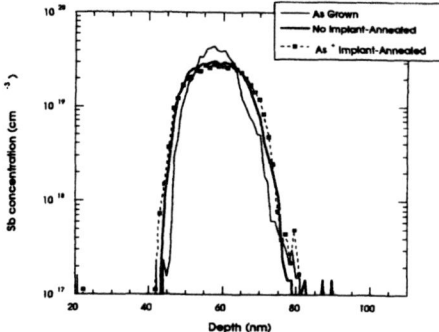

Figure 3: SIMS antimony concentration profiles for the shallowest peak of the antimony DSL as grown and after anneal in N_2 at 800 °C for 1 hr for both the no implant and 3keV 5×10^{15} cm^{-2} arsenic implant samples. The implanted profile exhibits no enhanced diffusion.

The results of this experiment show that the antimony marker layers are unaffected by the clustering and/or precipitation process. There are a number of reasons for the experimentally obtained results. Due to the low implant energy, the arsenic profile is initially contained within the first 10nm of the surface. It is possible that any vacancy flux is annihilated by recombination at the surface. Previous investigations of MeV and medium energy implantation into antimony structures have shown that vacancies formed in the near surface region do indeed diffuse into the bulk in concentrations significant enough to cause antimony diffusion.[24,25] Therefore, vacancies can travel into the bulk from the implanted region, but whether there is a sufficient flux of vacancies to cause enhanced diffusion is unknown at this point. Another possibility is that clusters consisting of As_nV release m arsenic atoms, not vacancies, to form $As_{n-m}V$ complexes upon precipitation. Regardless of the exact explanation, the absence of enhanced diffusion of the marker layers indicates that vacancy injection into the bulk is immeasurable upon precipitation.

The results of this experiment also eliminate possible explanations used in other experiments. For instance, the possibility of vacancies eliminating the extended defects in the EOR region may be discounted. It appears that the interstitials that make up the extended defects prefer to go to arsenic atoms and form arsenic-interstitial complexes during the initial stages of the annealing cycle. Since no vacancy injection results from annealing, the arsenic-interstitial interaction seems to be the controlling factor.

CONCLUSIONS

Experiments were performed to determine the point defect injection from the formation of SiAs precipitates. It was found that the interstitials in the EOR region were the only source of the enhancement seen. However upon further annealing higher dose arsenic implants caused a reduction in the interstitial flux to the boron marker layer. The cause for this behavior was thought to be either a vacancy injection that allowed for I-V recombination or the formation of As-I complexes. An addition experiment using antimony doped superlattices showed that no vacancy injection resulted due to precipitation and therefore, the results indicate the formation of As-I complexes to be the cause.

REFERENCES

1   M. I. Current, in *Technology Roadmaps for Doping of Semiconductor Transistors*, 1997.
2   A. Armigiliato and A. Parsini, J. Mater. Res. **6**, 1701 (1991).
3   D. Nobili, A. Carabelas, G. Celotti, and S. Solmi, Journal of the Electrochemical Society **130**, 922 (1983).
4   P. M. Rousseau, P. B. Griffin, W. T. Fang, and J. D. Plummer, Journal of Applied Physics **84**, 3593-601 (1998).
5   K. S. Jones, D. Downey, H. Miller, J. Chow, J. Chen, M. Puga-Lambers, K. Moller, M. Wright, E. Heitman, J. Glassberg, M. E. Law, L. Robertson, and R. Brindos, , Kyoto, 1998.
6   P. M. Fahey, P. B. Griffin, and J. D. Plummer, Reviews of Modern Physics **61**, 289-384 (1989).
7   A. Armigliato and A. Parisini, J. Mater. Res. **6**, 1701 (1991).
8   A. Armigliato, A. Parisini, M. Derdour, P. Lazzari, L. Moro, D. Nobili, and S. Solmi, Solid State Phenomena **19&20**, 393 (1991).
9   A. Parisini, D. Nobili, A. Armigliato, M. Derdour, L. Moro, and S. Solmi, Appl. Phys. A **54**, 221 (1992).
10   R. Brindos, P. Keys, M. E. Law, and K. S. Jones, Appl. Phys. Lett. **75**, 229 (1999).
11   R. Brindos, M. E. Law, K. S. Jones, and E. Andideh, in *Arsenic Trapping and its Effect on Enhanced Diffusion*, San Francisco, 1999 (Materials Research Society), p. 169.
12   R. B. Fair and G. R. Weber, J. Appl. Phys. **44**, 273-279 (1973).
13   K. C. Pandey, A. Erbil, G. S. Cargill, R. F. Boehme, and D. Vanderbilt, Phys. Rev. Lett. **61**, 1282 (1988).
14   M. Ramamoorthy and S. T. Pantelides, Submitted to Physical Review Lett. (1996).
15   S. Solmi and D. Nobili, Appl. Phys. Lett. **83**, 2484 (1998).
16   D. Nobili, S. Solmi, A. Parsini, M. Derdour, A. Armigliato, and L. Moro, Phys. Rev. B **49**, 2477 (1994).
17   K. S. Jones, S. Prussin, and E. R. Weber, J. Appl. Phys. **62**, 4114 (1987).
18   S. N. Hsu and L. J. Chen, J. Appl. Phys. **55**, 2304 (1989).
19   S. N. Hsu and L. J. Chen, Nuc. Instr. Meth. Phys. Res. B **55**, 620 (1991).
20   V. Krishnamoorthy, D. Venables, K. Moeller, K. S. Jones, and J. Jackson, MRS Proceedings **469**, 401-406 (1991).
21   H. J. Gossmann, F. C. Unterwald, and H. S. Luftman, Journal of Applied Physics **73**, 8327 (1993).
22   H.-J. Gossmann, A. M. Vredenberg, C. S. Rafferty, H. S. Luftman, F. C. Unterwald, D. C. Jacobson, T. Boone, and J. M. Poate, J. Appl. Phys. **74**, 3150-3155 (1993).
23   D. Venables, V. Krishnamoorthy, H.-J. Gossman, A. Lilak, K. S. Jones, and D. C. Jacobson, in *The Role of Vacancies and Interstitials in Transient Enhanced Difussion of Arsenic Implanted into Silicon*, 1997, p. 315.
24   D. J. Eaglesham, T. E. Haynes, H.-J. Gossman, D. C. Jacobson, P. A. Stolk, and J. M. Poate, Appl. Phys. Lett. **70**, 3281-3283 (1997).
25   G. Lulli, M. Bianconi, S. Solmi, E. Napolitani, and A. Carnera, Appl. Phys. **87**, 8461 (2000).

Mat. Res. Soc. Symp. Proc. Vol. 669 © 2001 Materials Research Society

Determining the Ratio of the Precipitated versus Substituted Arsenic by XAFS and SIMS in Heavy Dose Arsenic Implants in Silicon

M. A. Sahiner[1], S. W. Novak[1], J. C. Woicik[2], J. Liu[3], and V. Krishnamoorty[4]
[1]Evans East, East Windsor, New Jersey 08520
[2]National Institute of Standards and Technology, Gaithersburg, Maryland 20899
[3]Varian Semiconductor Equipment Associates, Gloucester, Massachusetts 01930
[4]Department of Materials Science and Engineering, University of Florida, Gainesville, Florida 32611

ABSTRACT

Doping silicon with arsenic by ion implantation above the solid solubility level leads to As clusters and/or precipitates in the form of monoclinic SiAs causing electrical deactivation of the dopant. Information on the local structure around the As atom, and the As concentration depth profiles is important for the implantation and annealing process in order to reduce the precipitated As and maximize the electrically activated As. In this study, we determined the local As structure and the precipitated versus substituted As for As implants in CZ (001) Si wafers, with implant energies between 20 keV and 100 keV, and implant doses ranging from 1 x 10^{15}/cm^2 to 1 x 10^{18}/cm^2. The samples were subjected to different thermal annealing conditions. We used secondary ion mass spectrometry (SIMS) and UT-MARLOWE simulations to determine the region where the As-concentration is above the solid solubility level. By x-ray absorption fine structure spectroscopy (XAFS), we probed the structure of the local environment around As. XAFS being capable of probing the short-range order in crystalline and amorphous materials provides information on the number, distance and chemical identity of the neighbors of the main absorbing atom. Using Fourier analysis, the coordination numbers (N) and the nearest-neighbor distances (R) to As atoms in the first shell were extracted from the XAFS data. When As precipitates as monoclinic SiAs, the nearest-neighbor distances and coordination numbers are ~2.37 Å and ~3, as opposed to ~2.40 Å and ~4 when As is substitutional. Based on this information, the critical implant dose where the precipitation/clustering of As starts, and the ratio of the substitutional versus cluster/precipitate form As in the samples were determined.

INTRODUCTION

Electrical deactivation of the dopant due to clustering or precipitation for high dose implants in Si is one of the major problems in the semiconductor technology [1]. In order to probe this problem and search for solutions, the detailed local structural information around the impurity atom is required. One of the powerful tools in extracting local structural information is x-ray absorption fine structure spectroscopy (XAFS). The coordination numbers, nearest neighbor distances, and the bond angles around an impurity atom can be determined using XAFS for both amorphous and crystalline materials. In high dose implants, As starts to form precipitate monoclinic SiAs above a critical concentration instead of going to substitutional sites. These two forms of As have local structural differences which can be directly probed by XAFS. In the substitutional form, As has Si atoms as the nearest neighbors at a distance about 2.40 Å and the coordination numbers (N) is four [2,3,4]. For the monoclinic As, the coordination number and the nearest neighbor (Si) distances are 3 and 2.37 Å, respectively [2,3,4].

In this study, by using XAFS analysis, we probed these differences in the coordination numbers and the nearest-neighbor distances to detemine the amount of precipitated versus substitutional As. The SIMS data provided the information on the As concentration versus depth so that the critical concentration where the precipitation starts was determined. As-bulk doped Si (with As concentration of 1.80×10^{19} atoms/cm^3) was used as a standard for pure substitutional form of As. Implant with the dose of 1×10^{18} atoms/cm^3 was our XAFS standard for the pure monoclinic SiAs form.

EXPERIMENT

Czochralski (001) Si wafers were implanted with As$^+$ first at 100 keV at a dose of 1×10^{15} atoms/cm^2. Then a second As implant at 20 keV or 30 keV was performed at a range of doses between 1×10^{15}-1×10^{18} atoms/cm^2. An arsenic bulk doped Si wafer with As concentration of 1.80×10^{19} atoms/cm^3 was also used in this study. The implanted samples were annealed at 800°C for 1 hour in nitrogen-controlled environment. The secondary ion mass spectroscopy results were performed at Evans East using a Physical Electronics Quadrupole instrument. Arsenic K-edge x-ray absorption fine structure spectroscopy (XAFS) experiments were performed at the National Institute of Standards and Technology's (NIST) beamline (X23A2) at National Synchrotron Light Source (NSLS) at Brookhaven National Laboratory (BNL). XAFS data were acquired in the fluorescence detection mode in using a Lytle fluorescence detector. The x-ray angle of incidence was set to 5° during the measurements.

RESULTS

SIMS and UT-MARLOWE Calculations

Fig. 1 shows SIMS depth profile and the UT-MARLOWE[5] simulation for the double As implanted Si wafer. The first implant is at 100 keV as with a dose of 1×10^{15} atoms/cm^2 and the second implant is at 30 keV at 2×10^{15} atoms/cm^2. The solid line is the SIMS profile and the dotted line is the calculated profile. The calculated profiles for the individual implants are also shown in the plot. There is a good agreement between the experimental and the calculated depth profiles. Also shown in the figure are the calculated individual profiles for the 100 keV and the 30 keV As implants. The intersitial data obtained from the UT-MARLOWE calculations indicate that the thickness of the amorphous layer formed by the first implant is about 120 nm. The peak of the 30 keV implant is around 30 nm and the As concentration at the peak is about 8×10^{20} atoms/cm^3.

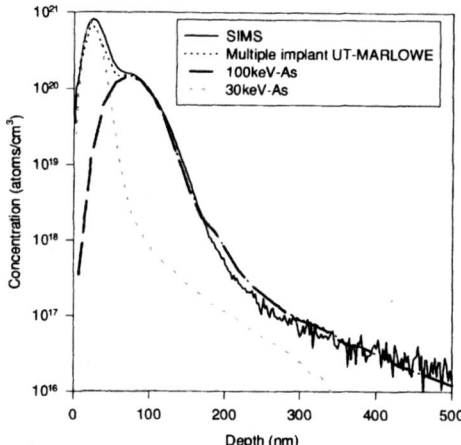

Figure 1. SIMS depth profiles and the UT-MARLOWE simulations for the double arsenic implants in silicon.

Fig. 2 shows a comparison plot of the SIMS profiles for the As implants with a second implant dose (30 keV) of 2×10^{15} atoms/cm^2 and 1×10^{16} atoms/cm^2 both for the as-implanted and the annealed samples at 800°C for 1hour in nitrogen-controlled environment.

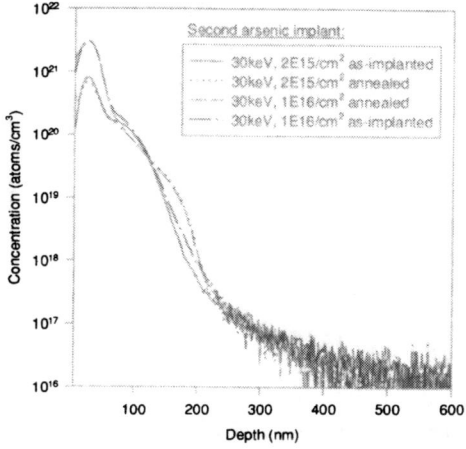

Figure 2. SIMS depth profiles for the as-implanted and the annealed samples for the low and high dose implants.

Comparing the as-implanted profiles with the annealed ones, it is observed that the As concentration in the range of the low energy (30 keV) implant remains constant. Upon annealing, arsenic in the range of the 100 keV implant is redistributed for both high dose and the low dose implants. Therefore all the clustering or the precipitation events are expected to occur within the implant range of the 30 keV implant.

XAFS Results

Arsenic implanted Si wafers with implant doses between 1×10^{15} atoms/cm^2 to 1×10^{18} atoms/cm^2 have also been used for the XAFS measurements. The reason for using very large dose implants (1×10^{18}atoms/cm^2) is to establish an XAFS standard in which arsenic is assumed to be in pure precipitated monoclinic SiAs form within the probing depth. At the other end, as an XAFS standard for the pure substitutional form of the As, the bulk doped sample was used.

In XAFS data analysis, the atomic absorption was subtracted from the energy dependent absorption coefficient, and then normalized to the absorption edge [6]. By Fourier transforming the resulting XAFS function, χ(k), the pseudo radial distribution function around the main absorbing As atom was obtained. Fig. 3 is a plot of the Fourier transformed XAFS data for the double As implanted Si samples. Fourier transformed data of the XAFS functions peak at the nearest-neighbor distances to the main absorbing atom As. Therefore, the first peak in the Fourier transformed data is due to the scattering from the nearest Si atoms. The peak distances are shifted from the actual As-Si distances because of the additional phase shift due to backscattering from the Si atoms. In Fig. 3, as-implanted samples exhibit no peaks beyond the first shell indicating an amorphous structure. For the annealed samples, the peaks between 3 to 4 Å are due to higher order near-neighbors confirming long-range order beyond the first shell.

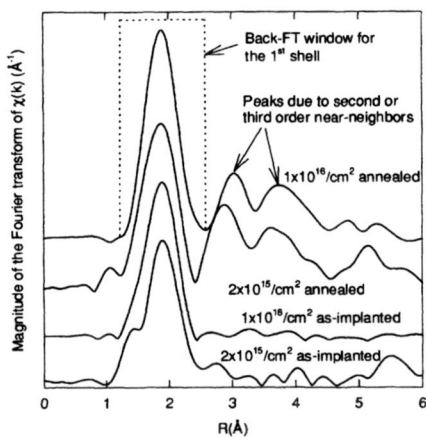

Figure 3. Fourier transformed data for the as-implanted and annealed double As implants in Si.

The Fourier transformed data show well-isolated first shell peaks for all the As implanted Si samples, allowing us to perform a back-Fourier analysis in order to extract the actual As-Si distances and the coordination numbers around the As atom. The first shell peaks are back-Fourier transformed to the momentum space using a transform window. Fig. 4 shows the back-Fourier transformed data (solid-lines) for the As-bulk sample and the annealed (800°C for 1hr) As implants with doses between 1×10^{15}-1×10^{18} atoms/cm^2. Also shown in Fig. 4, are the non-linear least squares fits (dotted lines) to the back-Fourier transformed data.

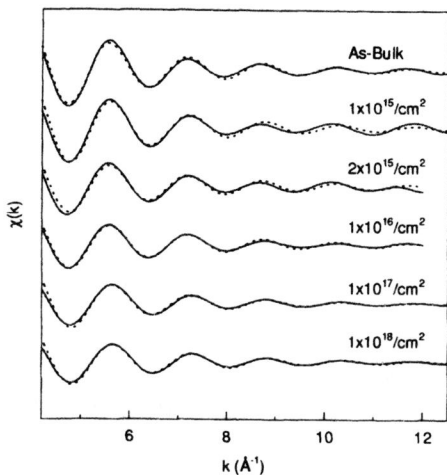

Figure 4. Back-Fourier transformed data and the non-linear least squares fits for the As-bulk and annealed As implants in Si.

In performing these fits, the University of Washington's XAFS simulation program (FEFF6.1) [7] has been used to calculate the back scattering amplitudes for the monoclinic SiAs and the substitutional form of As in a Si crystal. The fits indicate a good agreement with the back-Fourier transformed experimental data. The calculated nearest neighbor distances and the coordination numbers from these non-linear least squares fits are listed on Table I below.

Table I. The results of the non-linear least squares fits to the back-Fourier transformed data.

As Conc./Dose	R(Å)	N	Structure of As
Bulk-$1.8 \times 10^{19}/cm^3$	2.40 ±0.005	3.98±0.01	Substitutional
$1 \times 10^{15}/cm^2$	2.39 ±0.005	3.61±0.01	Substitutional
$2 \times 10^{15}/cm^2$	2.39 ±0.005	3.50±0.01	Cluster/precipitate
$1 \times 10^{16}/cm^2$	2.38 ±0.005	3.27±0.01	Cluster/precipitate
$1 \times 10^{17}/cm^2$	2.37 ±0.005	3.13±0.01	Cluster/precipitate
$1 \times 10^{18}/cm^2$	2.36 ±0.005	3.01±0.01	Cluster/precipitate

Since the bulk doped sample has an average As concentration of 1.80×10^{19} atoms/cm^3, which is low enough to prevent any clustering or precipitation, it is assumed to serve as a good standard for the pure substitutional form of As. For the bulk doped sample, the As-Si distance obtained from the fit is 2.40 Å and the coordination number around the As atom is 3.98 which is in agreement with the literature values for the substitutional form of As [2,3,4] confirming the validity of this assumption. As shown in Table I, the As-Si distance shows a steady decrease from 2.40 Å to 2.36 Å with the increase in the implant dose from 1×10^{15}-1×10^{18} atoms/cm^2. This decrease is an indicator of the formation of the SiAs precipitates with increasing As dose in Si. Similarly, a decrease in the coordination number from 3.98 to 3.01 is observed going from a dose of 1×10^{15} to 1×10^{18} atoms/cm^2. At an implant dose of 1×10^{18} atoms/cm^2 the nearest neighbor distances and the coordination numbers overlap with those of pure monoclinic SiAs so the assumption of considering 1×10^{18} atoms/cm^2 As implant as a standard for pure precipitated form of As is equally valid. By using the EXAFS functions of these two samples i.e., the bulk doped As and the 1×10^{18} atoms/cm^2 implants, the amount of the precipitated form of As in a specific sample can be determined.

In Fig. 5 the $\chi(k)$'s of the bulk doped As and the 1×10^{18} atoms/cm^2 are shown on the top of the figure. These two functions have quite different forms. The experimental $\chi(k)$'s of the 1×10^{15}, 1×10^{16}, and 1×10^{17} atoms/cm^2 implants are also plotted in solid lines in Fig. 5. Linear combinations of these two standards, substitutional (bulk doped As) and precipitated form (dose of 1×10^{18} atoms/cm^2) of As, are used to reproduce the $\chi(k)$'s of the 1×10^{15}, 1×10^{16}, and 1×10^{17} atoms/cm^2 implants and are plotted in dotted lines. As it can be seen from the overlay plots the agreement between the experimental $\chi(k)$'s and the linear combinations is very good.

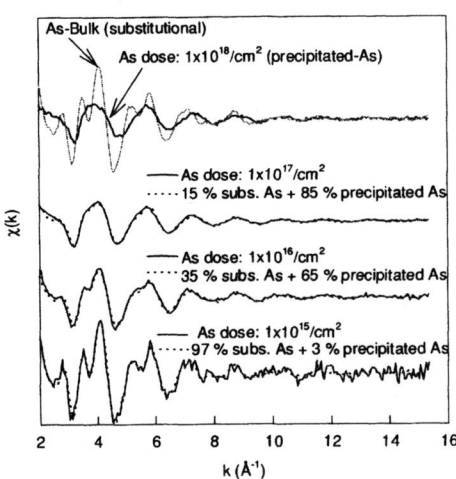

Figure 5. XAFS Functions $\chi(k)$ of Bulk doped and the 1×10^{18} atoms/cm^2 implants. By using the linear combinations of these two functions, the $\chi(k)$'s of the other implants are reproduced.

In this way, the ratios of the precipitated As to the substitutional As were determined for the implanted doses of 1×10^{15}, 1×10^{16}, and 1×10^{17} atoms/cm^2. These ratios are noted on the plot. According to the results of the linear combinations, the percentage of the precipitated As increases from 3 % to 85 % upon increasing the As dose from 1×10^{15} to 1×10^{17} atoms/cm^2. This method is a very efficient way of determining the percentage of the precipitated As for any implant dose and annealing conditions. The key to the problem is to be able to obtain the XAFS functions from well-established standards (bulk doped As and 1×10^{18} atoms/cm^2 implant) as used in this work.

CONCLUSIONS

The results of this work clearly indicate that XAFS can directly probe the precipitation and substitution problem in ion-implanted Si crystals. Combining the concentration depth profile information from SIMS, and the local structural information from XAFS, the critical concentration values for the precipitate formation under different annealing conditions can be determined. Furthermore using the pure substitutional and precipitated forms of As as XAFS standards, the amount of the substitutional versus precipitated As can be determined in an implant. In future, we will extend the XAFS analysis beyond the first shell and using multiple scattering analysis we will attempt to extract structural information on the nature of clusters and determine the relative concentrations of substitutional, clustered and precipitated forms of arsenic in these implants.

REFERENCES

1. K. S. Jones, S. Prussin, and E. R. Weber, *Appl. Phys. A* **45**, 1 (1988).
2. A. Erbil, W. Weber, G. S. Cargill III, and R. F. Boehme, *Phys. Rev. B* **34**, 1392 (1986).
3. J. L. Allain, J. R. Regnard, A. Bourret, A. Parisini, A. Armigliato, G. Tourillon, and S. Pizzini, *Phys. Rev. B* **46**, 9434 (1992).
4. A. Terrasi, E. Rimini, V. Raineri, F. Iacona, F. La Via, S. Colonna, and S. Mobilio, *Appl. Phys. Lett.*, **73**, 2633 (1998).
5. S.–H. Yang, D. Lim, S.J. Morris, A.F. Tasch, R. B. Simonton, D. Kamenitsa, C. Magee, and G. Lux, *J. Elec. Mat.* **23**, 801 (1994).
6. A. Bianconi, *X-ray Absorption: Principles, Applications, Techniques of EXAFS, SEXAFS, and XANES.* John Wiley & Sons, New York, 1988.
7. S. I. Zabinsky, A. Ankudinov, J. J. Rehr, and R. C. Albers, *Phys. Rev. B* **52** 2995 (1995).

Mat. Res. Soc. Symp. Proc. Vol. 669 © 2001 Materials Research Society

Modeling of Annealing of High Concentration Arsenic Profiles

Pavel Fastenko[1], Scott T. Dunham[1] and Graeme Henkelman[2]

[1] Department of Electrical Engineering, University of Washington,
Seattle, WA 98195, U.S.A.

[2] Department of Chemistry, University of Washington,
Seattle, WA 98195, U.S.A.

ABSTRACT

Understanding the diffusion and activation of arsenic is critical for the formation of low resistance ultra-shallow junctions as required for nanoscale MOS devices. In this work, we use results of *ab-initio* calculations in order to gain insight into the fundamental processes involved in arsenic activation/deactivation. Utilizing continuum modeling, we find it is possible to account for both the very rapid initial deactivation of arsenic as well as the strongly superlinear dependence of interstitial supersaturation on doping level which accompanies deactivation. The critical process is the rearrangement of As atoms via interstitial mediated diffusion leading to ejection of silicon atoms from arsenic complexes and formation of arsenic-vacancy clusters.

INTRODUCTION

There is a strong attractive interaction between arsenic and vacancies due to the combination of strain compensation and valence [1]. As a result, under most conditions deactivation of arsenic occurs primarily via the formation of arsenic vacancy complexes. *Ab-initio* calculations [2,1] find that As_4V complexes (4 substitutional As atoms surrounding an empty lattice site) are the most energetically favorable, which seems reasonable given that this configuration allows each valence 5 As to have 3 nearest neighbors. It has been observed experimentally for high active arsenic concentrations, initial deactivation is very rapid (within 15 seconds at 750°C) [3]. Arsenic also shows strong deactivation for temperatures as low as 400°C [4].

It has been observed that deactivation of high concentration arsenic layer injects interstitials into the substrate [5]. It was proposed, based on XSW (X-ray standing waves) [6] and positron annihilation spectroscopy (PAS) experiments [7] that several second nearest neighbor As atoms (two or more) may kick-out adjacent Si atom, forming an arsenic-vacancy cluster and a self-interstitial. *Ab-initio* calculations suggest [8] that energetically the most favorable reaction of this kind is: $As_4Si \rightarrow As_4V + I$. The strong binding energy between arsenic atoms and a vacancies dramatically reduces the energy associated with Frenkel pair formation at sites surrounded by arsenic.

We have investigated these processes via *ab-initio* and continuum simulations. By including both interstitial and vacancy-mediated diffusion processes, we find it is possible to account for both the very rapid initial deactivation of arsenic as well as the strongly superlinear dependence of interstitial supersaturation on doping level. The critical process is the rearrangement of As atoms via AsI pair diffusion leading to formation of arsenic clusters which are favorable for vacancy incorporation and interstitial ejection.

MODELING

Ab-initio calculations

The energy barriers for ejection of a silicon atom from an As_4Si tetrahedral cluster was calculated via density functional theory (DFT). The initial state for the process consisted of a 64 atom silicon lattice in which four silicon atoms were replaced with arsenic atoms. The As_4Si cluster forms a tetrahedron with a silicon atom at the center of the cluster. The arsenic atoms are thought to be electronically active in this configuration [2]. During the deactivation process the central silicon atom is ejected from the cluster, becoming an interstitial (I) and leaving behind a vacancy (V). A final state was chosen with the interstitial silicon atom as far from the deactivated As_4V cluster as possible in a hexagonal interstitial site. The minimum energy path for the process was calculated with the nudged elastic band (NEB) [9] method in which 8 images were used to connect the initial and final states. The DFT calculations were done with the VASP [10] code using the PW91 functional [11] and ultrasoft pseudopotentials [12]. Plane waves up to a 200 eV energy cutoff were used to represent the wavefunction in the unit cell. The calculations were done with eight points in the k-point mesh.

The results of the NEB calculation are shown in Fig. 1. There is a small initial barrier of 0.2 eV in which the silicon atom moves from the center of the As_4Si tetrahedron (a) to a stable site at the center of one of the tetrahedron faces (b) (e.g. in the center of the triangle formed by 3 of the 4 As atoms, which have relaxed apart). The dimer method [13] was used to search for low energy saddle points leading from the initial state. Five independent dimer searches found the process to (b) to have the lowest barrier. From (b) to (c) the mobile silicon atom continues in the same direction away from the cluster through a tetrahedral site to a hexagonal interstitial site over a barrier of 1.4 eV. At this point the arsenic cluster is thought to be inactive as the interstitial silicon atom is removed from the cluster. This is supported by the final process in which the interstitial silicon atom diffuses to another slightly lower energy hexagonal site (d) because the barrier of 0.3 eV is nearly the same as the bulk hexagonal-tetrahedral-hexagonal diffusion barrier. The overall process is particularly interesting because the presence of the arsenic atoms allow for the creation of an interstitial vacancy pair with a barrier of 1.4 eV which is much less than the Frankel pair formation DFT barrier of more than 7 eV. Because there does not appear to be any barrier higher than the normal interstitial diffusion barrier, we consider the reverse process to be diffusion limited in the continuum simulations described below.

Continuum simulations

To simulate arsenic deactivation we used four discrete arsenic-vacancy clusters, assumed to be neutral: $(As_2V)^0$, $(As_3V)^0$, and $(As_4V)^0$. Since clustering involves reactions of point defects and defect-dopant pairs of different charge, we include charge transfer reactions, and simulate all possible pathways of clustering reaction. The binding energies for these reactions were fitted to equilibrium activation levels from experimental data [14,15]. The comparison of simulation data to experiments is plotted in Fig. 2.

To model interstitial ejection during As deactivation, we include arsenic-assisted Frenkel pair generation in addition to bulk generation/recombination. Since *ab-initio* calculations described above find no additional barrier beyond that required for the I to diffuse away, for I ejection reactions we simply reduce the normal Frenkel pair energy by

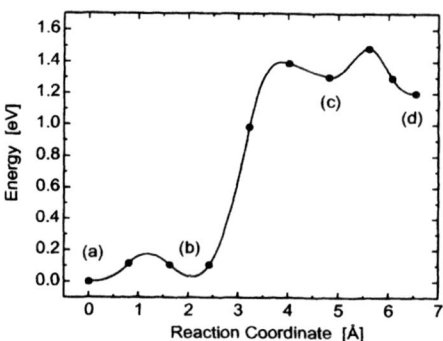

Figure 1. Energy versus distance for kick-out of a silicon atom from between 4 substitutional As atoms.

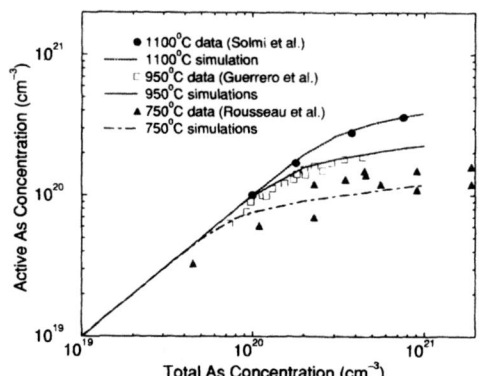

Figure 2. Comparison of arsenic-vacancy clustering model to experimental measurements of active versus total arsenic concentration at long times. The model was fitted to data from Solmi *et al.* [14] at 1100°C and Guerrero *et al.* [15] at 950°C and extrapolated to lower temperature. Extrapolation to 750°C temperature is compared to data from Rousseau *et al.* [16]

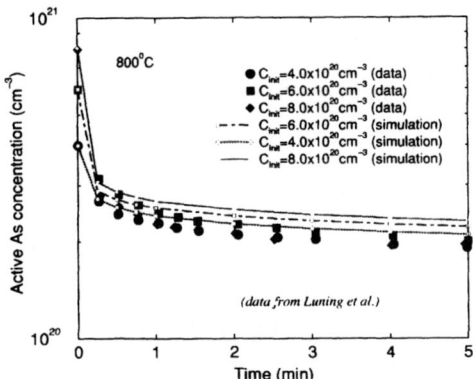

Figure 3. Comparison of simulation results to experimental observations of deactivation kinetics at 800°C for laser annealed As layers [3]. Note that initial deactivation is very rapid and different initial arsenic concentrations quickly reach similar activation levels. Initially As layer was fully activated by laser melt.

the same binding energies that were found from comparison to equilibrium activation. Our analysis shows that dominant reaction for high concentration As deactivation is:

$$(As_4Si)^{4+} + 4e^- \rightarrow (As_4V)^0 + I^0. \tag{1}$$

The initial number of As_4Si complexes after ion implantation and regrowth is estimated based on random distribution of dopants and the corresponding occupational probabilities,

$$C_{As_4Si} = C_{Si} \left(C_{As}/C_{Si} \right)^4 \left(1 - C_{As}/C_{Si} \right). \tag{2}$$

During deactivation anneal As_4Si complexes are assumed to be formed by AsI pair migration, again based on random reconfiguration. The strong super-linear dependence of deactivation reaction on As concentration gives a sharp onset of deactivation as a function of As concentration. Since a large number of interstitials are released during As deactivation, a corresponding large interstitial supersaturation is present throughout the process. We used a moment-based {311}/loop model to account for extended defect formation [17]. High interstitial supersaturation significantly increases As diffusivity via AsI pairs which helps maintain sufficient concentration of As_4Si complexes during the deactivation process.

COMPARISON TO EXPERIMENTS

It has been observed in isothermal anneal experiments [3], that for high arsenic active concentration, initial deactivation is very rapid (within 15 seconds). After this initial deactivation, activation level is similar for all doses. After initial deactivation (15 seconds), deactivation continues but on much slower time scale. At lower temperature the deactivation kinetics are substantially slower but show same basic behavior. Fig. 3 and Fig 4 shows comparison of our model to experimental data. It has also been observed via buried B marker layers that deactivation of high concentration arsenic layers injects

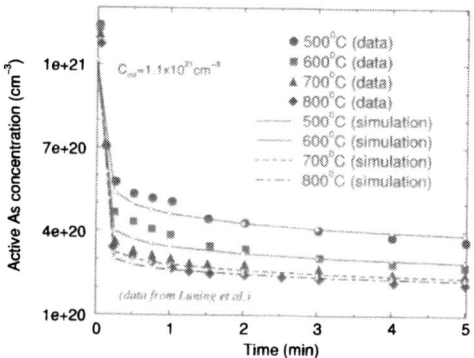

Figure 4. Deactivation kinetics for laser annealed As layers, with comparison of simulations to experimental data from Luning [3]. Even though equilibrium solubility is higher for higher T, initial deactivation is limited by kinetics, so the lower T anneals give higher active concentrations.

interstitials into the substrate [5]. Initial rapid deactivation is due to interstitial ejection from As clusters which form due to random dopant motions. Figure 5 shows comparison of simulation results to experimental data of Rousseau for arsenic layers which were initially fully-activated via laser anneal.

CONCLUSIONS

We utilized a combination of *ab-initio* calculations (DFT) and continuum modeling in order to gain understanding of arsenic diffusion and activation/deactivation. The results highlighted the importance of ejection of silicon atoms from arsenic clusters. The barrier for Frenkel pair generation is reduced by the strong binding of V to As clusters. These As clusters form via random interstitial-mediated diffusion. We find it is possible to account for both the very rapid initial deactivation of arsenic as well as the strongly superlinear dependence of interstitial supersaturation on doping level which accompanies deactivation.

ACKNOWLEDGMENTS

This research was supported by the Semiconductor Research Corporation (SRC). The authors would like to thank Marcie Berding, Jianjun Xie and Shao-Ping Chen for discussion regarding their *ab-initio* calculations.

REFERENCES

1. M. Ramamoorthy and S. Pantelides, *Phys. Rev. Lett.* **75**, 4753 (1996).
2. M. Berding and A. Sher, *Phys. Rev. B* **58**, 3853 (1998).
3. Scott Luning, PhD thesis, Stanford University (1996).
4. D. Nobili, A. Carabelas, G. Celotti, and S. Solmi, *J. Electrochem. Soc.* **130**, 922 (1983).
5. Paul Rousseau, PhD thesis, Stanford University (1996).

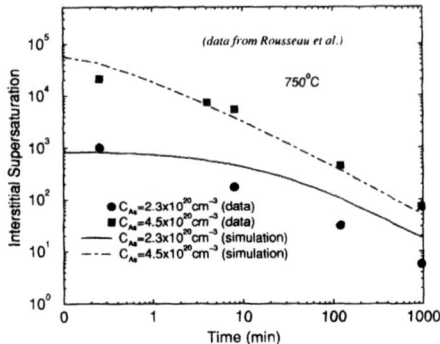

Figure 5. Comparison of simulation results to experimental observations of interstitial supersaturation versus time for different arsenic concentrations, measured via buried marker layer [16]. Note order of magnitude difference in supersaturation with factor of 2 change in initial As concentration.

6. A. Herrera-Gomez, P.M. Rousseau, G. Materlik, T. Kendelewicz, J.C. Woicik, P.B. Griffin, J. Plummer, and W.S. Spicer, *Appl. Phys. Lett.* **68**, 3090 (1996).
7. D.W. Lawther, U. Myler, P.J. Simpson, P.M. Rousseau, P.B. Griffin, W. T. Fang, and J. D. Plummer, *Appl. Phys. Lett.* **67**, 3575 (1995).
8. Jianjun Xie and S.P. Chen, *J. Appl. Phys.* **87**, 4160 (2000).
9. H. Jónsson, G. Mills, and K.W. Jacobsen. *Classical and Quantum Dynamics in Condensed Phase Simulations.* World Scientific (1998). ed. B. J. Berne, G. Ciccotti, and D. F. Coker, page 385.
10. G. Kresse and J. Hafner, Phys. Rev. B **47**, 558 (1993); **49**, 14251 (1994); G. Kresse and J. Furthmüller, *Comput. Mater. Sci.* **6**, 16 (1996); Phys. Rev. B **55**, 11169 (1996).
11. J.P. Perdew. *Electronic Structure of Solids.* (1991). eds. P. Ziesche and H. Eschrig.
12. D. Vanderbilt, *Phys. Rev. B* **41**, 7892 (1990).
13. G. Henkelman and H. Jónsson, *J. Chem. Phys.* **111**, 7010 (1999).
14. S. Solmi, D. Nobili, and J. Shao, *J. Appl. Phys.* **87**, 658 (2000).
15. E.Guerrero, H.Potzl, R.Tielert, M.Grasserbauer, and G.Stingeder, *J. Electrochem. Soc.* **129**, 1826 (1982).
16. P.M. Rousseau, P.B. Griffin, W. T. Fang, and J. D. Plummer, *J. Appl. Phys.* **84**, 3593 (1998).
17. A.H. Gencer, S. Chakravarthi, I. Clejan, and S.T. Dunham, in **Defects and Diffusion in Silicon Processing**, 359 (1997).

Mat. Res. Soc. Symp. Proc. Vol. 669 © 2001 Materials Research Society

Defect Evolution from Low Energy, Amorphizing, Germanium Implants on Silicon

Andres F. Gutierrez[1], Kevin S. Jones[2] and Daniel F. Downey[3]
[1]Intel Corporation, 5000 W. Chandler Blvd., Mailstop CH5-263, Chandler, AZ 85226, U.S.A.
[2]Materials Science and Engineering, Bldg #33 Rm 538, University of Florida, Gainesville, FL 32611, U.S.A.
[3]Varian Semiconductor Equipment Associates, 35 Dory Road, Gloucester, MA 01930, U.S.A.

ABSTRACT

Plan-view transmission electron microscopy (PTEM) was used to characterize defect evolution upon annealing of low-to-medium energy, 5-30 keV, germanium implants into silicon. The implant dose was 1×10^{15} ions/cm^2, sufficient for surface amorphization. Annealing of the samples was done at 750 °C in nitrogen ambient by both rapid thermal annealing (RTA) and conventional furnace, and the time was varied from 10 seconds to 360 minutes. Results indicate that as the energy drops from 30 keV to 5 keV, an alternate path of excess interstitials evolution may exist. For higher implant energies, the interstitials evolve from clusters to {311}'s to loops as has been previously reported. However, as the energy drops to 5 keV, the interstitials evolve from clusters to small, unstable dislocation loops which dissolve and disappear within a narrow time window, with no {311}'s forming. These results imply there is an alternate evolutionary pathway for {311} dissolution during transient enhanced diffusion (TED) for these ultra-low energy implants.

INTRODUCTION

Preamorphization is commonly used in the formation of ultra-shallow junctions for Si-based microelectronic devices [1-3]. However, this step alone will not achieve the necessary highly doped, low resistivity structures required for next generation devices since the process itself introduces end-of-range (EOR) defects that influence diffusion and dopant activation mechanisms upon annealing [4]. Knowing that excess interstitials provide a source for TED [5], it becomes necessary to understand all sources and conditions for interstitial formation and evolution from preamorphizing implants, especially at the technologically important low-energy regimes i.e., less than 5 keV. Low to high-energy Si$^+$ implants ($\sim$ 1-100 keV) have been previously studied [6-9], and a consensus regarding the path of interstitial evolution, as it affects TED, has emerged [10]. In essence, this path includes the formation of small precursor clusters, which coalesce into {311} type defects. With time, these EOR defects switch in function from interstitial sink to source, and dissolve, either driving dislocation loop formation or TED. At longer times, these loops either dissolve and drive TED, or become stable and trap interstitials. The objective of this work is to enhance the understanding of defect (and consequently interstitial) evolution from low energy, germanium implants on silicon, since it has not been reported in the literature.

EXPERIMENTAL DETAILS

Czochralski (CZ) grown (100) Si wafers were implanted with germanium ions at 5, 10 and 30 keV implant energies at a 7° tilt. These energies produced an amorphous/crystalline (α/c)

interface at 120, 220 and 480 A below the surface, respectively with increasing energy, as characterized by ellipsometry and cross-sectional TEM (XTEM). Dose was held constant at 1 x 10^{15} ions/cm^2 and implantation was carried out at room temperature. Samples were annealed at 750 °C using both a rapid thermal annealing system (AG Associates Heatpulse 410) and a conventional tube furnace under a nitrogen ambient. For anneals up to 120 seconds, the RTA was employed, while anneals greater than 5 minutes, up to 360 minutes, were performed in the tube furnace. Subsequently, samples were mechanically thinned and chemically etched using a standard HF:HNO$_3$ drip etch. A JEOL 200CX TEM under g$_{220}$ weak beam dark field conditions (WBDF) was used to analyze the annealed samples. Defect evolution was quantified from enlarged PTEM micrographs using the tracing method, [11], with and without image processing, and defect density as well as interstitial density counts were extracted.

RESULTS AND DISCUSSION

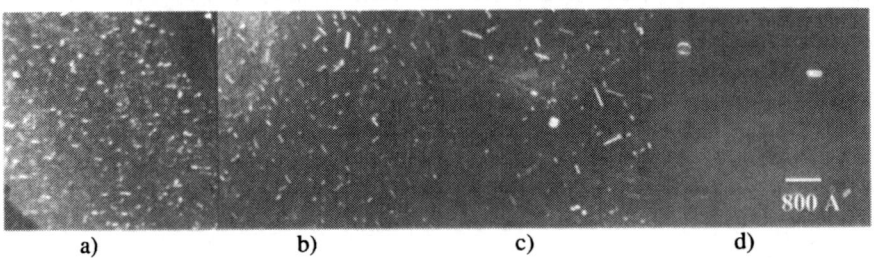

Figure 1. PTEM micrographs showing defect evolution for 30 keV, 1 x 10^{15} ions/cm^2 Ge$^+$ annealed at 750 °C for a) 2 min. b) 15 min. c) 45 min. and d) 360 min.

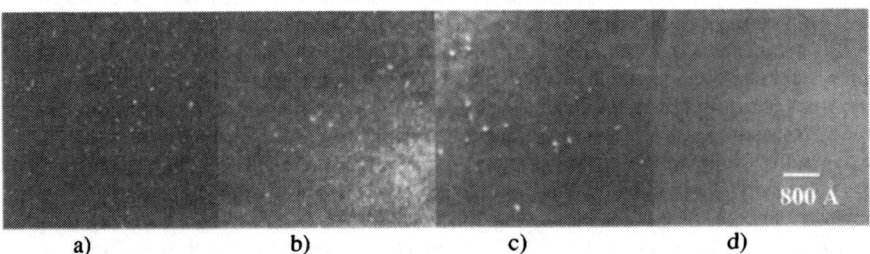

Figure 2. PTEM micrographs showing defect evolution for 5 keV, 1 x 10^{15} ions/cm^2 Ge$^+$ annealed at 750 °C for a) 5 min. b) 15 min. c) 45 min. and d) 360 min.

Qualitatively comparing figures 1 and 2, obvious differences are observed for the defect evolution from the 30 keV and 5 keV implants, respectively. The figures show the expected decreasing trend in defect density with time, but at the higher energy, the defects resulting from implantation evolve very differently from the low energy. Beginning with the nucleation of visible defects which grow into {311}'s with time, these dissolve, releasing the trapped interstitials for dislocation loop growth, and by 360 minutes only large, stable dislocation loops are observed. Figure 3-a illustrates this defect behavior by plotting the total defect density, and when visibly feasible, plotting the evolution of {311}'s and loops individually. The complete dissolution of {311}'s is represented by the points on the TEM detection limit line [12]. Hence, there is qualitative agreement between the evolution of defects from Si+ and Ge+ implants at medium energies, where {311}'s and loops coexist.

At the lowest energy, several noteworthy effects are observed. The two most striking differences between figure 1 and figure 2 are the absence of {311} type defects and the complete dissolution of all defects for the longest anneal time, both in the latter figure. Indeed, only small, unstable dislocation loops are formed for the 5 keV, which dissolve rapidly, as seen in figure 3-b. For the intermediate energy implant, i.e., 10keV, the defect evolution behavior is similar to the 5 keV in that only dislocation loops are formed, but these coarsen and remain even after annealing for 360 minutes. Apparently, the decrease in energy greatly affects the type and stability of the EOR defects that form, and evolve upon annealing, beyond the α/c interface.

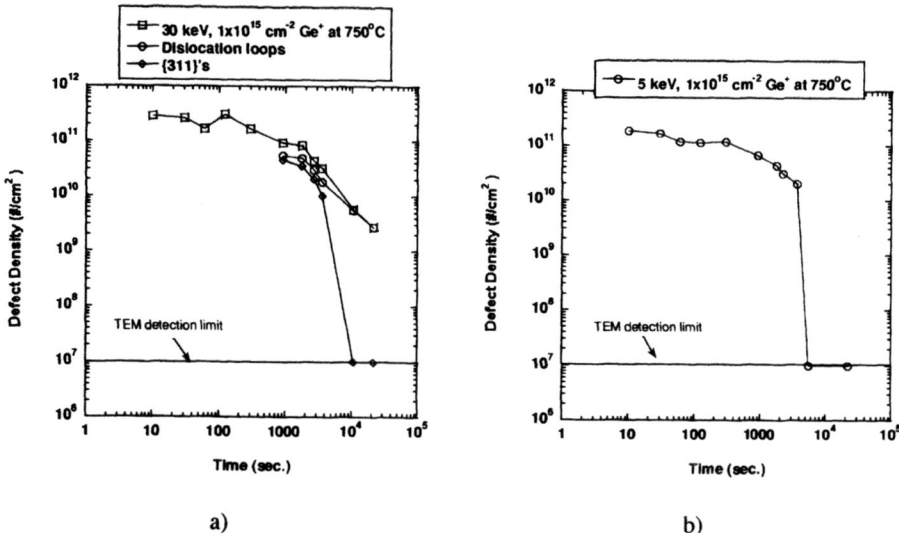

Figure 3. Defect evolution from a) 30 keV and b) 5 keV, $1x10^{15}$ ions/cm^2 Ge+ ions in silicon.

Figure 4 illustrates the interstitial behavior for all three energies following 750 °C anneals of $1x10^{15}$ ions/cm^2 Ge+ implants. With increasing time, the interstitial population is surprisingly stable up to 60 minutes, after which all energies exhibit a decrease in the number of interstitials. Moreover, the sharp decline for the 5 keV is unprecedented and mirrors the defect evolution behavior observed in figure 3-b. Consequently, within the scope of this experiment, a threshold for defect stability exists between the 5 and 10 keV energy levels.

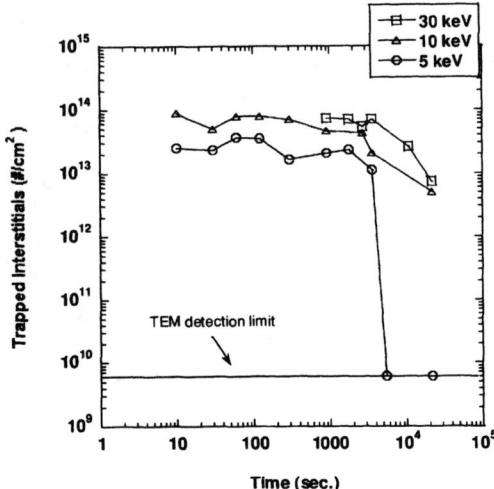

Figure 4. Interstitial evolution with time of the 30, 10 and 5 keV, 1×10^{15} ions/cm^2 Ge$^+$ annealed at 750 °C.

Lastly, the results for the lowest energy differ when compared to previous studies with similar experimental conditions [7, 8], which reported the presence of {311} type defects for 2 to 5 keV Si$^+$ implants, albeit, at a slightly lower dose. This contrast suggests a dependence on implant mass, since both Si$^+$ and Ge$^+$ are isoelectronic species. The much heavier germanium ion results in altogether different implantation kinetics and most of the energy loss is due to nuclear collisions, inducing a greater degree of disorder in the silicon lattice at lower doses [2]. Furthermore, upon annealing, defect behavior could have a strong dependence on implant depth, or surface proximity. For the lowest energy, the closeness to the surface (~120 Å) could adversely impact the stability of the Si interstitials, which make up the small dislocation loops observed, when compared to the implant depth achieved by the 10 and 30 keV implants (220 Å and 480 Å, respectively).

CONCLUSIONS

The defect evolution behavior of amorphizing, 1×10^{15} ions/cm^2, Ge$^+$ implants varies greatly with energy for anneals at 750 °C. For the highest energy studied (30 keV), the defects nucleate and evolve in a manner consistent with previously reported behavior for similar energy, Si$^+$ implants. However, with decreasing energy, the Ge$^+$ implants differ markedly in evolution behavior from Si$^+$. No {311} formation was observed for the 10 or 5 keV implants, and at the lowest energy, only small, unstable dislocation loops formed and quickly dissolved. In order to further the understanding of this unusual and, as of yet, unreported behavior, the effect of surface proximity needs to be elucidated so as to obtain corroborating conclusions.

ACKNOWLEDGEMENTS

This work was made possible through funding from the Semiconductor Research Corporation (task id 677.003) under the Master's Scholar Program.

REFERENCES

1. E. Myers, G. A. Rozgonyi, D. K. Sadana, W. Maszara, J. J. Wortman and J. Narayan, Mat. Res. Soc. Fall Meeting, Symposium B, 1985.
2. Mehmet C. Ozturk, Jimmie J. Wortman, Carlton M. Osburn, Atul Ajmera, George A. Rozgonyi, Eric Frey, Wei-Kan Chu and Clinton Lee, IEEE Transactions on Electron Devices, 1988. **35**(5): p. 659-667.
3. Shin Nam Hong, Gary A. Ruggles, Jimmie J. Wortman and Mehmet C. Ozturk, IEEE Transactions on Electron Devices, 1991. **38**(3): p. 476-486.
4. A. J. Murrell, E. J. Collart and M. A. Foad, J. Vac. Sci. Technol. B, 2000. **18**(1): p. 462-467.
5. D. J. Eaglesham, P. A. Stolk, H.-J. Gossmann and J. M. Poate, Appl. Phys. Lett., 1994. **65**(18): p. 2305-2307.
6. K. S. Jones, J. Liu, L. Zhang, V. Krishnamoorthy and R. T. DeHoff, Nucl. Instr. and Meth. in Phys. Res. B, 1995. **106**: p. 227-232.
7. D. J. Eaglesham, A. Agarwal, T. E. Haynes, H.-J. Gossmann, D. C. Jacobson and J. M. Poate, Nucl. Instr. and Meth. in Phys Res. B, 1996. **120**: p. 1-4.
8. Aditya Agarwal, Tony E. Haynes, David J. Eaglesham, Hans-J. Gossmann, Dale C. Jacobson, John M. Poate and Yu E. Erokhin, Appl. Phys. Lett., 1997. **70**(23): p. 3332-3334.
9. Jinghong Li and Kevin S. Jones, Appl. Phys. Lett., 1998. **73**(25): p. 3748-3750.
10. N. E. B. Cowern, G. Mannino, F. Roozeboom, P. A. Stolk, H. G. A. Huizing, J. G. M. van Berkum, N. N. Toan and P. H. Woerlee. in *195th ECS Meeting*. 1999. Seattle, WA.
11. S. Bharatan, in *Material and Process Characterization of Ion Implantation*, M.I. Current and C.B. Yarling, Eds. 1997, Ion Beam Press: Austin, TX. p. 224-243.
12. Jinning Liu, Kevin S. Jones, Daniel F. Downey and Sandeep Mehta, Mat. Res. Soc. Symp. Proc., 1999. **568**: p. 9-14.

Dopant Impurity Effects

Mat. Res. Soc. Symp. Proc. Vol. 669 © 2001 Materials Research Society

DIFFUSION AND DEFECT STRUCTURE IN NITROGEN IMPLANTED SILICON

Omer Dokumaci, Richard Kaplan, Mukesh Khare, Paul Ronsheim, Jay Burnham*, Anthony Domenicucci, Jinghong Li, Robert Fleming, Lahir S. Adam**, and Mark E. Law**

IBM SRDC, Hopewell Junction, NY 12533
* IBM Microelectronics, Burlington, VT.
* Electrical Engineering Dept. , University of Florida, Gainesville, FL 32611

ABSTRACT

Nitrogen diffusion and defect structure were investigated after medium to high dose nitrogen implantation and anneal. 11 keV N_2^+ was implanted into silicon at doses ranging from 2×10^{14} to 2×10^{15} cm^{-2}. The samples were annealed with an RTA system from 750°C to 900°C in a nitrogen atmosphere or at 1000°C in an oxidizing ambient. Nitrogen profiles were obtained by SIMS, and cross-section TEM was done on selected samples. TOF-SIMS was carried out in the oxidized samples. For lower doses, most of the nitrogen diffuses out of silicon into the silicon/oxide interface as expected. For the highest dose, a significant portion of the nitrogen still remains in silicon even after the highest thermal budget. This is attributed to the finite capacity of the silicon/oxide interface to trap nitrogen. When the interface gets saturated by nitrogen atoms, nitrogen in silicon can not escape into the interface. Implant doses above 7×10^{14} create continuous amorphous layers from the surface. For the 2×10^{15} case, there is residual amorphous silicon at the surface even after a 750°C 2 min anneal. After the 900°C 2 min anneal, the silicon fully recrystallizes leaving behind stacking faults at the surface and residual end of range damage.

INTRODUCTION

Nitrogen implant into silicon has been observed to retard the rate of oxide growth [1]. Nitrogen implant has been utilized to obtain thinner and more uniform oxides. It has also been investigated to obtain multiple gate oxide thicknesses on the same chip for System-On-A-Chip applications [2]. Nitrogen implant before oxide formation also helps suppress boron diffusion through the oxide [3]. Implanted nitrogen diffuses to the oxide/silicon interface during annealing and piles up around the interface [4]. The same experiments also show that diffusivity of nitrogen is quite high even at 750°C. The diffusion of nitrogen at low doses ($<1 \times 10^{15}$ cm^{-2}) has been modeled successfully with a point defect based model [5] . The resulting model gives excellent fit to oxide thicknesses for various oxidation temperatures (800°C-1050°C) and nitrogen implant doses (1×10^{14}-1×10^{15} cm^{-2}).

Nitrogen implants significantly enhance buried layer boron diffusion due to the implant damage [6]. On the other hand, nitrogen pre-implants at high doses decrease enhanced diffusion of boron due to amorphization and EOR loops [7]. Nitrogen co-implants into p+-gates have been found to reduce boron penetration through gate oxide. This has been attributed to the suppression of boron diffusion in poly due to the presence of nitrogen [8].

As CMOS gate dimensions scale down, thinner gate oxides are needed to obtain a larger inversion capacitance per unit area. As a result, the tunneling current through the oxide increases every generation. In order to prevent excess tunneling currents, dielectric constant of the oxide should be increased. A common way of achieving this is to incorporate more nitrogen into the oxide. High dose nitrogen implant is an option to incorporate a large amount of nitrogen into the oxide. This work is aimed at understanding diffusion and defect structure of high dose nitrogen implants in silicon. Diffusion of nitrogen is an important factor in the determination of the oxide thickness and the amount of nitrogen in the oxide, whereas the defect structure can affect the reliability of the oxide. Nitrogen implants also have an effect on the dopant profile and carrier mobility in the channel as well as on the subsequent processing steps such as extension formation.

EXPERIMENT

The experiments were carried out on <100> epi silicon wafers. After a 6 nm sacrificial oxide growth, N_2^+ was implanted at 11 keV between 2×10^{14} and 2×10^{15} cm^{-2}. Part of the wafers were annealed in a rapid thermal anneal (RTA) system in nitrogen atmosphere between 750°C and 900°C for 2 minutes. The remaining wafers were oxidized at 1000°C in an RTA system in a nitrogen plus oxygen ambient after the sacrificial oxide was stripped and the surface was cleaned. Nitrogen SIMS profiles were obtained from the samples with the sacrificial oxide intact. Cross Sectional TEM was carried out on selected samples. In the oxidized samples, nitrogen and oxygen profiles were acquired by TOF-SIMS with a 450 eV Cs+ primary beam.

RESULTS AND DISCUSSION

Fig. 1(a)-(c) show the nitrogen SIMS profiles in samples that have received the inert anneals. The sacrificial oxide was not removed before SIMS to avoid nitrogen loss around the silicon/oxide interface. The interface is approximately where the peak of the nitrogen profile is. The 2×10^{14} implant has the typical characteristics of low dose nitrogen diffusion [4]. Upon annealing, nitrogen profile shifts towards the surface. Higher temperatures cause more nitrogen to diffuse to the silicon/oxide interface. Nitrogen is captured and piles up at the interface as it reaches the interface. The diffusion of nitrogen is so rapid that only a small amount of nitrogen remains in silicon after the 900°C 2 min anneal. The 7×10^{14} implant is like the 2×10^{14} implant except that there is now a bump in the nitrogen profile centered around 17 nm in silicon. This bump is due to the segregation of nitrogen into the end-of-range dislocation loops. Also, more nitrogen is inside silicon than the lower dose case. In the 2×10^{15} case, nitrogen still diffuses towards the surface. However, the diffusion has slowed down considerably with respect to the lower doses, especially at 750°C and 850°C. After the 900°C anneal, a significant portion of nitrogen still remains in silicon. This is in contrast to the 2×10^{14} case where the amount of nitrogen in silicon is negligible.

Fig. 2 shows the percentage of remaining nitrogen in silicon and the amount of nitrogen movement. In order to calculate the percentage, first a dose is calculated for each profile by integrating the nitrogen profile from 10 nm onward. Then, the dose of the diffused profile is normalized to the similarly calculated dose of the corresponding as-implanted profile. The amount of movement is the depth difference between the as-implanted profile and the diffused profile at a fixed concentration of 1×10^{19} cm^{-3}. It is representative of the nitrogen diffusivity. In terms of percentage dose, 2×10^{14} and 7×10^{14} implants show similar behavior whereas the 2×10^{15} implant is completely different. For the two smaller doses, 50% of nitrogen stays in silicon after the 750°C anneal whereas only 15% remains after the 900°C anneal. For the 2×10^{15} implant,

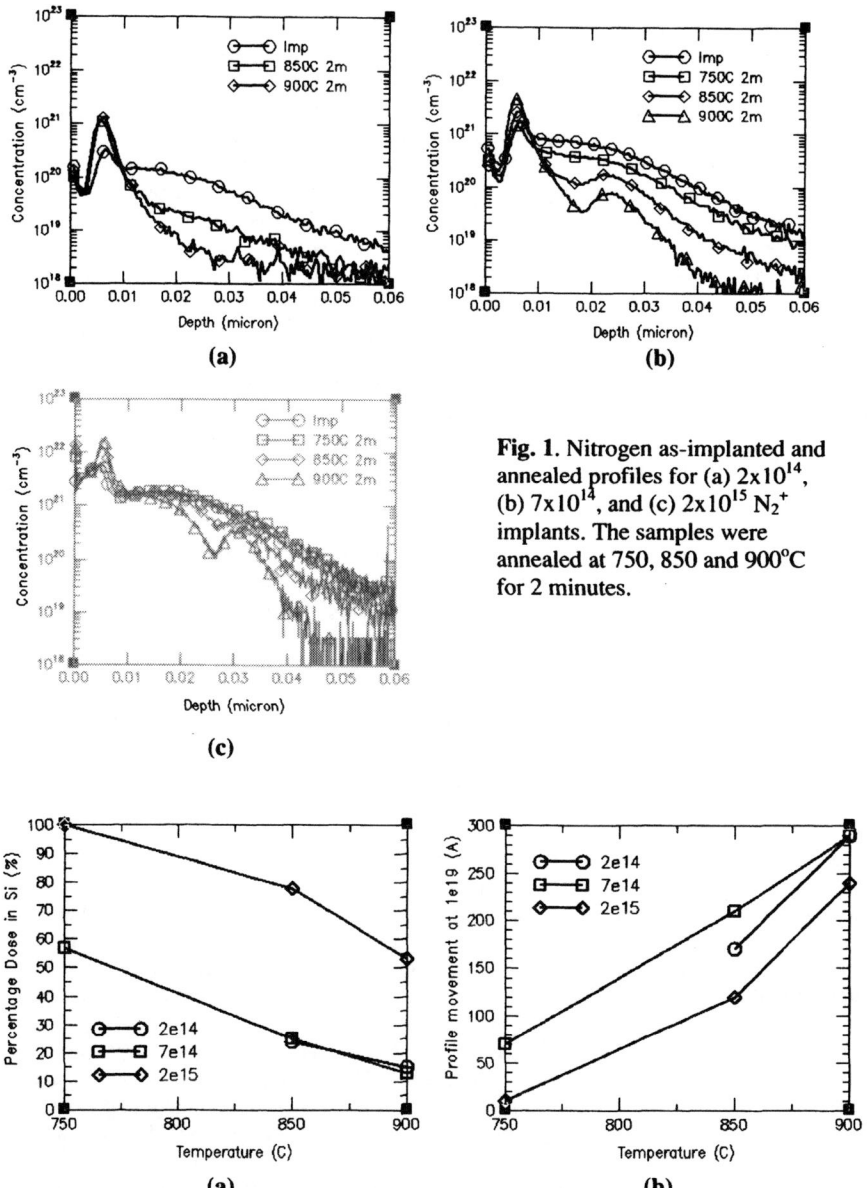

Fig. 1. Nitrogen as-implanted and annealed profiles for (a) 2×10^{14}, (b) 7×10^{14}, and (c) 2×10^{15} N_2^+ implants. The samples were annealed at 750, 850 and 900°C for 2 minutes.

Fig. 2. (a) Percentage of nitrogen remaining in silicon and (b) nitrogen profile movement as a function of annealing temperature.

approximately 100% is in silicon after 750°C, and 50% after 900°C. So, significantly more nitrogen remains in silicon at the highest dose for all annealing conditions.

The movements of the profiles suggest that diffusivity of nitrogen is similar for the $2x10^{14}$ and $7x10^{14}$ implants, At 750°C, the diffusivity is much larger for the lower dose implants than for the $2x10^{15}$ implant . For this temperature, the higher remaining dose in the $2x10^{15}$ case can be explained by a much lower diffusivity. When the diffusivity decreases, less nitrogen is able to reach the surface and more nitrogen remains in silicon. However, in the 900°C case, there is not much difference between the diffusivities, but still 50% of nitrogen is in silicon for the $2x10^{15}$ case. A likely explanation is that the interface traps are geeting saturated by nitrogen at the highest dose case. When the interface is saturated with nitrogen, the amount of nitrogen at the interface will not increase further. At that point, no more nitrogen will be removed from silicon into the interface.

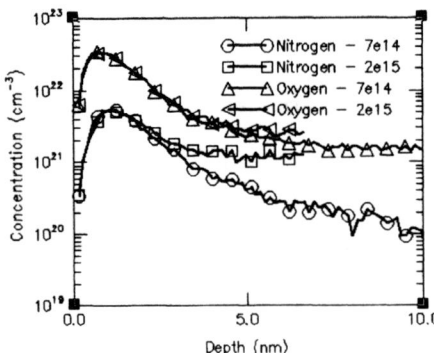

Fig. 3. TOF-SIMS profiles of nitrogen and oxygen in the thin oxide.

Another evidence for the saturation of the interface with nitrogen comes from TOF-SIMS measurements as shown in Fig. 3. In this case, an oxide has been grown in the rapid thermal anneal equipment at 1000°C, after the nitrogen implant, sacrificial oxide removal and surface clean. Nitrogen distribution and dose are similar around the oxide/silicon interface for both $7x10^{14}$ and $2x10^{15}$ although there is more nitrogen in the bulk in the higher dose case. These profiles suggest that the amount of nitrogen around the oxide/silicon interface is not increasing even if the implant dose is significantly increased. The interface is completely saturated with nitrogen and, just like the solubility effect, no more nitrogen can pile up at the interface.

Fig. 4 shows the cross-section TEM pictures of the samples implanted with the $7x10^{14}$ dose. After implantation, a continuous amorphous layer forms that has a thickness of 11 nm. The amorphous layer completely regrows during the 750°C anneal. End-of-range loops form at a depth of approximately 16 nm. There are also stacking faults intersecting the surface. After the 900°C anneal, there are still some loops and stacking faults remaining in the sample. The density of the loops goes down and their size becomes bigger during the higher temperature anneal due to the Ostwald ripening effect.

Fig. 4 also shows the cross-section TEM images of the samples with the $2x10^{15}$ implant. The thickness of the amorphous layer after the implant is 22 nm. Surprisingly, recrystallization is not complete after the 750°C 2 min anneal with still some amorphous layer remaining close to the surface. In undoped silicon, an amorphous layer of the same thickness should grow within a couple tenths of a second [9]. But, it has also been observed that nitrogen retards the recyrstallization rate [10]. In our experiment, the $2x10^{15}$ implant retards the regrowth rate by a factor of around 1000 with respect to an undoped sample. The a/c interface is very rough during the regrowth. This can explain the formation of stacking faults at the surface. As this rough a/c

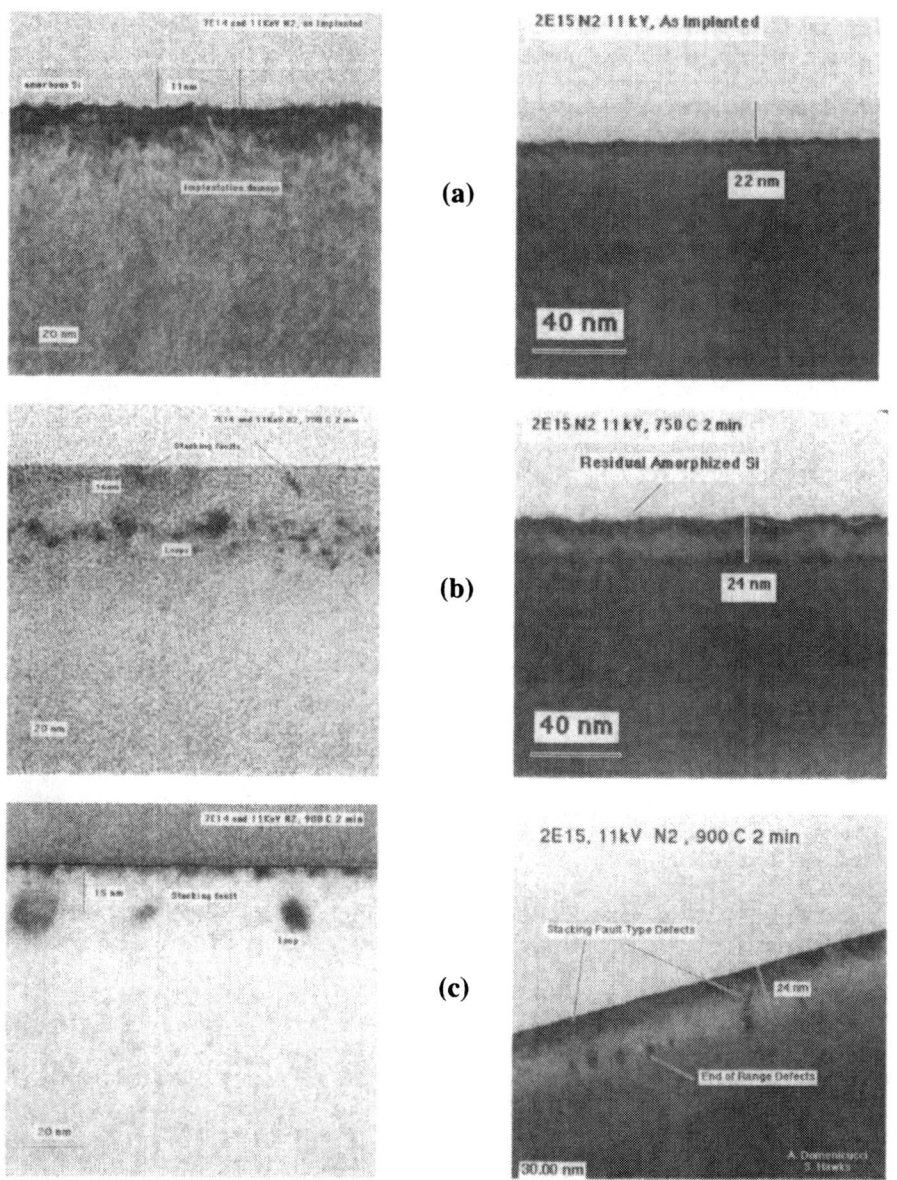

Fig.4. XTEM images of, (a) as-implanted, (b) 750°C 2 min annealed, and (c) 900°C 2 min annealed samples. Left-hand side images are from the 7×10^{14} sample, whereas right-hand side images are from the 2×10^{15} sample.

front intersects the surface, stacking faults can form as a result of the misalignment between the a/c interface and the surface. Similar surface defects have also been observed in BF_2^+ implanted samples and have been explained by the adverse effect of fluorine on amorphous layer regrowth [11]. After the 900°C anneal, a higher density of stacking faults appear in the $2x10^{15}$ sample than in the $7x10^{14}$ sample, confirming the correlation between the stacking faults and the nitrogen content.

CONCLUSIONS

High dose nitrogen implant, diffusion and defect structure were investigated. The amount of nitrogen that can be put at the silicon/oxide interface saturates as the dose of the nitrogen implant is increased. N_2^+ doses higher than $7x10^{14}$ will not likely improve the dielectric constant of the gate oxide much further. The amount of nitrogen remaining in silicon after the anneals increases significantly above a dose of $1x10^{15}$ N_2^+ because of lower nitrogen diffusivity and saturation of interface traps with nitrogen. High amounts of nitrogen in silicon can adversely affect the carrier mobility in the channel. At high doses, stacking faults form at the interface due to the very rough a/c growth front. These faults can cause problems with the oxide quality and leakage. End-of-range loops also form at high doses, and their density decreases at higher temperatures. The stacking faults and loops are not desirable in devices. At high doses, higher annealing temperatures (>900°C) and longer times will be needed to obtain a more defect and nitrogen-free silicon crystal under the gate oxide.

ACKNOWLEDGMENTS
The authors would like acknowledge ASTC for the processing of the wafers.

REFERENCES
1. C.T. Liu, Y. Ma, J. Becerro, S. Nakahara, D.J. Eaglesham, and S. J. Hillenius, IEEE Electron Device Lett. 18, 105 (1997).
2. C.T. Liu, Y. Ma, M. Oh, P.W. Diodato, K.R. Stiles, J.R. McMacken, F. Li, C.P. Chang, K.P. Cheung, J.I. Colonell, W.Y.C. Lai, R. Liu, E.J. Lloyd, J.F. Miner, C.S. Pai, H. Vaidya, J. Frackoviak, A. Timko, F. Klemens, H. Maynard, and J.T. Clemens, IEDM Tech. Dig., 589 (1998).
3. C.T. Liu, Y. Ma, H. Luftman, and S.J. Hillenius, IEEE Electron Device Lett. 18, 212 (1997).
4. L.S. Adam, M.E. Law, K.S. Jones, O. Dokumaci, C.S. Murthy, and S. Hegde, J. Appl. Phys. 87, 2282 (2000).
5. L.S. Adam, M.E. Law, O. Dokumaci, and S. Hegde, IEDM Tech. Dig., (2000).
6. O. Dokumaci, P. Ronsheim, S. Hegde, D. Chidambarrao, L.S. Adam, and M.E. Law, Mat. Res. Soc. Symp. Vol. 610, B.5.9.1 (2000).
7. T. Murakami, T. Kuroi, Y. Kawasaki, M. Inuishi, Y. Matsui, and A. Yasuoka, Nucl. Instr. and Meth. in Phys. Res. B 121, 257 (1997).
8. S. Hakayama, and T. Sakai, J. Electrochem. Soc. 144, 4326 (1997).
9. G.L. Olson, and J.A. Roth, Mater. Sci. Rep. 3, 1 (1988).
10. E.F. Kennedy, L. Csepregi, J.W. Mayer, and T.W. Sigmon, J. Appl. Phys. 48, 4241 (1977).
11. C. Carter, W. Maszara, D.K. Sadana, G.A. Rozgonyi, J. Liu, and J. Wortman, Appl. Phys. Lett. 44, 459 (1984).

Mat. Res. Soc. Symp. Proc. Vol. 669 © 2001 Materials Research Society

Modeling Boron and Indium Electrical Activities in Silicon in the Presence of Nitrogen

Vladimir Zubkov, Sheldon Aronowitz, Helmut Puchner and Juan P. Senosiain[1]
LSI Logic Corporation, 3115 Alfred Street,
Santa Clara, CA 95054, U.S.A.
[1]Department of Materials Science and Engineering, Stanford Unversity
Stanford, CA 94305, U.S.A.

ABSTRACT

The ab initio pseudopotential code (VASP) was employed to explore indium and boron electrical activities in silicon in the presence of nitrogen. Electrical activities for the combinations B+N, In+N, and In+B+N were explored. Formation energy of a negatively charged supercell, $(E^-)_f$, and a band gap, E_g, from calculations with one k point were chosen as indicators of acceptor activity. For separate dopants the calculated $(E^-)_f$ and E_g values indicate that substitutional B and In are effective acceptors and N is an extremely weak donor. When nitrogen is adjacent to, or separated 3 - 5 bonds from B or In, it suppresses acceptor activity. Binding is greater for In+N than for B+N in agreement with secondary ion mass spectroscopy (SIMS) data that demonstrates a greater retention of N by In. This should lead to a greater drop in activity for In+N combination versus B+N one, in agreement with spreading resistance profiling (SRP) experiments. Loss of activity in In+B+N combination might be due to long range interactions between dopants.

INTRODUCTION

Indium has been used to create super-steep retrograde channels in n-channel devices in silicon. This permits undesired effects on device performance such as reverse short channel effects to be significantly reduced. The limitations to the use of In are that it displays a low solid solubility [1] and electrical activation appears to be inefficient. Implanting high doses of nitrogen to modify gate oxide growth further complicates the picture by reducing the activation to a level that excludes In use [2]. (Implant dose for In: 1×10^{13} [115]In^+/cm^2 at 180 keV, 40 times less than that for N; more details in [2,3]). The threshold voltage was minimally reduced with channels formed with a boron implant and the same pre-gate nitrogen implant dose. The effects of combinations of indium and boron when nitrogen is present or absent have been explored by (SIMS) and SRP [2,3]. According to SRP data boron is much more active than indium and activity of In+B is close to that of B. The presence of nitrogen reduces the concentration of active carriers. This loss is quite substantial in the case of In and In+B: 28 and 54% , respectively, but rather small for B: only 7% [2,3].

In this work, an ab initio method was employed to explore if the calculations could capture the major experimental observations. We assumed that it would be relatively simple, once plausible criteria for electronic activation were established, to study the relative activity of any combination of dopants. That is, we were not interested in diffusion mechanisms but in possible configurations after an anneal was performed.

METHOD

The atomic level calculations were performed using the Vienna Ab initio Simulation Package (VASP) [4], using Vanderbilt-type pseudopotentials [4,5]. All calculations were performed for one k point ($\frac{1}{4},\frac{1}{4},\frac{1}{4}$) in a Si64 supercell. It was found that calculations with a greatly extended 4x4x4 Monkhorst and Pack k-sampling [6] gave virtually the same values for the total energies of the supercells.

Two parameters were adopted as indicators of electrical activity of the acceptors In and B. One of them is the energy of formation of a negatively charged cell:

$$(E^-)_f = (E^-)_{total} - E_0 - 5.28 \text{ eV} \tag{1}$$

where $(E^-)_{total}$ is the total energy of the negatively charged cell, E_0 is the total energy of the neutral cell and 5.28 eV is the middle of the gap for a cell composed of 64 silicon atoms. Values of $(E^-)_f$ were assumed to correlate with the acceptor electron affinity. The other indicator chosen is the energy gap, E_g, between the singly occupied orbital and the highest doubly occupied valence band orbital for the individual acceptor dopants or nitrogen. In the case when there are two dopants present and an even number of electrons, E_g is the difference between the lowest vacant and the highest occupied orbitals. Interaction between dopants was characterized by a binding energy, E_{bind}. This energy is defined with respect to a cell composed of sixty-four silicon atoms. For example, the net binding energy between dopants A and B equals

$$E_{bind} = E(AB) - E(A) - E(B) + E(Si64). \tag{2}$$

Here, $E(AB)$ is the total energy of the cell containing dopants A and B; $E(A)$ and $E(B)$ are the total energy of the cells containing only dopant A or B but in their same configuration as in the cell containing both dopants. $E(Si64)$ is the total energy of the sixty-four silicon atom cell.

Besides interactions of adjacent dopants in substitutional and hexagonal positions we also considered interactions between dopants separated by several bonds. In one set of calculations two substitutional dopants, e.g. In_s and N_s, were separated by a mimimum of five bonds within the supercell. But due to periodic boundary conditions (PBC), there will be a In_s atom in a neighboring cell separated by four bonds from the N_s in the original cell. In_s in other cells will be at larger distances from the original N_s dopant, and thus affect it to a lesser degree. Such configurations of separated dopants were designated as set 5/4. In another set of calculations the separation of N_s from two In_s atoms in the original and neighboring cells were three and four bonds, respectively (set 3/4). Due to PBC effects interactions between separated dopants might be overestimated, but it remains of interest whether they are present at all or not. These interactions are referred to as "long range interactions" throughout this text.

RESULTS

Before dealing with electrical activities of dopant combinations we tested our approach to estimate qualitatively dopant electrical activity via $(E^-)_f$ and E_g values on single dopant species incorporated into the silicon lattice. It appears to work quite satisfactorily. Thus, $(E^-)_f$ values were negative (several tenths of an eV, Table 1) for substitutional (s) acceptors and positive for interstitials, both hexagonal (H) and tetrahedral (T), indicating that substitutional acceptors

Table 1. Calculated activities and bindings for A and A + N (A = In, B; values in eV)

B + N	E_g	$(E^-)_f$	E_{bind}	In + N	E_g	$(E^-)_f$	E_{bind}
B_s	$\cong 0$	-0.51		In_s	$\cong 0$	-0.27	
B^H	0.39	0.08		In^H	0.7	0.38	
B_s-B_s	0.04	-0.59	1.98	In_s-In_s	0.46		-0.72
B_s-B^H	0.41^b		0.69	In_s-In^H	0.58		-2.04
B_s-N_s	0.67	-0.06	0.71	In_s-N_s	0.46	0.34	-1.54
B_s-N^H	0.92	0.30	-2.34	In_s-N^H	0.82		-2.62
B_s-N^T	0.66		0.16	In_s-N^T	0.61		0.73
B^H-N_s	0.68		0.16	In^H-N_s	0.82		-4.36
B_s...N_s^a	0.57	-0.19	0. 82 (0.96)^a	In_s....N_s^a	0.32	-0.08	-0.13 (-0.17)^a
B_s....N^H	0.69		-1.68 (-1.82)	In_s....N^H	0.76		-3.45 (-2.83)
B^H...N_s	0.61		-2.32 (-0.61)	In^H.....N_s	0.52		-2.91 (-2.25)

a. N_s is five/four bonds away; E_{bind} in brackets for the set 3 /4 (see text)

readily acquire an electron contrary to acceptors in interstitial positions. Similar conclusions were obtained from E_g values which were virtually zero for acceptors in substitutional positions and several tenths of an eV for acceptors in interstitial positions. The latter represents a very large gap for acceptors to be active. ($(E^-)_f$ values were calculated only for some cases with $E_g >$ 0.3 eV, since such a large E_g means no activity).

As was mentioned above, In is less active than B at the same concentration. This may be due to various reasons, e.g. some In atoms remain interstitial after anneal and/or cluster as inactive pairs In_s-In_s and In_s-In^H (Table1). Formation of In pairs is favorable, contrary to that of B-B pairs. For N_s, N^H, and N^T E_g = 0.47, 0.38, 0.71 eV, respectively. This means that N is an extremely weak donor and its contribution to electrical activity is negligible.

While exploring combinations In + N and B + N we examined both adjacent, A-N, and separated, A...N, pairs (A = B, In). In all cases that were examined, both calculated values of energy gaps and the energy of formation of negatively charged supercells indicated that electrical activity disappears when any A+N pair is formed (Table 1). Consequently, activity is absent for all adjacent A-N as well as for A...N pairs separated by several bonds and activity of the whole Si + dopants system decreases according to the number of A-N and A...N pairs that are formed. (Since T interstitials are less favorable than H ones, only H ones were considered).

Some drastic structural changes were found in some cases. In the pair B_s-N^H boron and nitrogen form a covalent bond and both atoms become trivalent (see figure 1a) which is their most favorable valence state. The similar local structures with trivalent In and N were found for In_s-N^H and In^H-N_s. However, in the case of B^H-N_s no bond was formed between B^H and N_s, B^H being repulsed from N_s and forming bonds with Si only (figure 1b).

Before suggesting an interpetation of experimental data on loss activity of B and In in the presence of N we look into nitrogen retention by indium and boron as revealed by SIMS experiments. The SIMS experiments [2,3] revealed that the nitrogen profile in the In+N

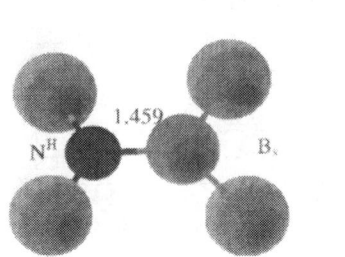

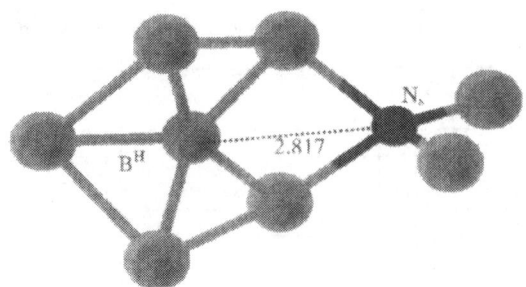

Figure 1a. Atoms B_s and N^H form a chemical bond. (Bonds in Å).

Figure 1b. B^H is more attracted to Si than to N_s. There is no bonding between B^H and N_s.

combination is shifted significantly towards the interface. The shift of the N peak corresponds to the region where the In peak is located. The presence of B does not alter this behavior. Boron by itself does not produce the N shift but the N concentration is reduced. Thus the ordering of nitrogen retention in the presence of the other dopant species is N{In or In+B} > N{B}. There is no noticeable interaction between In and B, probably because their concentrations are so small. The chemical profiles do not give any indication of a special redistribution of species that would result in the diversity observed in the net electrical activities of the various combinations of dopants.

Atomistic modeling offers an explanation of differences in N retention by different acceptor species. E_{bind} values (Table 1) were used as a measure of attraction or repulsion between dopants. Note that all but one entries for In+N interactions are attractive, i.e. $E_{bind} < 0$. In the case of B+N combinnations only three of the seven E_{bind} values are attractive. Thus, the calculations show that ordering of attraction should be In+N > B+N similar to experiment, i.e. on average nitrogen is more readily attracted to indium than to boron. Consequently, there is a greater loss of activity in In than in B due to the presence of N.

It should be noted that E_{bind} dependence on distance is different in B_s+N^H and B^H+N_s combinations compared with similar combinations involving In: in the B_s+N^H combination attraction decreases with distance contrary to the results for In_s+N^H. In the case of B^H+N_s the reason might be that adjacent B^H-N_s dopants are repulsive. It also may be an artifact of periodic boundary conditions.

Concluding that strong interactions exist between separated dopants seems unambiguous. Their presence is evident in figures 2a and 2b. Nitrogen, by itself, in the hexagonal interstitial position does not form bonds with adjacent silicon atoms (figure 2a). However, the presence of an indium atom separated from the nitrogen by several bonds alters the nitrogen interactions with adjacent atoms and results in the nitrogen forming chemical bonds with three silicon atoms (figure 2b).

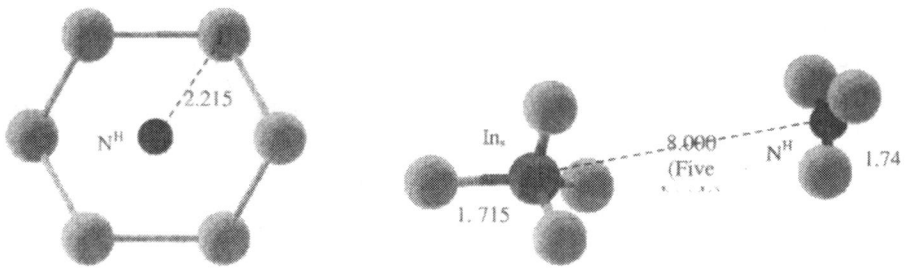

Figure 2a. Hexagonal bonding environment around N^H, no bonds formed.(Distance in Å)

Figure 2b. When In_s is five bonds away, N^H becomes trivalent. (Bonds in Å)

Experimentally, the In+B+N combination is much less active than In+B or B+N. Because indium is less active than boron, the reduction of activity of In+B+N versus B+N then should be due to the reduction of the activity of B arising from some effect introduced by the combined presence of In+N. Experimental results and modeling indicate that In interacts with N more strongly than B. We explored the possibility that In+N pairs formed in In+N+B attract B atom. If B becomes adjacent to N of In+N and interacts with it, this should reduce activity since, according to our calculations, adjacent B-N pairs are always inactive. Values of E_{bind} for B interaction with In+N pairs were calculated. For comparison we also explored how the N of the B+N pair interacts with another B. Attraction was found only for the cases where B^H was adjacent to the N_s of a separated $A...N_s$ pair (Table 2). This is denoted as $A...N_s \leftarrow B^H$.

Table 2. E_{bind}(eV) for B interaction with $A...N_s$ separated 5/4 bonds

Combination	A = In	A = B
$A_s...N_s \leftarrow B^H$	-0.42 (-0.12)[a]	0.22 (0.82)
$A^H...N_s \leftarrow B^H$	-2.21 (-2.91)	-1.6 (-2.32)

a In brackets E_{bind} for the A...N pair.

Attraction of B^H to $A...N_s$ suggests that during anneal some B^H might remain interstitial and therefore inactive. In the absence of In, the adjacent N_s-B^H pair is repulsive (Table 1). Thus the presence of a remote In can facilitate the creation of an inactive N_s+B^H pair. (It should be mentioned that in the combination $A^H...N_s \leftarrow B^H$ atoms N_s and B^H form bonds only with Si atoms). It should be noted that the calculated E_{bind} for the separated A...N pairs (numbers in brackets in Table 2) supports the possibility that In...N pairs are formed and that these might attract B^H. Thus, the modeling indicates that differences in activities in the In+B+N combination compared with the B+N one might be due to long range interactions. As for the B...N$\leftarrow$B cases it is reasonable to suggest that although there is an attraction for the $B^H...N_s \leftarrow B^H$, the concentration of B^H during an anneal is less than that of In^H and this pair is less probable than $In^H...Ns$. This agrees with a small loss of activity for the B+N combination.

CONCLUSIONS

The ab initio pseudopotential calculations have demostrated that indicators of dopant electrical activities suggested in this work: energy of formation of a negatively charged supercell and energy gap at one k point correlate with experimental observations. Calculations show that B and In are electrically active only when they are substitutional and N is a weak donor. Conclusion about deactivation of acceptors by adjacent or separated nitrogen seems unambigious. Another important result is the significant role played by long rang interactions. In the case of the Si64 supercell, this means a separation of dopants by 3- 5 bonds although their quantitative effect might be overestimated as a result of the PBC. Calculations suggest an interpretation of relative reduction in acceptor activity in the presence of nitrogen that would not be possible from SIMS profiles alone. This interpretation uses calculated of electrical activity indicators in combination with estimation of acceptor – nitrogen binding which showed a greater attraction in the In+N combination than in the B+N one. Activity of the In+N+B combination is less than that of B+N pair because formation of inactive B-N pairs might become more favorable in the presence of a remote In atom.

It appears that macroscopic expresssions of net activities can be directly correlated with atomistic arrangements. Understanding dopant electrical activities in various regions of a semiconductor device might be helpful in creating designer devices involving a comparatively small number of lattice atoms.

References

1. O. Dokumaci et al., *Mat. Res. Soc. Symp. Proc.* **568**, 205 (1999).
2. S. Aronowitz, H. Puchner, and V. Zubkov, *2000 International Conf. on Simul. of Semicond.Processes and Dev.*, 159 (2000); H. Puchner, S. Aronowitz, and V. Zubkov, *Proc. 30th European. Sol.-State Dev. Res. Conf.*, 104 (2000).
3. S. Aronowitz, V. Zubkov, H. Puchner, and J.Kimball, *J. Appl. Phys.* (submitted)
4. G. Kresse and J. Hafner, *Phys. Rev.* **B47**, 558 (1993); ibid **49**, 14251,1994; G.Kresse and J. Furthmuller, *Comput. Mater. Sci.* **6**, 15 (1996); G. Kresse and J. Furthmuller, *Phys. Rev.* **B54**, 11169 (1996); G. Kresse and J. Hafner, *J. Phys. Condens. Matte* **6**, 8245 (1994).
5. D. Vanderbilt, *Phys. Rev.* **B41**, 7892 (1990).
6. H. J. Monkhorst and J. D. Pack, *Phys. Rev.* **B13**, 5188 (1976).

Mat. Res. Soc. Symp. Proc. Vol. 669 © 2001 Materials Research Society

Silicon Interstitial Driven Loss of Substitutional Carbon from SiGeC Structures

M. S. Carroll and J. C. Sturm, Dept. of Electrical Engineering, Princeton University, Princeton, NJ; E. Napolitani, D. De Salvador and M. Berti, INFM and Dept. of Physics, University of Padova, Padova, Italy; J. Stangl and G. Bauer, Institute for Semiconductor Physics, Johannes-Kepler University Linz, Linz, Austria; D. J. Tweet SHARP Laboratories of America Inc., Camas WA

Abstract

The effect of annealing silicon capped pseudomorphic $Si_{0.7865}Ge_{0.21}C_{0.0035}$ or $Si_{0.998}C_{0.002}$ layers on silicon substrates in nitrogen or oxygen at 850°C was examined using x-ray diffraction (XRD) and secondary ion mass spectrometry (SIMS). Most substitutional carbon is lost from the alloy layers due to carbon out-diffusion rather than from precipitation. The carbon is found to diffuse more rapidly out of the SiGeC layer than the SiC layer after nitrogen and the carbon is found to leave the sample entirely, an effect that is enhanced by oxidation and thin cap layers. All substitutional carbon can be removed from the sample in some cases implying negligible formation of silicon-carbon complexes. Furthermore, it is found that each injected silicon interstitial atom due to oxidation causes the removal of one additional carbon atom for the SiGeC layer.

Introduction

Substitutional carbon incorporation in silicon and SiGe has drawn significant attention because of reduced boron diffusion in carbon's presence, due to its ability to consume silicon interstitials, which mediate boron diffusion [1]. However, the potential that the interstitial-carbon product is a defect (i.e ß-SiC precipitation or carbon clusters [2,3]) may limit the usefulness of carbon for diffusion engineering. Previous studies of carbon thermal stability in SiGeC confirm that carbon can precipitate in SiGeC [2,4]. In this letter carbon out-diffusion from thin SiGeC layers is examined and carbon out-diffusion is found to be the dominant mechanism of carbon loss for samples close to the surface, even in the regime of carbon concentration far above solid solubility or in the presence of excess interstitials injected during oxidation.

Experiment

Two test structures with 25 nm thick $Si_{0.7865}Ge_{0.21}C_{0.0035}$ layers capped by 50 or 280 nm of silicon and one 150 nm thick $Si_{0.998}C_{0.002}$ layer capped by a 45 nm silicon layer were grown on <100> silicon substrates by rapid thermal chemical vapor deposition (RTCVD) at temperatures between 550°C and 750°C using dichlorosilane, disilane, germane, and methylsilane as the silicon, germanium, and carbon sources respectively [5]. Samples of the as-grown and annealed structures were examined using secondary ion mass spectrometry (SIMS), which were sputtered using 1-2 keV Cs^+ (Fig. 1) or O^+ (Fig. 2) ions. Depths were determined using standard profilometry of the sputtered craters leading to a 5% uncertainty in depths, a 20% (Fig. 1) or 10% (Fig. 2) uncertainty in carbon concentrations and approximately a 2% uncertainty in the absolute germanium fraction.

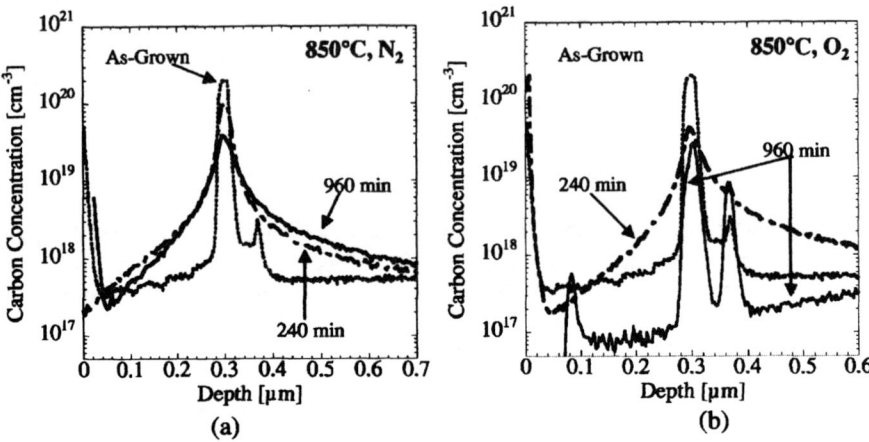

Figure 1. Carbon concentration profiles of the 280 nm silicon capped SiGeC layers before and after annealing at 850°C in nitrogen or oxygen ambient for 240-960 minutes overlaid on the as-grown carbon profile, (a) and (b) respectively.

All as-grown samples were examined by x-ray diffraction (XRD) using a double crystal rocking curve geometry around the (004) Bragg reflection to determine the amount of carbon that was substitutional in the buried alloy layers. The as-grown carbon was 100% substitutional in all three samples [6]. The substitutional carbon in the SiGeC layers after oxidation was also examined using XRD. Rocking curves of as-grown and oxidized samples of the SiGeC sample with a 50 nm cap were fit by simulations and agreed well with germanium and carbon profiles obtained by SIMS indicating that the carbon in the SiGeC layer compensated the strain as if it was all substitutional. As-grown and 960 minute nitrogen or oxygen annealed samples of the 280 nm silicon capped SiGeC layer were also examined for relaxation in the plane parallel to the growth surface by scanning around the (224) Bragg reflection. No relaxation was observed.

Carbon profiles from the SiGeC samples with a 280 nm silicon cap, annealed in nitrogen ambient for 240 or 960 minutes are overlaid on the as-grown carbon profile, Fig. 1 (a). Carbon diffusion in silicon is believed mediated by an interstitial kick-out [7],

$$C_s + I \rightarrow C_i \qquad (1)$$

where C_s is a carbon atom in a substitutional site, I is the silicon interstitial C_i, is the mobile interstitial carbon defect. The distinct non-gaussian broadening of the carbon profiles after nitrogen annealing has been explained as a result of a depleted interstitial concentration in the carbon-rich region. The carbon reaction with the self-interstitial is believed to proceed so rapidly that it produces an undersaturation of interstitials locally [7]. Because the interstitial population is depleted in the carbon region, a "quasi-stationary" profile is observed surrounded by tails of carbon "kicked-out" by the transport-limited diffusion of interstitials from the surrounding silicon [7].

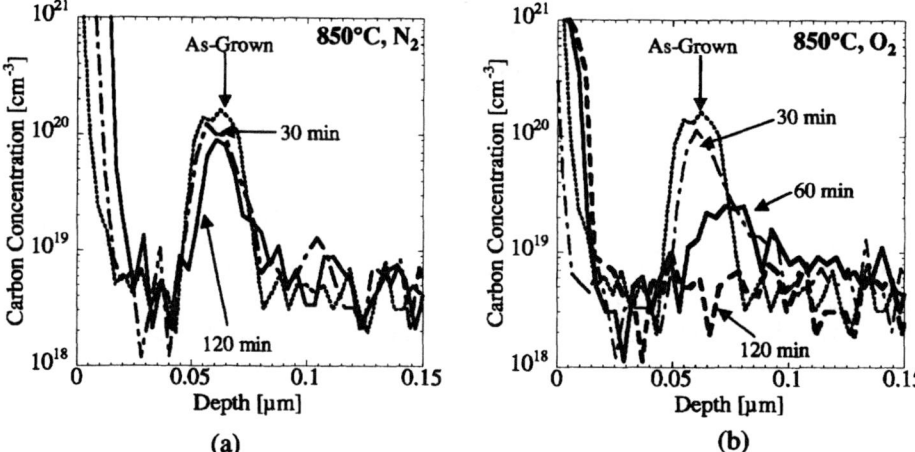

Figure 2 Carbon concentration profiles from 50 nm silicon capped SiGeC layers before and after annealing at 850°C in nitrogen or oxygen ambient for 30-120 minutes overlaid on the as-grown carbon profile, (a) and (b) respectively.

The carbon concentration of the out-diffusing tail on the surface side is notably less than that on the substrate side of the SiGeC layer. No sign of carbon build-up on the surface side combined with a relatively constant concentration gradient towards the surface over the entire 16 hour anneal, indicates that carbon is diffusing towards and out the surface from the SiGeC layer. Previous studies have also reported loss of carbon from slightly carbon enriched (8×10^{17} cm^{-3}) crystalline silicon out the surface after annealing in either oxygen or nitrogen ambient [8].

The rate of carbon loss from the SiGeC layer is enhanced by oxidation, Fig. 1 (b), and the carbon is reduced well below the as-grown background concentration (3×10^{17} cm^{-3}) except two spikes of immobile carbon, which persist after 960 minutes of oxidation located at 300 and 370 nm. The oxide-silicon interface after 960 minutes of oxidation is indicated by the carbon spike located at a depth of 100 nm in this sample.

Carbon concentration profiles of the SiGeC layer with a 50 nm Si cap after annealing for 30 or 120 minutes in nitrogen ambient are overlain on the as-grown profile, Fig. 2 (a). A clear reduction of the total carbon concentration is observed although very little broadening of the profile is observable. Presumably, the primary mechanism of loss is diffusion to the surface and the carbon tails are obscured by the higher carbon detection limits (3×10^{18} cm^{-3}). The carbon concentration profiles after oxidation of 30 to 120 minutes, Fig. 2 (b), shows a more rapid decrease of the carbon concentration resulting in no detectable carbon in this sample after 120 minutes of oxidation.

Immobile carbon (density of ~ 2×10^{19} cm^{-3}) is observed in the SiGeC layer with a 300 nm silicon cap after oxidation, but is not observed in the SiGeC layer capped with a 50 nm silicon cap. The formation of immobile carbon in the SiGeC layer structure with the 50 nm cap is likely prevented by the rapid carbon out-diffusion to the surface. The carbon concentration is above the solid solubility in the SiGe layer for a shorter time in the 50 nm capped layer not allowing time for carbon to condense into its immobile form. Previous reports of carbon precipitation or immobile carbon are typically from much thicker carbon layers in $Si_{1-x}C_x$ or $Si_{1-x-y}Ge_xC_y$ [2,3,4]. In these cases the carbon concentration in the middle of the layer remains near the as-grown value longer (typically far above the solid solubility) because the carbon out-diffusion in nitrogen is slower and the carbon concentration decreases only near the edges of the layers. Indeed, in the thick (150 nm) $Si_{1-x}C_x$ layer in this study SIMS does show immobile carbon formation after 8 hours of oxidation, but only after the top 100 nm of carbon layer has eroded away with no sign of immobile carbon.

The carbon detected by SIMS in the alloy layers after annealing in either nitrogen (solid) or oxygen (hollow) was integrated over the layer thickness for all annealing times. The change in carbon content in the 50 nm Si capped $Si_{1-x-y}Ge_xC_y$ and 45 nm Si capped $Si_{1-x}C_x$ layer is shown in Fig 3 (a) and (b), respectively. After 120 minutes of annealing the 50 nm Si capped SiGeC layer $\sim 1.5 \times 10^{14}$ cm^{-2} carbon diffuses out due to the intrinsic carbon diffusion mechanism in SiGeC. The carbon out-diffusion from the SiC layer is, however, much slower and even after 8 hours of annealing in nitrogen only $\sim 1 \times 10^{14}$ cm^{-2} is lost from the layer. Furthermore, the carbon in the layer during the nitrogen anneal remains substitutional, measured by XRD. The source of the extra carbon out-diffusion from the SiGeC layer remains to be identified. The extra carbon lost due to oxidation (carbon in alloy layer after nitrogen anneal subtracted from that after oxidation) is due to the extra silicon self-interstitials injected by the oxidation process and increases with decreasing silicon cap thickness, Fig. 4.

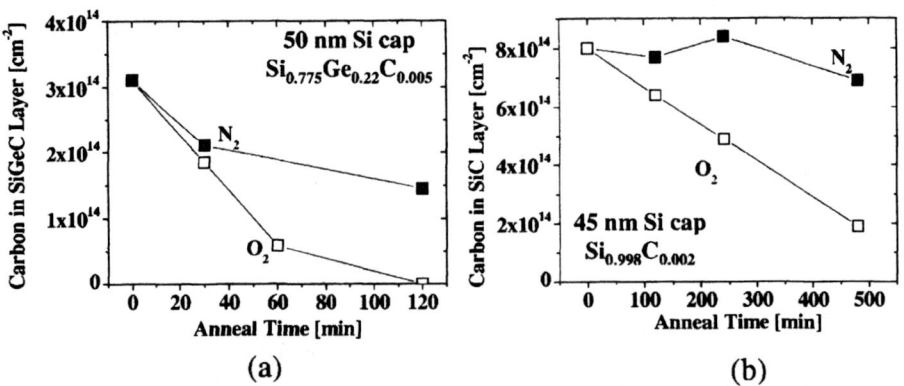

Figure 3. Total carbon in (a) SiGeC layers and (b) SiC layer after annealing in either nitrogen (solid symbol) or oxygen (hollow symbol) ambient at 850°C. Carbon concentrations measured by SIMS.

Interstitial Flux

Oxidation is known to inject interstitials into the silicon bulk at the surface and the enhanced carbon diffusion and carbon loss from the SiGeC layers after oxidation qualitatively can be explained by an increase in mobile carbon, C_i, due to the injected interstitials. For high carbon concentrations in buried SiGeC layers, similar atomic compositions as those in this work, SiGeC layers are near perfect interstitial sinks for all interstitials injected during oxidation [9]. Therefore the additional injected interstitials are expected to "kick-out" a similar number of carbon, since all injected interstitials are consumed at the SiGeC layer.

During oxidation the surface concentration of interstitials is constant [10], resulting in a linearly decaying interstitial profile from the surface supersaturation concentration to approximately zero at the SiC or SiGeC layer [11, 12]. The interstitial flux into the silicon during oxidation can therefore be calculated as:

$$J_I = -D_I \frac{dI}{dx} = n_{surf} \times \frac{D_I I^*}{\Delta x}$$

(2)

where, n_{surf} is the ratio of the interstitial surface concentration to the bulk interstitial concentration ($I/I^* = 12.7$) at 850°C, $D_I I^*$ is the interstitial transport product measured by metal tracer diffusion (1×10^4 cm^{-1}s^{-1}) [11,13], and Δx is the depth of the SiGeC layer.

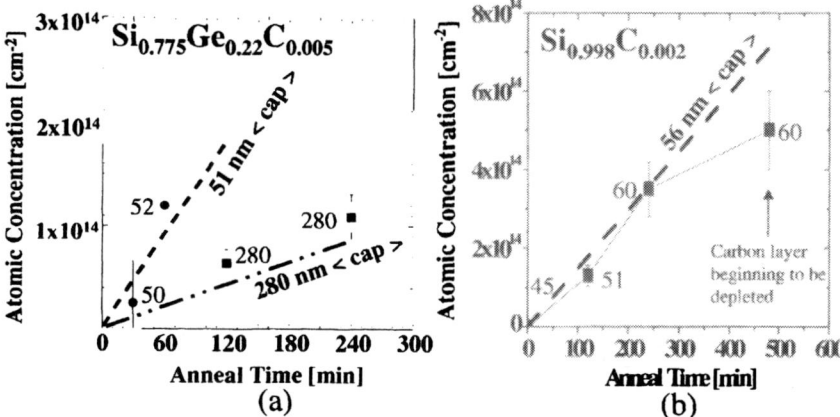

Figure 4. Summary of oxidation enhanced carbon loss from the (a) 50 nm and 280 nm silicon capped SiGeC layers and (b) the 45 nm silicon capped SiC layer. The number of injected interstitial silicon atoms after oxidation is calculated for each test structure using the average silicon cap thickness over the entire oxidation time. The silicon cap thicknesses are found next to each of the carbon loss data in nanometers for each oxidation condition, respectively.

A crude estimate of the total number of injected interstitials is made using an average silicon cap thickness for the entire oxidation time. The silicon cap thickness is, however, not constant during the oxidation. The cap layer thickness depends on two competing processes, the erosion of the carbon layer (increasing the cap) and the consumption of surface silicon by the oxidation reaction (decreasing the cap). This effect is greatest in the thick SiC layer that undergoes the longest oxidation times (8 hrs.). However, even after eroding approximately 100 nm of carbon from the SiC layer, the silicon cap thickness (the total silicon between the edge of the carbon layer and the silicon-oxide interface) remains nearly constant for the entire oxidation time (see cap thickness, Fig. 4, determined by SIMS). Apparently, the two competing processes naturally reach a temporary steady-state increasing or decreasing the total silicon cap until the carbon layer erosion rate matches the silicon consumption rate due to oxidation.

The calculated number of interstitials injected into each carbon layer is compared to the extra carbon removed due to oxidation, Fig. 4. The extra carbon out-diffusion and the calculated number of injected interstitials both scale with the silicon cap thickness and agree reasonably well for all three oxidation cases considering the uncertainties involved indicating that nearly every injected self-interstitial removes one carbon from the carbon alloy layer. Note: the extra carbon loss after the longest oxidation time may deviate from the calculated number of injected interstitials because by this time the carbon layer is beginning to become completely depleted of carbon. For this reason the carbon loss from the SiGeC layers after the longest oxidation times is omitted because SIMS clearly shows that the SiGeC layers are nearly completely depleted of carbon.

Conclusion

Carbon out-diffusion from $Si_{0.998}C_{0.002}$ or $Si_{0.7865}Ge_{0.21}C_{0.0035}$ layers has been examined after annealing in nitrogen or oxygen ambient at 850°C. Carbon out-diffusion from carbon alloy layers is the dominant mechanism of carbon loss, not precipitation. Carbon is found to diffuse out the surface and the carbon diffusion from the SiC or SiGeC layer is enhanced by oxidation. Each injected interstitial leads to the removal of one carbon from the SiC or SiGeC layer. Finally, the rapid and complete loss of carbon from the 50 nm silicon capped SiGeC layer after oxidation clearly indicates that the end-product of the reaction between the injected silicon self-interstitials and the substitutional carbon is mobile carbon. No evidence of immobile carbon formation or precipitation is observed in this important case.

Acknowledgements
This work was supported by ARO and DARPA.

References
[1] P. A. Stolk, H.-J. Gossmann, D. J. Eaglesham, D. C. Jacobson, J. M. Poate, and H. S. Luftmann, Appl. Phys. Lett. **66**, 568 (1995)
[2] P. Warren, J. Mi, F. Overney, and M. Dutoit, J. of Crystal Growth **157**, 414-419, (1995)
[3] J. W. Strane, H. J. Stein, S. R. Lee, S. T. Picraux, J. K. Watanabe, J. W. Mayer, J. Appl. Phys. **76**, 3656 (1994)
[4] L. V. Kulik, D. A. Hits, M. W. Dashiell, J. Kolodzey, Appl. Phys. Lett. **72**, 1972, (1998)
[5] J. C. Sturm, P. V. Schwartz, E. J. Prinz, H. Manoharan, J. Vac. Sci. Tech. B **9**, 2011 (1991)
[6] D. De. Salvador, M. Petrovich, M. Berti, F. Romanato, E. Napolitani, A. Drigo, J. Stangl, S. Zerlauth, M. Muehlberger, F. Schaeffler, G. Bauer, P. C. Kelires, Phys. Rev. B. **61**, 13005 (2000)
[7] R. F. Scholz, P. Werner, U. Gosele, T. Y. Tan, Appl. Phys. Lett. **74**, 392 (1999)
[8] L. A. Ladd and J. P. Kalejs in *Oxygen, Carbon, Hydrogen, and Nitrogen in Crystalline Silicon*, Editors: Mikkelsen, Pearton, Corbett, Pennycook. Vol. **59**, p 445. Materials Research Symposium Proceedings. Pittsburgh PA (1986)
[9] M. S. Carroll, C-L. Chang, J. C. Sturm, T. Buyuklimanli, Appl. Phys. Lett. **73**, 3695 (1998)
[10] S. T. Dunham, J. Appl. Phys. **71**, 685 (1992)
[11] M. S. Carroll and J. C. Sturm in *Silicon Front-End Processing- Physics and Technology of Dopant-Defect Interactions II*, Vol. **620**. Material Research Symposium Proceedings. (to be published)
[12] H. Ruecker, B. Heinemann, W. Roepke, R. Kurps, D. Krueger, G. Lippert, H. J. Osten, Appl. Phys. Lett. **73**, 1682 (1998)
[13] H. Bracht, N. A. Stolwijk, H. Mehrer, Phys. Rev. B 52, 16542 (1995)
[14] "Carbon in monocrystalline Silicon", G. Davies and R. C. Newman Handbook on Semiconductors, Editor: T. S. Moss. Elsevier 1558 (1994)

Mat. Res. Soc. Symp. Proc. Vol. 669 © 2001 Materials Research Society

Carbon Diffusion and Clustering in SiGeC Layers Under Thermal Oxidation

D. De Salvador[1], E. Napolitani[1], A. Coati[1], M. Berti[1], A.V. Drigo[1], M. Carroll[2], J.C. Sturm[2], J. Stangl[3], G. Bauer[3], L. Lazzarini[4]

[1]Dept. of Physics, University of Padova and INFM, Padova, ITALY
[2] Dept. of Electrical Engineering, Princeton University, Princeton, USA
[3]Inst. For Semiconductor Physics, J. Kepler University of Linz, Linz, AUSTRIA
[4] CNR-MASPEC, Parma, ITALY

ABSTRACT

In this work we investigated the diffusion and clustering of supersaturated substitutional carbon 200nm thick SiGeC layers buried under a silicon cap layer of 40nm. The samples were annealed in inert (N_2) or oxidizing (O_2) ambient at 850°C for times ranging from 2 to 10 hours. The silicon self-interstitial (I) flux coming from the surface under oxidation enhances the C diffusion with respect to the N_2 annealed samples. In the early stages of the oxidation process, carbon escape by diffusion across the layer/cap interface dominates. This phenomenon saturates after an initial period (2-4h) which depends on the C concentration. This saturation is due to the formation and growth of C containing precipitates which are promoted by the I injection and act as a sink for mobile C atoms. The competition between clustering and diffusion is discussed for two different C concentrations.

INTRODUCTION

In the last years strong efforts have been devoted to the investigation of the structural properties of SiGeC alloy, due to its potential use as a Si-based material with band-gap [1] and lattice parameter [2] tailoring properties. Recently, the role of C in Si and SiGe alloys as a Si self-interstitial (I) trap was evidenced by the reduction of Boron diffusion [3] and by the suppression of both B transient enhanced diffusion (TED) and B oxidation enhanced diffusion (OED) [4,5]. This pushed up a renewed interest in the use of SiGeC layers to control the diffusion of dopants in silicon devices.

Therefore, understanding the behavior of C in SiGeC/Si heterostructures under I supersaturation is of crucial importance. It is known that C diffusion is strongly enhanced by I supersaturation [6], as the silicon self-interstitials promote the formation of mobile C interstitial atoms via the kick-out or Frank-Turnbull mechanisms. Recently, a C-diffusion enhancement by interstitials injection was observed also in SiGeC alloys with high C concentration, above 10^{20} cm^{-3} [7]. Nevertheless, even in the absence of an external I injection (i.e. after inert thermal annealing), it has been reported the tendency of C to precipitate and to form β-SiC clusters [8].

All these facts suggest that the control of the C behavior in Si-based materials under thermal treatments is challenging for technological applications. In this work we report on the diffusion and clustering behavior of C in buried SiGeC layers under supersaturated I non-equilibrium conditions, induced by thermal oxidation of the silicon cap. We'll describe the presence of a complex competitive mechanism between C diffusion and C accumulation in clusters.

EXPERIMENT

Two nominally 200nm thick $Si_{0.926}Ge_{0.07}C_{0.004}$ and $Si_{0.922}Ge_{0.07}C_{0.008}$ layers covered by a 40nm Si-cap were grown over a 200nm thick silicon buffer layer by rapid thermal chemical vapor deposition (RTCVD) at temperatures between 625°C and 750°C on a p-type Czochralski (100) silicon wafer. The samples were cut in several pieces that were thermally treated in furnace at 850°C under O_2 or N_2 fluxes for time from 2 to 10 hours. The N_2 annealing experiments were performed in order to distinguish the pure thermal effects from those produced by the I injection under oxidation.

The C and Ge chemical concentration depth profiles were obtained by Secondary Ions Mass Spectroscopy (SIMS) on a CAMECA IMS-4f spectrometer, while using Cs^+ or O_2^+ primary beam at impact energy below 2 keV and at glancing incidence, in order to improve the depth resolution. The C concentration was calibrated using the C total dose obtained by means of resonant Backscattering Spectrometry (rBS) technique [9], while the depth scale was calibrated by measuring the crater depth with a profilometer. High Resolution X-Ray Diffraction (HRXRD) measurements were performed by a Philips MRD diffractometer in standard setup (Ge(220) Bartels monochromator); (004) rocking curves were taken using a detector aperture of about 0.5 degrees.

RESULTS AND DISCUSSION

Fig. 1a and 1b report the C profiles of all the samples annealed in O_2, and of the sample annealed for 10 h in N_2. In order to consistently compare the profiles all the depth scales where shifted so as to have the Ge cap/layer interface at the same position, which has been conventionally assumed to be at zero (dashed vertical line in Fig. 1). In fact, after annealing, the cap/layer interface of the Ge signal results to be at different depths due to the volume expansion produced by the silicon cap oxidation. It is worth noting that SIMS analysis (not shown) reported negligible Ge diffusion in all the processed samples either at the cap/layer and at the

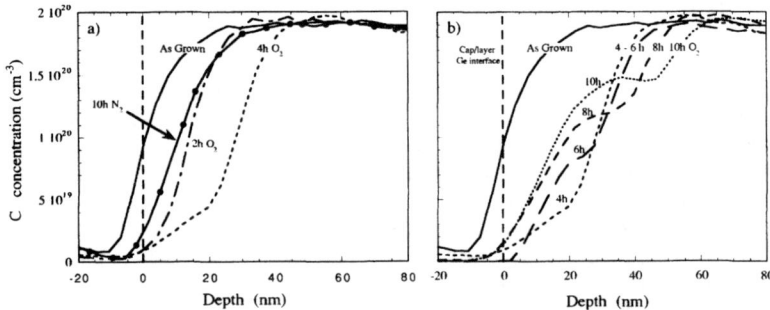

Figure 1 *Carbon concentration profile of the0.35at% C samples relative to: (a) as grown (continuous line), 10 hours annealed in inert atmosphere (marked by the arrow) and 2h and 4h oxidation; (b) as grown (continuous line), and 4h, 6h, 8h, and 10 h oxidation . The vertical line marks the presence of the Ge cap/layer interface.*

layer/substrate interfaces. Moreover, the C diffusion is limited to the portion of the layer closer to the silicon cap, whether no diffusion occurs at the deeper interface. This is the reason why Fig. 1 reports only the region close to the cap/layer interface.

By comparing the C concentration profiles of the processed samples with that of the as grown (Fig.1a) a clear diffusion effect is evidenced by a shift of the C interface with respect to that of Ge. This effect is much more remarkable in the O_2 annealed samples with respect to the N_2 samples. The I flux evidently enhances the formation of mobile C which can diffuse out from the layer. The C profile evolution for annealing times longer than 4h in O_2 (Fig. 1b) is the following: the interface moves deeper into the sample as the annealing time increases and, at the same time, an accumulation kink grows up in the region between 20 and 45nm in depth.

The C diffusion is not the single mechanism induced in the layer by the I flux. In Fig.2 a cross sectional TEM image of the 0.35 at% C sample annealed for 10h in O_2 atmosphere is reported. The formation of precipitates in the SiGeC layers is clearly visible. Such precipitates have a diameter of about 3–5nm and are not revealed in the N_2 annealed samples. The clusters are not homogeneously distributed in the layer being located between 20 and 120nm in depth with a maximum density at the same depth as the accumulation kink of the C profile. XTEM on other samples (not reported) shows that both the total number of clusters and their maximum distribution depth increase with oxidation time. Such observations indicate a strong correlation between the clusters location and the C accumulation kink present in the SIMS profiles, suggesting that the clusters contain C and act as a sink for the mobile C produced by the I injection.

HRXRD analyses confirm and give further insight to the above process. Fig. 3a reports the (004) rocking curve of the sample annealed for 10 h in O_2. Simulations of the rocking curves based on dynamical scattering theory were attempted by using the C and Ge SIMS profiles, and considering the Ge and C effect on strain as described in Ref.[2].

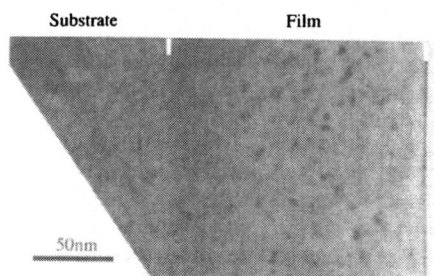

Substrate   Film

50nm

Figure 2 *TEM cross section of the 0.35at% C sample annealed in O_2 atmosphere for 10 hours. Three zones are visible in the picture. The left zone corresponds to the substrate, the central zone corresponds to the layer while the white right zone corresponds to the glue. The cap is not visible because almost completely oxidized and then removed by HF treatment. The precipitates are distributed inside the layer.*

While this approach provides a successful fit of the as grown sample, in the case of the oxidized samples, there is no agreement between the experimental HRXRD data and the dynamical simulations (Fig.3a, dashed line). However, the SIMS and TEM data suggest that not all the C atoms present in the layer are in substitutional sites and hence producing strain, but part of the C atoms is contained in the precipitates. We suppose that in the first 40nm of the layer (kink zone) C is fully precipitated while in the region between 40 and 100nm its substitutional fraction varies linearly from 0 to 1. The resulting substitutional-C profile is shown in Fig.3b as the solid line. This profile allows to produce the rocking curve reported in Fig.3a (solid line). The good agreement with the experimental data conclusively

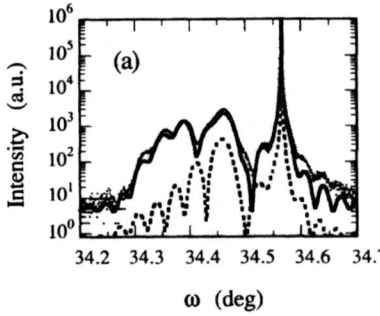 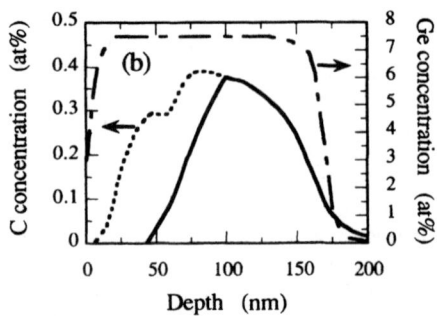

Figure 3 *(a) HRXRD (004) rocking curve of 0.35at% C 10 h oxidized sample (dots) compared to different simulations. (b) Concentration profiles used for the simulations. The Ge SIMS profile (dot-dashed line) was used for both the simulations. The C SIMS profile (dashed line) was used for the dashed line simulation in (a), assuming C and Ge to be fully substitutional. Solid line is the substitutional C profile used to generate the continuous line simulation in (a).*

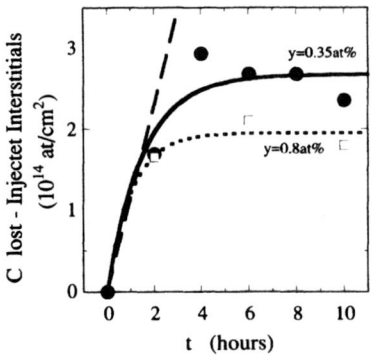

Figure 4 *C dose lost by the layer due to the oxidation process as a function of the annealing time. The C loss is evaluated by making the difference between the N_2 annealed and the O_2 annealed C dose. Data of both 0.35at% C series (full circles) and 0.8at% C series (open squares) are reported. The dashed line represents the I-injected during the thermal oxidation. Saturating exponential fits are reported to guide the eye.*

demonstrates the presence of a large zone of non substitutional C atoms in coincidence with the clusters revealed by TEM.

All the experimental observations indicate that the surface injection of I produces a strong structural change of the SiGeC layer. The change proceed with the annealing time from the cap/layer interface and involves about 100nm of the layer after 10h of annealing in O_2 when the C concentration is of 0.35at%. The main physical processes causing the structural modifications are C diffusion and precipitation.

During the early stages of the oxidation, the amount of C lost by diffusion is comparable to the I injected in the layer computed as in Ref.[10] (dashed straight line in Fig.4). This is similar to what observed in thinner samples of identical C concentration [7], where the complete loss of C from the layer was observed. On the contrary, in our samples there is only a partial loss of C atoms which saturates after an initial transient.

The saturation of the C loss can be understood on the basis of the observed clustering phenomenon. It is quite reasonable that the mobile C atoms promoted by the I injection can diffuse both inside the layer and towards the surface. During the first stages of the annealing, the C atoms moving towards the surface have an increased mobility, because they move inside a region which is both richer in I [10] and poorer in traps (as C itself is a trap for mobile C), so they can easily leave the film through the cap. Instead, C atoms moving inside the layer can start nucleating clusters by reacting with other C atoms. As a matter of fact, while after 2h of annealing the net result appears to be a simple shift of the cap/layer C interface, after 4h a slight slope change in the bottom part of the interface shows the formation of the first clusters. These clusters are revealed by TEM in the first part of the layer. The growth of the kink from 4 to 10 h of annealing assesses that the clusters are able to capture the mobile C atoms moving towards the surface. The C loss saturation observed after 4 hours demonstrated that at this stage of the process the cluster density is sufficiently high to trap all C atoms diffusing from the layer.

It is quite reasonable that the clustering and diffusion processes will change by changing the C concentration. Results concerning the 0.8at% C samples confirm this guess. The clustering probability is higher with respect to the low C concentration case as demonstrated by the cross sectional TEM image of the 2h O_2 annealed sample in Fig.5. As can be seen, small clusters just below the cap/layer interface appear. On the contrary, no evidence of nucleation of such precipitates is visible in the equivalent samples with 0.35at% C (not shown). The increase of the clustering probability causes a reduction of the diffusion effects. As a matter of fact, the total amount of C lost by the layer is lower at higher concentration and the saturation of the loss process is reached in a shorter time (see Fig.4). Furthermore, the redistribution of C inside the layer is also reduced being the evolution of an accumulation kink not present in the SIMS analyses (not shown) as in the case of low C content. In other words the clusters efficiency in trapping the mobile C increases with the concentration or equivalently the mean free-path of mobile C before being trapped decreases.

CONCLUSIONS

In this work the evolution of structural properties of SiGeC layers buried under a Si cap under oxidation was investigated. Clear evidences of both C diffusion and clustering are reported. It is experimentally demonstrated that the oxidation-induced I flux strongly enhances both

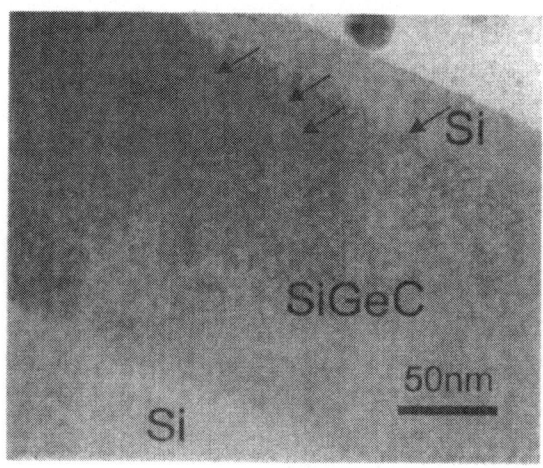

Figure 5 *Cross sectional TEM image of the 0.8at% C sample annealed for 2 hours in oxygen atmosphere. The arrows indicate the presence of small clusters just bellow the cap layer interface. Such clusters are not revealed in the equivalent sample with 0.35at% C composition.*

phenomena with respect to the annealing in inert atmosphere. A strong competition between the diffusion and clustering processes is observed. In the early stages of oxidation out-diffusion from the cap/layer dominates, whereas, when clustering begins to take place the C loss is progressively suppressed. Clusters are demonstrated to be full efficient in trapping the mobile C after an initial transient of C out diffusion. The transient duration and the total amount of C loss depend on the C concentration. Indeed, we have demonstrated that the higher is the C concentration the higher is the clustering probability and efficiency in trapping mobile C.

ACKNOWLEDGEMENTS

The authors wish to thank C. Spinella and S. Pannitteri for TEM analyses. This work was partially supported by SIGENET, the EC program for improving Human Potential-Research Training Network, Contract No. HPRN-CT-2000-00123.

REFERENCES

1.   K. Eberl, K. Brunner and O. G. Schmidt, *Germanium Silicon, Physics and Materials*, edited by R. Hull and J. C. Bean, Vol. **56** of Semiconductors and Semimetals (Academic, San Diego, 1999).
2.   D. De Salvador, M. Petrovich, M. Berti, F. Romanato, E. Napolitani, A. V. Drigo, J. Stangl, S. Zerlauth, M. Mühlberger, F. Schäffler, G. Bauer and P. C. Kelires, Phys. Rev. B **61**, 13005 (2000).
3.   H. Rücker, B. Heinemann, W. Röpke, R. Kurps, D. Krüger, G. Lippert and H. J. Osten, Appl. Phys. Lett. **73**, 1682 (1998).
4.   P. A. Stolk, D. J. Eaglesham , H. - J. Gossmann, and J. M. Poate, Appl. Phys. Lett. **66**, 1370 (1995).
5.   M. S. Carrol, C.L. Chang, J. C. Sturm and T. Büyüklimanli, Appl. Phys. Lett. **73**, 3695 (1998).
6.   U. Gösele, P. Laveant, R. Scholz, N. Engler and P. Werner, Mat. Res. Soc: Symp. Proc. **610**, B7.1.1 (2000).
7.   M. S. Carrol, J. C. Sturm, D. De Salvador, E. Napolitani, M. Berti, J. Stangl and G. Bauer, presented at the 2001 MRS Spring Meeting, San Francisco, CA, 2001 (unpublished).
8.   J. W. Strane, H. J. Stein, S. R. Lee, S. T. Picraux, J. K. Watanabe and J. W. Mayer, J. Appl. Phys. **76**, 3656 (1994); A. R. Powell, F. K. LeGoues, and S. S. Iyer, Appl. Phys. Lett. **64**, 324 (1994); G. G. Fischer, P. Zaumseil, E. Bugiel, and H. J. Osten, J. Appl. Phys. **77**, 1934 (1995); L.V. Kulik, D.A. Hits, M.W. Dashiell, J. Kolodzey, Appl. Phys. Lett. **72**, 1972 (1998).
9.   M.Berti, D. De Salvador, A. V. Drigo, F. Romanato, A. Sambo, S. Zerlauth, J. Stangl, F. Schäffler and G. Bauer, Nucl. Instr. and Meth. **B 143**, 357 (1998).
10.   M. S. Carrol and J. C. Sturm, Mat. Res. Soc. Symp. Proc. **610**, B4.10.1 (2000).

Mat. Res. Soc. Symp. Proc. Vol. 669 © 2001 Materials Research Society

Boron Segregation and Electrical Properties in Polycrystalline SiGeC

E. J. Stewart, M. S. Carroll*, and J.C. Sturm
Center for Photonics and Optoelectronic Materials, Department of Electrical Engineering
Princeton University, Princeton NJ
* present address: Agere, Murray Hill, NJ

ABSTRACT

Previously, it has been reported that PMOS capacitors with heavily boron-doped polycrystalline SiGeC gates are less susceptible to boron penetration than those with poly Si gates [1]. Boron appears to accumulate in the poly SiGeC layers during anneals, reducing boron outdiffusion from the gate despite high boron levels in the poly SiGeC at the gate/oxide interface. In this abstract, we report clear evidence of strong boron segregation to polycrystalline SiGeC layers from poly Si, with boron concentration in poly SiGeC (Ge=25%, C=1.5%) increasing to four times that of adjacent poly Si layers. A separate experiment confirms that this result is not due to any SIMS artifacts. Electrical measurements of heavily in-situ doped single layer samples show that the conductivity of poly SiGeC is similar to poly Si and remains roughly constant with annealing at 800^0C. However, in a two-layer sample where the poly SiGeC is initially lightly doped and subsequently heavily doped by diffusion by from an adjacent poly Si layer, conductivity appears lower than in poly Si.

INTRODUCTION

Heavily boron-doped polysilicon is typically used as the gate material for modern p-channel MOSFETs. Doping the polysilicon gate is usually achieved by ion-implantation using B or BF_2^+, followed by an anneal to activate the dopant and drive it throughout the gate. For devices with very thin gate oxides, boron can actually diffuse through the gate oxide and into the substrate during this activation anneal. This boron penetration into the substrate changes the doping in the channel, causing undesirable positive threshold voltage shifts in the device [2].

Previously, it has been shown that, by placing a thin layer of polycrystalline SiGeC at the bottom of the polysilicon gate, boron penetration can be greatly suppressed for both PMOS capacitors and p-channel MOSFETs [1,3]. It is well known that boron diffusion in single-crystal SiGeC can be over an order of magnitude slower than in single-crystal Si [4]. However, the polycrystalline SiGeC gate layers do not appear to be acting as boron diffusion barriers (grain boundary diffusion may be enhancing diffusion in poly vs single crystal SiGeC). Instead, boron diffuses through and appears to accumulate in the poly SiGeC layers during the anneal, giving a high dopant level at the gate/oxide interface [1,3]. This tendency for boron to segregate to poly SiGeC layers may be responsible for the reduced boron penetration.

In this work, we present clear independent evidence of strong boron segregation to polycrystalline SiGeC layers. We also examine the electrical properties of boron in polycrystalline SiGeC layers subjected to annealing.

EXPERIMENT

Thermal oxides (~200 nm) were first grown on n-type substrates. Polycrystalline layers were then deposited by Rapid Thermal Chemical Vapor Deposition at ~575°C for poly SiGe and poly SiGeC and 700°C for poly Si. SiH_4, GeH_4, $SiCH_6$, and B_2H_6 were used as source gases for silicon, germanium, carbon, and boron, respectively. For the SiGe or SiGeC layers, Ge concentrations were ~25% and C concentrations were ~1.5%. Previous work shows that in single crystal SiGeC layers grown under similar conditions, most of the carbon is substitutional [5]. To study segregation, a two layer structure was grown, consisting of a ~100 nm lightly (in-situ) doped polycrystalline SiGeC layer underneath a 300 nm heavily (in-situ) doped poly Si layer. This sample was annealed at 800°C for long times to allow boron to move from the heavily doped poly Si layer to the lightly doped poly SiGeC layer. SIMS profiles, sheet resistance, and spreading resistance measurements were taken before and after the anneal to examine boron diffusion and electrical activity in the sample.

To study the electrical properties of isolated poly SiGeC layers subjected to annealing, single layers of heavily in-situ doped ($[B] \sim 1 \times 10^{20}$ cm^{-3}) poly Si (~300 nm), SiGe (~100 nm), and SiGeC(~100nm) were grown and annealed under the same conditions as the two-layer sample. Along with SIMS and sheet resistance measurements, Hall measurements were also taken for these samples.

BORON SEGREGATION TO POLYCRYSTALLINE SIGEC

Boron segregation to poly SiGeC is clearly demonstrated in the two-layer structure. SIMS profiles of this sample are plotted in figure 1. Before the anneal (as-grown sample), the poly Si layer is heavily doped at ~ 4×10^{19} cm^{-3}, while the poly SiGeC layer is lightly doped at ~ 2×10^{18} cm^{-3}. If there were no segregation between the layers, this profile would be expected to flatten out during the anneal. However, after an 800°C,

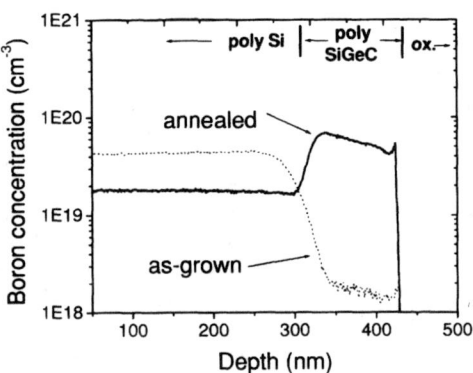

Figure 1: SIMS profiles of boron concentration in two-layer structure before and after annealing at 800°C for 44 hours in N_2.

44 hour anneal in N_2, boron levels have risen in the poly SiGeC layer to ~4-6x10^{19} cm^{-3}, while boron concentration has decreased in the poly Si to ~ 1.5x10^{19} cm^{-3}, demonstrating strong segregation to the poly SiGeC. Defining m as the ratio of boron concentration in the poly SiGeC to the poly Si at the interface, we find m ~ 4 for these parameters (Ge = 25%, C=1.5%) and annealing conditions. A separate sample, consisting of several layers of poly SiGeC (Ge ~ 20%, C levels varied from 0% to 1%) sandwiched between poly Si layers, revealed that boron segregation increases steadily with C concentration [3]. This sample, also annealed in N_2 at 800^{0}C, gave a segregation coefficient m~4.3 for a 1% carbon level, roughly consistent with these results.

   A separate experiment was carried out to confirm that the observed segregation is not attributable to a SIMS artifact associated with measuring boron in SiGeC layers with high C levels. Three samples were used for this experiment. The first was just a lightly doped (~1x10^{15} cm^{-3}) n-type substrate. The second sample consisted of an n-type substrate, on top of which was grown a ~80 nm SiGe layer, followed by a ~50 nm Si cap (all undoped). The third sample was similar to the second, except a SiGeC layer was used in place of the SiGe. Ge and C levels were 20% and 1%, respectively. All 3 samples were then ion-implanted with boron at several doses and energies to create a roughly flat boron profile in each sample from a depth of about 50 nm to 300 nm. The implant conditions were identical for the 3 samples. No anneals were performed. SIMS measurements of boron were then taken for all 3 samples, expecting to see similar boron profiles in each (perhaps slightly different profiles are expected in the samples with SiGe and SiGeC layers vs. the all Si sample, due to a different stopping coefficient for boron in SiGe vs. Si). If SIMS measurements exaggerate boron levels in the SiGeC layers, then an anomalous spike in the measured SIMS profile might be expected in this sample. No such spike is observed, however. Figure 2 shows the SIMS plots of boron concentration vs. depth for the 3 samples. All samples have almost indistinguishable profiles, with boron levels rising quickly to ~ 2x10^{19} cm^{-3} at 50 nm, leveling off, and then dropping at ~300 nm. In particular, no significant difference is seen between the samples with a SiGe

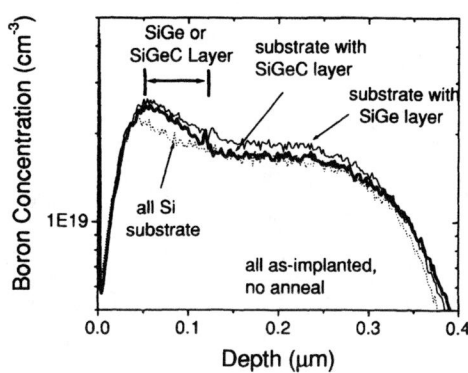

Figure 2: SIMS profiles of boron concentration for implanted samples.

layer vs. a SiGeC layer, indicating that SIMS measurements are not exaggerating boron levels in SiGe layers with high C concentrations. This confirms that the segregation measured by SIMS on the two-layer poly sample is not a measurement artifact.

The mechanism driving the segregation is not known at this point. Weak boron segregation to strained single crystal SiGe layers vs. Si (m~1.3) has been previously reported [6]. This has been explained by smaller boron atoms relieving strain in the strained SiGe, and by the smaller bandgap of SiGe vs. Si [7]. However, SiGeC has less strain than SiGe, and a bandgap closer to that of Si, both of which predict less segregation to SiGeC vs. SiGe. This is opposite of what is observed. One possibility is the formation of B-C related defects, which is discussed in the next section.

ELECTRICAL PROPERTIES

Electrical measurements indicate that boron incorporated by in-situ doping in polycrystalline SiGeC remains electrically active when subjected to furnace anneals. Figure 3 shows plots of resistivity, Hall mobility, and Hall carrier concentration vs. furnace anneal time at 800^{0}C for the single layer in-situ doped samples. The as-grown

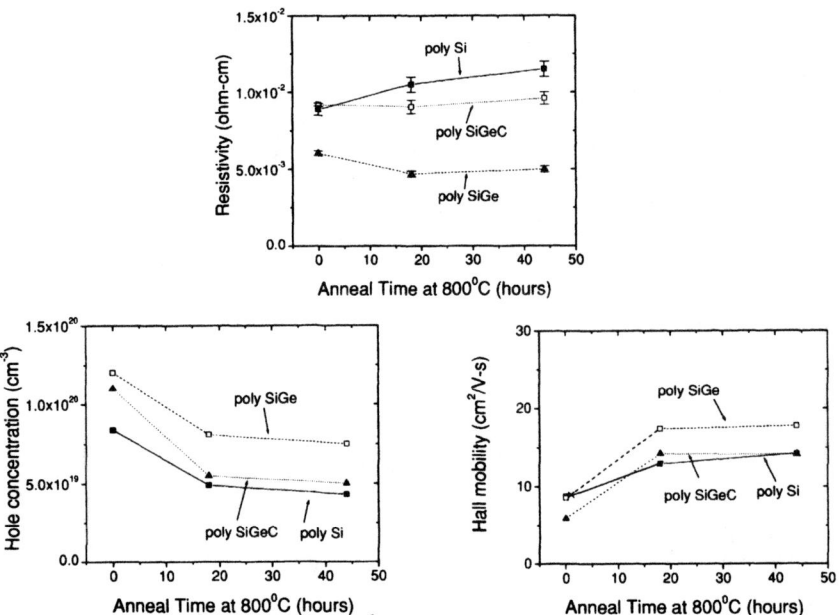

Figure 3: Resistivity, hall mobility, and hall concentration vs. anneal time at 800^{0}C for single in-situ doped layers ([B]~10^{20} atoms/cm^{3}) of poly Si, SiGe, and SiGeC.

poly SiGeC sample has a resistivity equal to the poly Si sample, both being higher than the poly SiGe sample. As all samples are annealed, the resistivity of the poly SiGeC sample remains roughly constant and similar to the poly Si sample. This indicates that large amounts of boron are not becoming inactive due to the formation of some inactive carbon-related defect. This is not due to the lack of enough carbon atoms – even with $\sim 10^{20}$ cm^{-3} boron doping, there are still ~ 7 carbon atoms for every boron atom.

Hall measurements show that the carrier concentrations in all 3 samples decrease quickly and level off. This is qualitatively to be expected, since all samples are doped with boron above solid solubility limits at this temperature ($\sim 6 \times 10^{19}$ cm^{-3} for poly Si [8]) and therefore should see a loss in substitutional boron with annealing. Hall mobilities increase with annealing however, offsetting the loss in conductivity caused by the reduced carrier concentration. Increased mobility in all samples may be due to an increase in grain size during the anneal.

Polycrystalline SiGeC appears less conductive than poly Si, however, when boron diffuses and segregates into the layer rather than being incorporated in-situ during growth. Figure 4 shows sheet resistance and spreading resistance measurements of the two-layer sample used above to demonstrate segregation. Initially, as shown in figure 1, all of the dopant is in the poly Si layer, and thus the resistivity of the sample is dominated by the poly Si layer. As boron moves into the poly SiGeC layer during the anneal (also figure 1), the sheet resistance of the sample becomes more and more determined by the electrical properties of the poly SiGeC layer. As shown in figure 4, the sheet resistance of the sample goes up for longer anneal times, and, therefore, with more boron in the poly SiGeC layer. This suggests that the boron in the poly SiGeC is less conductive than in the poly Si. Spreading resistance plots of this sample confirm this result. After the anneal, even though the doping level in the poly SiGeC is at least 3 times as higher than in the poly Si, the resistivity is only about equal, indicating $\sim 3X$ higher resistance for a given doping level. Whether this is due to a lower hole mobility or a lower level of boron activation in the poly SiGeC in not known.

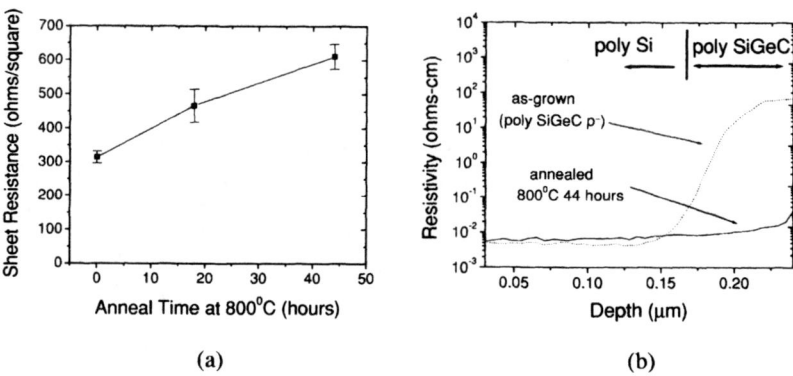

Figure 4: (a) Sheet resistance and (b) spreading resistance vs. anneal time at 800°C for the two layer poly Si/poly SiGeC sample.

This higher resistance in the poly SiGeC is inconsistent with the single layer results (figure 3), which showed that poly Si and poly SiGeC have similar resistivities after being subjected to the same annealing procedure. There are several possibilities for this difference. First, some boron may be associated with carbon-related defects in both samples (and hence be electrically inactive), but this is not detected in the heavily doped single-layer poly SiGeC sample because only a fraction of the boron is conductive anyway (total boron concentration in the single-layer sample was $\sim 10^{20}$ cm^{-3}, electrically active boron was only $\sim 5 \times 10^{19}$ cm^{-3}, see figure 3). Second, the strain in the poly Si or poly SiGeC in the two-layer structure may be different than in the single layer samples, particularly if there was a columnar grain structure and the epitaxy within a grain were pseudomorphic. The large difference in strain could cause a difference in conductivities of the two samples. Third, in the multilayer case, the boron presumably entered the layer via grain boundary diffusion, instead of being grown in-situ, as with the single layer case. Experiments to elucidate the implications of the difference between the single and multilayer samples are in progress.

CONCLUSIONS

Boron-doped poly SiGeC is of great interest as a gate material for p-channel MOSFETs. Boron segregates to polycrystalline SiGeC layers with high levels of carbon. For Ge=25% and C=1.5%, the concentration ratio is ~4 between poly SiGeC and poly Si after annealing for 44 hours at 800^{0}C in N$_2$. Heavily in-situ doped poly SiGeC does not lose conductivity during long anneals at 800^{0}C, indicating that boron is not being deactivated by forming C-related defects. These layers have a resisitivity that closely tracks that of similarly doped poly Si layers. However, when boron is diffused into an initially lightly doped poly SiGeC layer, the resulting conductivity of the poly SiGeC appears less than when doped in-situ. The cause of this discrepancy is under further investigation.

ACKNOWLEDGEMENTS

This work was supported by DARPA and ARO.

REFERENCES

1. C.L.Chang and J.C. Sturm, *Applied Physics Letters* **74,** 2501 (1999).
2. J.R. Pfiester, F.K. Baker, T.C. Mele, H. Tseng, P.J. Tobin, J.D. Hayden, J.W. Miller, C.D. Gunderson, and L.C. Parrillo, *IEEE Trans. Electron Devices* **ED-37**, 1842 (1990).
3. E. J. Stewart, M S. Carroll, and J.C. Sturm, ECS 199th meeting (March 2001).
4. M.S. Carroll, L.D. Lanzerotti, and J.C. Sturm, *MRS Symp. Proc.* **527**, 417 (1998).
5. C.W.Liu, A. St. Amour, J.C. Sturm, Y.R.J. Lacroix, M.L.W. Thewalt, C.W. Magee, and D. Eaglesham, *Journal of Applied Physics* **80**, 3043 (1996).
6. S.M. Hu, D.C Ahlgren, P.A. Ronsheim, and J.O. Chu, *Physical Review Letters* **67**, 1450 (1991).
7. S.M. Hu, *Physical Review Letters,* **63**, 2492 (1989).
8. K. Suzuki, N. Miyata, and K. Kawamura, *Jpn. J. of Appl. Phys.* **34**, 1748 (1998).

Mat. Res. Soc. Symp. Proc. Vol. 669 © 2001 Materials Research Society

A Comprehensive Model for Carbon Suppression of Boron Transient Enhanced Diffusion

Julie L. Ngau, Peter B. Griffin, and James D. Plummer
Center for Integrated Systems, Stanford University,
Stanford, CA 94305, U.S.A.

ABSTRACT

In this work, the time evolution of B transient enhanced diffusion (TED) suppression due to the incorporation of 0.018% substitutional carbon in silicon was studied. The combination of having low C concentrations, which reduce B TED without completely eliminating it, and having diffused B profiles for several times at a single temperature provides much data upon which various models for the suppression of B TED can be tested. Recent work in the literature has indicated that the suppression of B TED in C-rich Si is caused by non-equilibrium Si point defect concentrations, specifically the undersaturation of Si self-interstitials, that result from the coupled out-diffusion of carbon interstitials via the kick-out and Frank-Turnbull reactions. Attempts to model our data with these two reactions revealed that the time evolved diffusion behavior of B was not accurately simulated and that an additional reaction that further reduces the Si self-interstitial concentration was necessary. In this work, we incorporate a carbon interstitial, carbon substitutional (C_iC_s) pairing mechanism into a comprehensive model that includes the C kick-out reaction, C Frank-Turnbull reaction, {311} defects, and boron interstitial clusters (BICs) and demonstrate that this model successfully simulates C suppression of B TED at 750 °C for anneal times ranging from 10 s to 60 min.

INTRODUCTION

Recently, research has shown that the incorporation of substitutional carbon in silicon suppresses boron transient enhanced diffusion (TED). Out-diffusion experiments of highly supersaturated C have indicated that the depletion of Si self-interstitials (I) in C-rich Si, which in turn leads to the reduction of B TED, is caused by the coupled out-diffusion of carbon interstitials via the kick-out [1] and Frank-Turnbull reactions [2].

We have previously investigated the suppression of boron TED through localized, substitutional incorporation of C into $Si_{1-x}Ge_x$ for various times at 750 °C [3]. This work examines the phenomena of suppressed B TED for C-doped Si and eliminates any strain or chemical interactions between Ge and C. Simulating the time evolution of the B diffusion profiles has provided insight into the possible mechanisms through which C, B, and I interact to produce the observed experimental results. We have found that the kick-out and Frank-Turnbull reactions for C are insufficient for modeling the B and C profiles for all anneal times investigated. Therefore, we propose that an additional reaction that further reduces the Si I concentration is necessary.

EXPERIMENTAL PROCEDURE

The test structure used in these experiments consisted of 2 X 10^{18} cm^{-3} boron marker layers positioned above and below a pure Si or C-doped Si layer with [C]= 0.018% as illustrated in Fig. 1. The samples were grown by atmospheric pressure chemical vapor deposition at 700 °C for the carbon containing layer and 750 °C for boron and intrinsic silicon layers. Ethylene and

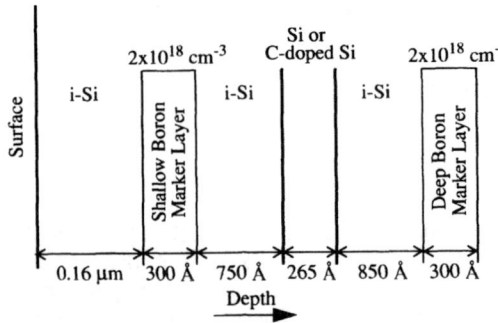

Figure 1. Schematic illustration of test structure. A pure Si or C-doped Si layer with [C]=0.018% lies in between a shallow B marker layer and a deep B marker layer.

dichlorosilane were used as the carbon and silicon sources, respectively.

In order to create interstitial damage, the samples were implanted with 40 keV Si ions to a dose of 2.5×10^{13} cm^{-2}. Inert nitrogen anneals of both as-grown samples and implanted samples were then done at 750 °C for 10 s and 4.25 min in a rapid thermal annealer and for 60 min in a furnace. The as-grown samples were annealed concurrently with the implanted samples to ensure that the difference in the resulting B profiles could clearly be attributed to the implant damage. Boron and carbon concentration profiles were measured using secondary ion mass spectroscopy (SIMS) with O_2 and Cs ions, respectively.

RESULTS

In this section, we make some qualitative and quantitative observations about how the presence of a C layer between shallow and deep B layers affects the TED of the B. As expected, the SIMS profiles of the 750 °C annealed, implanted pure Si control samples indicated that the shallow and deep marker layers undergo approximately the same time averaged B diffusivity enhancement. In contrast, analysis of the B SIMS profiles for the C-doped Si sample revealed that the presence of C caused the deep peak to experience ~2.7 times less diffusivity enhancement than the shallow peak. Fig. 2 gives the time averaged B diffusivity enhancements for the shallow and deep B peaks of the pure Si and the C-doped samples. Compared to the diffusivity enhancements

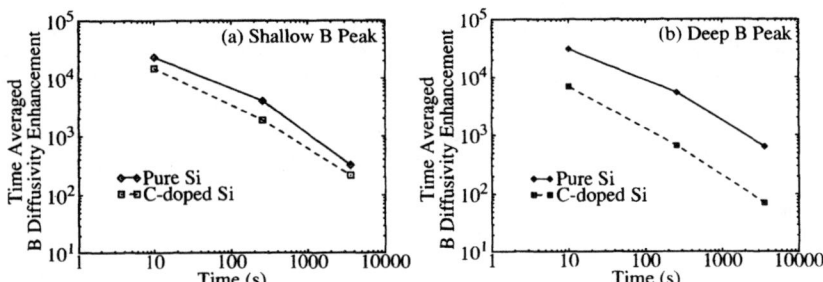

Figure 2. Time averaged B diffusivity enhancements of the (a) shallow B marker layers and (b) deep B marker layers for the pure Si and C-doped Si samples.

of the pure Si sample, the deep B peak enhancement of the C-doped sample was reduced by a factor of ~4.5 at 10 s and ~8.2-8.9 at longer times. The shallow B peak of the C-rich sample also experienced less TED, by a factor of ~1.6 at 10 s and ~1.5-2.2 at longer times, than that of the Si sample. The fact that the B in front of the C-doped layer experiences suppressed TED indicates that C need not physically block the path of interstitials to B to reduce TED.

MODELS AND DISCUSSION

The first step in modeling the diffusion data obtained in these experiments was to simulate the boron interstitial clustering (BIC) and the TED evident in the SIMS data of the pure Si sample. The BIC model from TSUPREM-4 [4] was implemented in *Avant!*'s Taurus Process & Device simulation software using the physical model and equation interface (PMEI) feature [5] for the B_3I cluster, proposed in the literature as one of the more dominant large BICs [6]. BICs, as well as {311} defects resulting from the ion implantation step, were then used to fit the implanted and annealed B SIMS data of the control Si sample. Good fits of the SIMS data for the pure Si samples were obtained using the BIC and {311} parameters from Ref. [3]. These parameters used to fit the pure Si data were then applied to the data from the C-rich samples.

The next step in modeling the diffusion data of this work was to incorporate the coupled diffusion of carbon and silicon point defects into the simulations [1][2][7]. Diffusion of C in Si occurs through a substitutional-interstitial mechanism [8][9]. Mobile interstitial carbon (C_i) are formed via the kick-out reaction (KO) between immobile substitutional carbon atoms (C_s) and I:

$$C_s + I \Leftrightarrow C_i. \tag{1}$$

The C_i can also interact with vacancies (V) through the Frank-Turnbull reaction (FT):

$$C_i + V \Leftrightarrow C_s. \tag{2}$$

These reactions reveal that the flux of C_i away from regions of high carbon concentration must be balanced by an opposite flux of I into this region as well as an accompanying flux of V away from the C-rich area. However, the flux of C_i is limited by the diffusion of the Si point defects and leads to an undersaturation of I and a supersaturation of V in the C-rich area. The diffusion of C thus results in the depletion of I, which would otherwise contribute to TED of B. In addition, the supersaturation of V due to C diffusion leads to enhanced diffusivity of dopants such as As and Sb which diffuse primarily via a V mechanism [7].

Equations (1) and (2) were implemented along with BICs and {311} defects using Taurus's PMEI module. The transport parameters used here to simulate the coupled diffusion of C are the same as those listed in Ref. [3] and are all consistent with values from the literature. The B and C diffusion fits that were obtained for the implanted and annealed C-doped Si samples are shown in Fig. 3. Note that the surface peak of C in the SIMS data is due to the knock-on of C atoms from hydrocarbon contamination on the sample surfaces and is therefore not included in the initial C profile used in these simulations. Instead, the relatively flat concentration level of C from ~0.1 μm to the beginning of the C-doped layer is extended towards the surface.

In Fig. 3, which uses only the KO and FT reactions, the fit of the diffused B profile at 10 s is good; however, the fits at 4.25 and 60 min show too much diffusion. The interstitial supersaturation in the regions outside of the carbon layer is too large to suppress the TED of B appropriately at longer times. If the forward reaction rate of the KO reaction is increased by many orders of magnitude so that more I are consumed by mobile carbon, or conversely the reverse reaction rate is reduced so that the release of free I is decreased, better fits of the B diffusion profiles can be obtained at either 4.25 or 60 min. Since a single set of parameters cannot fit the diffused B profiles for all times in these experiments, this study reveals that the KO and FT reactions for C are insufficient for the modeling of the time evolution of the suppression of B TED at 750 °C. The reactions can be used to match the B profiles only for a single time and temperature point. To the best of our knowledge, there has been no previous work in which a time evolution study of C suppression of B TED has been done at a given temperature. In addition, we were able to obtain these physical insights because the C concentration in our experiments was not large enough to eliminate the TED of B completely.

The inability of the KO and FT reactions for C to model the suppression of B TED for all times suggests that an additional reaction needs to be taken into account. The fact that increasing the forward reaction rate or decreasing the reverse reaction rate of the KO reaction improves the fits of the B profiles at longer times indicates that the I concentration must be reduced. Taking a carbon-related trap (T) for I, such as $I + T \Leftrightarrow IT$, into consideration is one way of achieving this effect. Another approach is the implementation of carbon clusters. One clustering reaction that would effectively reduce the Si self-interstitial concentration is the pairing of a C_i and a C_s to form a stable, immobile carbon complex [10][11][12]:

$$C_i + C_s \Leftrightarrow C_i C_s. \tag{3}$$

Reaction (3) ties up C_i in a $C_i C_s$ complex thereby reducing the reverse C kick-out reaction and the subsequent release of free Si self-interstitials. The $C_i C_s$ pairing reaction has also been implemented in simulations in the literature in order to study the effect these carbon complexes may have on Si self-interstitial diffusion and the role they may play in reconciling the discrepancies in the various published values for I diffusivity [10][13].

Fig. 4 shows the fits of the boron and carbon diffusion profiles for the implanted and annealed C-doped samples which were obtained by including the $C_i C_s$ pairing reaction along with BICs, {311} defects, the C kick-out reaction, and the C Frank-Turnbull reaction. The forward and reverse reaction rates for Eq. (3) were used as fitting parameters and were found to give good results when they were equal to 2×10^{-14} and 1×10^{-5}, respectively, at 750 °C. The simulated profiles for B are quite good overall for all times 10 s through 60 min.

As discussed above, a feature of the coupled diffusion of C is that a vacancy supersaturation, in addition to an interstitial undersaturation, is induced in the C-region [7]. In order to test the effect of the $C_i C_s$ pairing reaction upon the concentration of V, the simulation example of Ref. [7] was run with the $C_i C_s$ reaction included in the model. It was found that a V supersaturation still resulted, thus indicating that the $C_i C_s$ reaction is consistent with previous observations [2][7].

It should be noted that all of the findings for B TED suppression in C-rich silicon discussed above are consistent with those found previously for $Si_{1-x-y}Ge_x C_y$ in Ref. [3].

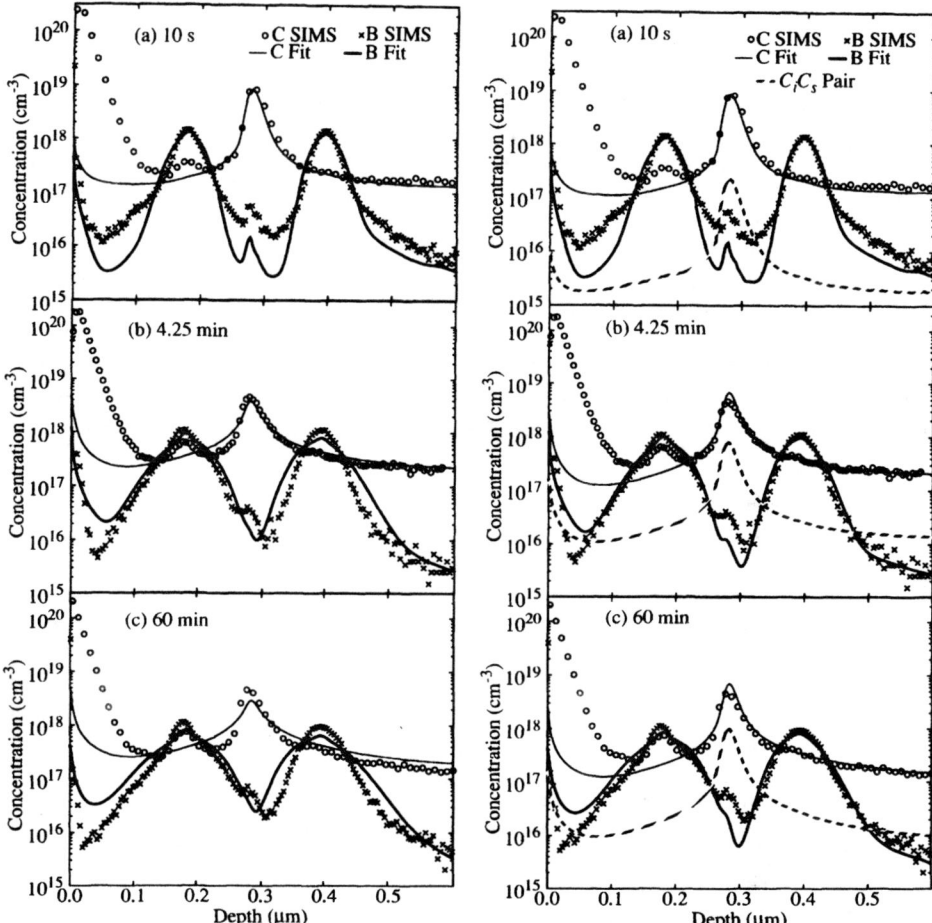

Figure 3. B and C profiles simulated using BICs, {311} defects, C kick-out, and C Frank-Turnbull reactions in C-doped Si samples for (a) 10 s, (b) 4.25 min, and (c) 60 min, 750 °C inert anneals.

Figure 4. B and C profiles simulated using BICs, {311} defects, C kick-out, C Frank-Turnbull, and C_iC_s pairing reactions in C-doped Si samples for (a) 10 s, (b) 4.25 min, and (c) 60 min, 750 °C inert anneals.

CONCLUSIONS

In these experiments, B marker layers exhibited suppressed TED in the presence of C-rich Si. Attempts to fit the resulting B profiles by including BICs, {311} defects, and the coupled diffusion of C via the kick-out and Frank-Turnbull reactions revealed that this model predicted too much B diffusion as the anneals progressed in time. This indicated that an additional reaction that further reduces the Si self-interstitial concentration, and thus preventing them from contributing to B diffusion, was necessary. We achieved this effect by implementing a pairing reaction between interstitial carbon and substitutional carbon to form a stable, immobile C_iC_s complex. The C_iC_s complex effectively ties up interstitial carbon causing the reverse C kick-out reaction, and therefore the Si self-interstitial concentration, to be reduced and permitting the correct amount of B TED suppression to occur over time. Excellent fits of the B diffusion profiles for all anneal times examined in this experiment were obtained upon using the C_iC_s pairing mechanism in conjunction with the C kick-out and Frank-Turnbull reactions. Furthermore, this pairing reaction was demonstrated to be consistent with reported observations of a vacancy supersaturation in C-rich regions of Si.

ACKNOWLEDGEMENTS

The authors are grateful to Dr. H. Rücker for supplying his computer script for the coupled diffusion of carbon as a starting point for the modeling of this work. We also offer many thanks to Dr. M. D. Johnson for his invaluable help with the implementation of Taurus's PMEI module.

REFERENCES

[1] R. Scholz, U. Gösele, J. -Y. Huh, and T. Y. Tan, Appl. Phys. Lett. **72**, 200 (1998).
[2] R. F. Scholz, P. Werner, U. Gösele, and T. Y. Tan, Appl. Phys. Lett. **74**, 392 (1999).
[3] J. L. Ngau, P. B. Griffin, and J. D. Plummer, submitted to J. Appl. Phys.
[4] TSUPREM-4 computer code from *Avant!* Corporation, Fremont, CA.
[5] Taurus computer code from *Avant!* Corporation, Fremont, CA.
[6] L. Pelaz, M. Jaraiz, G. H. Gilmer, H. -J. Gossmann, C. S. Rafferty, D. J. Eaglesham, and J. M. Poate, Appl. Phys. Lett. **70**, 2285 (1997).
[7] H. Rücker, B. Heinemann, D. Bolze, D. Knoll, D. Krüger, R. Kurps, H. J. Osten, P. Schley, B. Tillack, and P. Zaumseil, in *International Electron Devices Meeting Technical Digest*, IEDM 1999, Washington, DC, USA, p. 345 (1999).
[8] U. Gösele in *Oxygen, Carbon, Hydrogen, and Nitrogen in Crystalline Silicon*, edited by J. C. Mikkelsen, Jr., S. J. Pearton, J. W. Corbett, and S. J. Pennycook, MRS Proceedings Vol. 59 (Materials Research Society, Pittsburgh, 1986), p. 419.
[9] F. Rollert, N. A. Stolwijk, and H. Mehrer, Mater. Sci. Forum **38-41**, 753 (1989).
[10] P. A. Stolk, H. -J. Gossmann, D. J. Eaglesham, and J. M. Poate, Mater. Sci. Eng. B **36**, 275 (1996).
[11] L. W. Song, X. D. Zhan, B. W. Benson, and G. D. Watkins, Phys. Rev. B **42**, 5765 (1990).
[12] G. Davies, K. T. Kun, and T. Reade, Phys. Rev. B **44**, 12146 (1991).
[13] M. D. Johnson, M. -J. Caturla, and T. Diaz de la Rubia, J. Appl. Phys. **84**, 1963 (1998).

Mat. Res. Soc. Symp. Proc. Vol. 669 © 2001 Materials Research Society

A Quantitative Model of the Electrical Activity of Metal Silicide Precipitates in Silicon Based on the Schottky Effect

Teh Y. Tan and Pavel S. Plekhanov
Department of Mechanical Engineering and Materials Science, Duke University
Durham, NC 27708-0300

ABSTRACT

A quantitative model of the electrical activity of metallic precipitates in Si is presented. An emphasis is placed on the properties of the Schottky junction at the precipitate-Si interface, as well as the carrier diffusion and drift in the Si space charge region. Carrier recombination rate is found to be primarily determined by the thermionic emission charge transport process across the Schottky junction rather than the surface recombination process. It is shown that the precipitates can have a very large minority carrier capture cross-section.

INTRODUCTION

Metal and metal silicide precipitates are abundant in multicrystalline Si used for manufacturing of low cost solar cells. In order to evaluate the effect of such impurities on solar cell performance, it is necessary to know the minority carrier capture cross-section of the impurity atoms as well as that of the precipitates. For most transition metals, the capture cross-section of individual atoms dissolved in Si has been measured [1,2], but much less is known regarding the precipitates. These precipitates mainly form at crystal imperfections (dislocations and grain boundaries), act as very active recombination centers to reduce minority carrier lifetimes in Si [3-8]. In electron beam induced current (EBIC) experiments it was found that the recombination activity of metallic precipitates is very high [3,5,7]. This can be explained by the presence of an electric charge on the precipitates due to Schottky effect [9,11]. Deep level transient spectroscopy (DLTS) study of metallic precipitates confirms this conjecture [12] and also indicates that precipitates form band-like states in the semiconductor bandgap [4,12]. Moreover, the recombination activity due to precipitates decreases with the increase of the generation rate [10]. There exists several theoretical calculations of the recombination activity due to precipitates [13,14] in the formation of EBIC contrast, but there was no attempt to predict the precipitate capture cross-section (CCS) of charges based on the precipitate size and materials properties. Also, the issue of the recombination mechanism was not addressed. Understanding the mechanism of recombination at the precipitates and obtaining the associated CCS values are necessary to evaluate the effect of precipitated metallic impurities on the minority carrier lifetime.

THEORY

A metal or silicide precipitate embedded in the Si bulk forms Schottky junction with Si. The precipitate is charged and surrounded by a Si space charge region. When the semiconductor is illuminated, generation of non-equilibrium carriers occurs, and their distribution is characterized by quasi Fermi levels, separate for electrons and holes. These quasi Fermi levels are flat in the absence of recombination at the precipitate, which implies existence of an impermeable barrier at the precipitate-Si interface and an absence of interface states. In the other extreme case, if carrier recombination at the precipitate-Si interface is infinitely fast, the equilibrium concentration of carriers is restored at the interface, and the quasi Fermi levels for electrons and holes merge (Fig 1a). Away from the interface, the positions of the quasi Fermi levels are determined by the carrier flux density from the Si bulk to the interface. This flux, in turn, is determined by the carrier diffusion and drift in the electric field, which is due to the charge on the precipitate and the space charge region in Si. In the absence of electric field, minority carrier concentration is much less than that of majority carriers everywhere in the semiconductor, and the supply of minority carriers is the limiting factor of recombination process at the precipitate. However, the electric field attracts minority carriers to the precipitate and repulses majority carriers. Therefore, the supply of majority carriers is decreased, whereas the supply of minority carriers is increased and may even exceed that of majority carriers. However, it cannot be qualitatively judged which type of carriers will be the limiting factor of the recombination process. Hence, no a priori assumption should be made in this regard.

Having reached the precipitate, carriers can recombine either in Si at the precipitate interface states or inside the precipitate. In the former case, the local carrier recombination rate is determined by the surface recombination velocity due to interface states, while in the latter case it is governed by the thermionic emission mechanism for charge transport across the Schottky junction. Regardless of the mechanism, the limited recombination rate will result in the quasi Fermi levels' not completely merging at the interface, but only approaching each other. In the Schottky barrier mechanism, electrons and holes flow to the precipitate interior. Electrons entering the precipitate have energies above the metal Fermi level, but rapidly attain Fermi-Dirac distribution. Similarly, holes entering the precipitate create empty states below the metal Fermi level, and these states are very rapidly populated by electrons. Thus, upon entering the precipitate, an electron-hole pair will almost immediately recombine. If there is no barrier for the charge carrier transport across the precipitate-Si interface, the recombination rate would be sufficiently large, and the band diagram of the system would look like the one in Fig. 1a. However, the transport of carriers across the precipitate-Si interface gives rise to the thermionic emission potential at the interface. For a Schottky diode, the transport of only majority carriers across the junction is important. In the case of recombination current, in steady state, the absolute values of minority and majority carrier currents are equal. Therefore, transport of both types of carriers, and the appropriate thermionic emission potentials are essential. The influence of these potentials on the band structure is shown in Fig. 1b. Thermionic emission potentials for electrons and holes (V_{tn} and V_{tp}) have opposite signs, and the ratio of their absolute values depends on the recombination current and effective masses of electrons and holes in the semiconductor. The existence of non-zero V_{tn} and V_{tp} alters the charge on the precipitate, and hence, changes the position of the Fermi level in metal with respect to the semiconductor bands. Carrier concentrations in Si at the surface of the precipitate are determined by the Schottky barrier height φ_b, which is an intrinsic property of the material, together with the thermionic emission potentials, which vary with the recombination current. The carrier concentrations in Si away from the precipitate surface are controlled by diffusion, drift, generation, and background recombination due to centers other than precipitates, e.g. dissolved metal atoms.

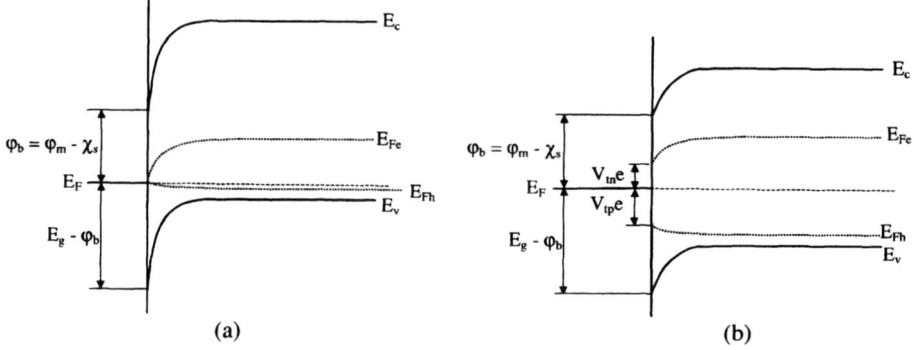

Fig. 1. Band structure of the precipitate with metallic properties in Si. (a) with infinite recombination rate at the interface; (b) with thermionic emission barrier for carrier transport at the interface.

To find out whether carrier recombination occurs primarily inside the precipitate or at the surface states, the interface recombination velocity of an ideal metal-semiconductor junction having no interface states was calculated based on the Schottky model. For p-type Si under the weak generation condition $n_o \ll \delta n \ll p_o$, we obtain that the surface recombination velocity $s = Am_e^* T^2 / (emN_c) = 5.0 \times 10^6$ cm/s at room temperature [15]. Here A is the Richardson constant, m_e^* is the effective mass of an electron in Si, m is electron mass, T is the absolute temperature, e is the elementary charge, and N_c is the density of states in the conduction band. Since the interface recombination velocity due to interface states is usually much less than 5.0×10^6 cm/s, it can be neglected. The effect of non-zero interface recombination velocity have also been considered.

RESULTS

Numerical modeling of recombination at precipitates was carried out using a variety of materials parameter values. Examples of electron and hole concentration profiles around the precipitate for p-type Si and FeSi precipitate are shown in Fig. 2. The CCS plots versus precipitate radius are shown in Figs. 3 and 4, along with the precipitate cross-section area and the total CCS of dissolved in Si metal atoms of the same amount as those in the precipitate. The precipitate CCS is roughly proportional to its diameter, as predicted by the diffusion-limited model [16]. In most cases, the precipitate CCS value lies between its geometric cross-section area and the total CCS of its constituent metal atoms dissolved in Si. Under certain conditions, the precipitate CCS can even exceed the latter value. This is more characteristic of smaller precipitates. A higher precipitate concentration leads to a larger CCS. The model also predicts that the CCS drops at higher generation rates, as observed in experiments [10]. The dependence of CCS on φ_b and background recombination constant is relatively weak. Under the conditions considered, the supply of minority carriers is the limiting factor. The supply of holes becomes influential only at lower p-doping levels and higher generation rates, for which the weak generation condition no longer holds.

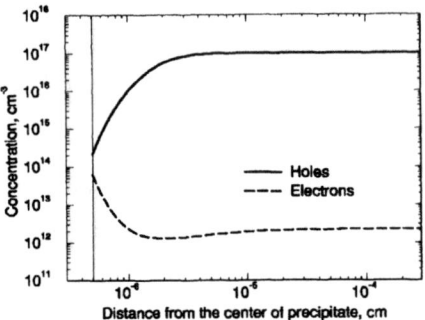

Fig. 2. Electron and hole concentration profiles around a precipitate. Parameters used: background recombination constant 10^5 s^{-1}, generation rate 10^{19} s^{-1} cm^{-3}, precipitate concentration 10^{10} cm^{-3}, precipitate diameter 10^{-6} cm, p-type doping level 10^{17} cm^{-3}, $\varphi_b = 0.68$ eV (FeSi)

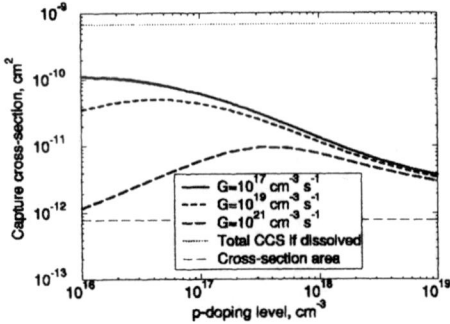

Fig. 3. Dependence of the precipitate capture cross-section on Si doping level at various generation rates. Other materials parameters used are the same as in Figure 2.

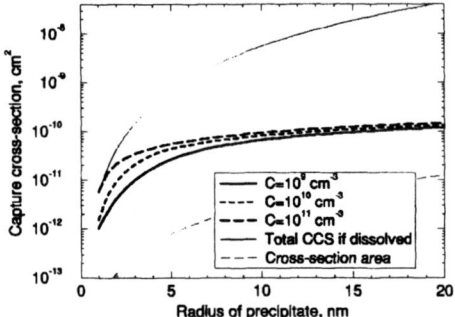

Fig. 4. Dependence of precipitate capture cross-section on its diameter at various precipitate concentrations. Other materials parameters used are the same as in Figure 2.

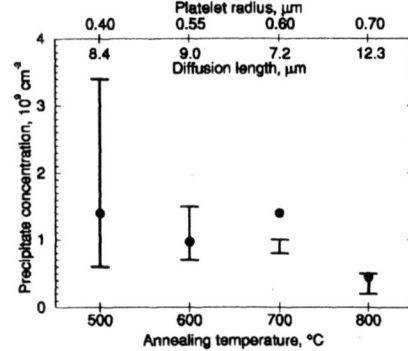

Fig. 5. Concentration of platelet-shaped NiSi$_2$ precipitates in Si, as measured by TEM and EBIC (bar) and calculated (circle). For each of 4 experimental samples, the minority carrier diffusion length and precipitate radius are shown.

In order to verify the model, available experimental data were fitted. The published measured parameters of platelet-shaped NiSi$_2$ precipitates in n-type Si were used [11]. For each of four samples examined, two values of precipitate concentration were reported, as measured by TEM and EBIC. The resulting ranges of values are shown in Fig. 5 as a bar, with TEM values being higher than respective EBIC values for each sample. Also published were precipitate dimensions and observed minority carrier diffusion length for each sample, which are also shown in Fig 5. Based on the experimental data and using the Schottky barrier model, the precipitate concentrations were calculated for each sample and compared to the experimentally measured values. The calculated concentrations are shown in Fig. 5 as black circles. The generation rate and the background recombination constant were not known and were used as fitting parameters, 10^{18} s^{-1} cm^{-3} and 10^5 s^{-1}, correspondingly. The same fitting values were used for all samples. Other parameters used are: n-type doping level 4×10^{14} cm^{-3}, $\varphi_b = 0.79$ eV (B-NiSi$_2$). The fit is good in three samples out of four. It can as well be considered satisfactory in the fourth case, since the experimental error span is not shown in the diagram, but is present and must be taken into account. The fit remains satisfactory even if the fitting parameter values are varied by up to one order of magnitude. It should be noted that the carrier generation in EBIC experiments is localized and dispersed in time, while in our modeling a uniform generation rate is assumed. Thus, the experimental generation rate value generally is not equal to the one used for fitting since the latter contains time and spatial averaging effects.

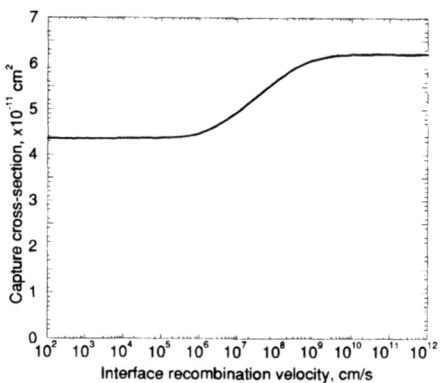

Fig. 6. Capture cross-section of a metallic precipitate as a function of the interface recombination velocity associated with interface states.

The dependence of a precipitate CCS on the interface recombination velocity is shown in Fig. 6. The curve has two horizontal sections and a transition region in between. The total recombination velocity at the precipitate-Si interface is the sum of the recombination velocity due to interface states s_i and effective recombination velocity due to recombination current through the interface s, $s_t = s_i + s$. If $s_i \ll s$, then s_i does not significantly affect the precipitate CCS, which corresponds to the left horizontal section of the curve in Fig. 6. If s_i is large enough, then the precipitate CCS becomes independent of the value of s_t and s_i, as is seen in the right horizontal section of the curve in Fig. 6. For very large s_t, the recombination becomes limited by carrier diffusion towards the precipitate. This condition has been used elsewhere [16] but without accounting for the effects of precipitate charge and Schottky current. Since in practice s_i is likely to be much less than s, and therefore, in most cases it can be assumed that $s_i = 0$.

CONCLUSIONS

A model of the carrier recombination process due to a precipitate with metallic properties in Si has been developed. Calculated results agree with experimental data. It was found that metal precipitates can serve as very efficient recombination centers with minority carrier CCS in great excess of their geometric cross-section area and, under certain conditions, as large as, or slightly larger than, the total CCS of precipitate's constituent metal atoms dissolved in Si. Under the weak generation conditions, the supply of minority carriers, via diffusion and drift, is the limiting factor of the recombination process.

ACKNOWLEDGMENT

This work is supported by US Department of Energy via National Renewable Energy Laboratory Subcontract No. XAF-7-17607-1. Computing facility support was provided by North Carolina Supercomputing Center.

REFERENCES

[1]   A. A. Istratov, H. Hieslmair, E. R. Weber, *Appl. Phys. A*, **69**, 13-44 (1999).
[2]   A. L. P. Rotondaro, T. Q. Hurd, A. Kaniava, J. Vanhellemont, E. Simoen, M. M. Heyns, C. Claeys, G. Brown, *J. Electrochem. Soc.*, **143**, 3014-3019 (1996).
[3]   M. Kittler, C. Ulhaq-Bouillet, V. Higgs, *J. Appl. Phys.*, **78**, 4573-83 (1995).
[4]   F. Riedel, J. Kronewitz, U. Gnauert, M. Seibt, W. Schroter, *Diffusion and Defect Data Part B (Solid State Phenomena)*, **47-48**, 359-64 (1996).
[5]   D. M. Lee, D. M. Maher, F. Shimura, G. A. Rozgonyi, in *"Semiconductor Silicon 1990"*, editors H. R. Huff and J. Chikawa (The Electrochemical Society, Pennington, NJ, 1990), p. 639-650.
[6]   J.-L. Maurice,C. Colliex, *Appl. Phys. Lett.*, **55**, 241-243 (1989).
[7]   M. Kittler, W. Seifert, Z. J. Radzimski, *Appl. Phys. Lett.*, **62**, 2513-15 (1993).
[8]   W. A. Brantley, O. G. Lorimor, P. D. Dapkus, S. E. Haszko, R. H. Saul, *J. Appl. Phys.*, **46**, 2629-2637 (1975).
[9]   H. C. Card, E. S. Yang, *IEEE Trans. Electron Devices*, **ED-24**, 397-402 (1977).
[10]   P. De Pauw, R. Mertens, R. Van Overstraeten, S. C. Jain, *Solid-State Electron.*, **27**, 573-87 (1984).
[11]   M. Kittler, J. Larz, W. Seifert, M. Seibt, W. Schroter, *Appl. Phys. Lett.*, **58**, 911-913 (1991).
[12]   A. A. Istratov, H. Hedemann, M. Seibt, O. F. Vyvenko, W. Schroter, T. Heiser, C. Flink, H. Hieslmair, E. R. Weber, *J. Electrochem. Soc.*, **145**, 3889-98 (1998).
[13]   C. Donolato, *Semicond. Sci. Technol.*, **7**, 37-43 (1992).
[14]   W. Seifert, M. Kittler, M. Seibt, A. Buczkowski, *Solid State Phenomena*, **47-48**, 365-370 (1996).
[15]   P. S. Plekhanov, T. Y. Tan, *Appl. Phys. Lett.*, **76**, 3777-3779 (2000).
[16]   P. S. Plekhanov, R. Gafiteanu, U. M. Gosele, T. Y. Tan, *J. Appl. Phys.*, **86**, 2453-2458 (1999).

Laser Annealing

Mat. Res. Soc. Symp. Proc. Vol. 669 © 2001 Materials Research Society

A Study of the Deactivation of High Concentration, Laser Annealed Dopant Profiles in Silicon

Yayoi Takamura, Sameer Jain, Peter B. Griffin, James D. Plummer
Center for Integrated Systems, Stanford University, Stanford, CA, 94305

ABSTRACT

As semiconductor device dimensions continue to decrease, the main challenge in the area of junction formation involves decreasing the junction depth while simultaneously increasing the active dopant concentration. Laser annealing is being investigated as an alternative to rapid thermal annealing (RTA) to repair the damage from ion implantation and to activate the dopants. With this technique, uniform, box-shaped profiles are obtained, with dopant concentrations that can exceed equilibrium solubility limits. Unfortunately, these super-saturated dopant concentrations exist in a metastable state and deactivate upon further thermal processing. In this work, a comprehensive study of the deactivation kinetics of common dopants (P, B, and Sb) was performed across a range of annealing conditions. For comparison, As deactivation data from the work of Rousseau et al.[1] is also presented. Each dopant exhibits different deactivation behavior, however, As and P can be classified as unstable species while B and Sb are stable against deactivation until higher temperatures of 700-800ºC. In addition, a means of maintaining these metastably doped layers is being investigated with the goal of meeting the International Technology Roadmap for Semiconductors (ITRS) requirements for ultrashallow junctions.

INTRODUCTION

The size of the Metal – Oxide – Semiconductor (MOS) transistor has been scaled aggressively as semiconductor manufacturers strive to create faster microprocessors with more functionality. In the area of front-end technology, this rapid scaling translates to major obstacles with the thin gate oxides and the formation of the highly doped junctions at the source and drain. In the latter category, the lateral abruptness of the dopant profiles needs to be increased with a simultaneous decrease in the junction depth and the sheet resistance of these doped regions. Current technology uses a two-step process of ion implantation followed by rapid thermal annealing (RTA) to introduce and electrically activate dopants. Electrical solubility of the dopants, which is on the order of $2\text{-}3\times10^{20}$ cm^{-3}, limits the active dopant concentrations. Once this limit is reached, a fundamental tradeoff exists between decreasing junction depth and increasing sheet resistance.

Laser annealing is being studied as a possible alternative to RTA. With laser annealing, a pulsed laser melts the near surface region of the silicon crystal for durations of about 70ns. The non-equilibrium nature of the liquid phase epitaxy provides for several advantages over conventional techniques. First, the dopant diffusivities in the liquid silicon are 10^8 times greater than that in solid silicon, allowing the dopants to diffuse throughout the molten region. Second, the liquid-to-solid regrowth front travels so fast that the dopant atoms are trapped on substitutional lattice positions and are electrically active. Finally, the segregation coefficient, the ratio of the dopant concentration in the solid to that in the liquid, approaches unity for dopants such as B, P, Sb and As. In the case of indium, however, the segregation coefficient remains well below one, resulting in the pileup of dopant at the wafer surface. For the other dopants, the results of these unique properties are uniform, abrupt, box-shaped dopant profiles with active dopant concentrations that can exceed the equilibrium solid solubility limit. One drawback of laser annealing is that these super-saturated dopant concentrations exist in a metastable state and deactivate upon subsequent thermal processing. Previously, Rousseau[1] studied the deactivation of As doped samples after laser annealing. In this work, we perform a comprehensive study of the deactivation kinetics of other common dopants (P, B, and Sb) across a range of annealing conditions.

Inactive dopants can take either the form of small clusters or precipitates, though the exact nature is often difficult to determine and is subject to debate. Clusters typically consist of a small number of dopant atoms and point defects, either interstitials (I) or vacancies (V). They are difficult to identify with analysis techniques such as Transmission Electron Microscopy (TEM) and Rutherford Backscattering Spectrometry (RBS) channeling analysis due to their small size and the fact that they only produce a small distortion of the host silicon lattice. Their formation can sometimes be detected due to the kick out of the opposite point defect than the one contained in the cluster. Precipitates, on the other hand, are a second phase of the dopant – Si system. They have their own crystal structure and are often a more macroscopic structure so that their presence is easier to detect through TEM images. Their formation requires the diffusion of many dopant atoms to their location, hence, they often are seen only after significant dopant diffusion. The predominant structure of the inactive dopant depends heavily on the point defect population within the crystal and on the annealing conditions.

EXPERIMENT

(100) Czochralski (CZ) silicon wafers were preamorphized with a 1×10^{15}cm^{-2} silicon implant at 55-keV. A second implant of the selected dopant was performed at 35-keV and 3.2×10^{16}cm^{-2} for P, 15-keV and 1×10^{16}cm^{-2} for B and 60-keV and 3.2×10^{15}cm^{-2} for Sb. For the As work[1], a single 1.6×10^{16}cm^{-2} implant of As was performed at 35-keV. The target dopant concentrations in the samples were an order of magnitude above the solid solubilities of the respective dopants[2,3,4,5]. Additional wafers were included containing a boron superlattice structure that was grown in an ASM Espilon II epitaxial reactor. The superlattice structure consists of three 2×10^{18}cm^{-3} boron spikes with FWHM ~30nm and a peak-to-peak separation of 180nm, capped with 300nm of undoped silicon. The laser annealing was performed at Verdant Technologies on a frequency doubled Nd:YAG laser operating at 532nm. A laser fluence of 0.9 J/cm^2 was used with ten shots at each location. The resulting melt depth is 180nm, which was sufficient to repair all of the damage from the previous implants. The wafers were diced into 5mm x 5mm squares and a low temperature oxide (LTO) was deposited at 300-350ºC. Inert thermal anneals were performed in a furnace or RTA at temperatures ranging from 500 to 900ºC and times between 15 seconds and 40 minutes. The oxide was etched off with dilute hydrofluoric acid (HF) and Hall Measurements were performed with the van der Pauw geometry to determine the electrical activation. Additionally, Secondary Ion Mass Spectroscopy (SIMS) was used to measure the chemical dopant profiles.

RESULTS

In order to compare the deactivation behavior for the different dopants, 40 minute isochronal furnace anneals were performed at temperatures between 500 and 900°C. Figure 1 plots the resulting active carrier dose as a function of temperature, normalized to the measured dose after laser annealing. The active carrier dose after ten minute anneals for As doped samples from the work done by Rousseau[1] is also included in the figure. From this plot, we can see that the number of active dopants for the Sb and B doped samples remain relatively stable against deactivation until 700-800°C, whereas the P and As doped samples deactivate substantially at temperatures as low as 500°C. For the P doped samples, nearly 50% deactivation was observed after the LTO deposition. At temperatures above 500°C, the P samples show a slow increase in active dose with temperature, known as the reverse annealing effect. While this effect was absent from Rousseau's work[1], Nobili et al.[6] saw the reverse annealing of As after anneals at higher temperatures. It is believed that this increase in active dose tracks the increase in electrical solubility with temperature. The B and Sb samples do not exhibit any reverse annealing, as the active dopant concentrations are still several times above equilibrium solubility values even after the 900°C anneal. This electrical data demonstrates that B and Sb can be classified as stable species and P and As are unstable species.

Isothermal RTA anneals for times between 15 seconds and five minutes were performed to study the time dependence of the deactivation process. Temperatures were chosen according to where deactivation was a maximum. Figure 2 shows the active carrier dose of the Sb and B samples after RTA anneals at 900°C and the As[1] and P samples after anneals at 600°C. The values have been normalized to that of the zero second time point. The P and As doped samples show a sharper initial decrease during the first 15 seconds of the anneal compared to the Sb and B doped samples, despite the lower anneal temperature. Not only do the As and P doped samples deactivate at lower temperatures, but the kinetics of the reaction are faster than in B and Sb doped samples.

The boron superlattice structures were used to detect the influx of point defects generated by the deactivation process. Because boron diffuses primarily through the interstitial mechanism, an enhancement in the measured B diffusivity can be correlated to a perturbation of the interstitial population above the equilibrium value by the same factor. The SIMS analysis showed that enhanced diffusion of the B superlattice structures occurred for all the dopants. The diffusivity enhancement factor was extracted

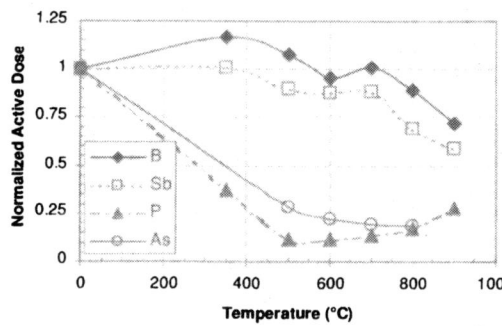

Figure 1: Normalized active dose is plotted against anneal temperature after 40 minute furnace anneals. Arsenic data[1] is from Rousseau after 10 minute anneals. P and As can be classified as unstable dopants while B and Sb are stable against deactivation.

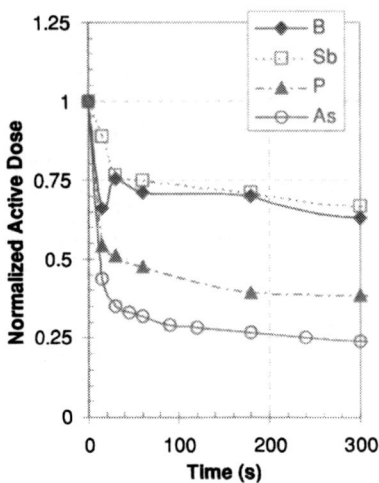

Figure 2: Normalized active dose is plotted against anneal time. B and Sb data is for 900°C anneals, while P and As data is at 600°C. As and P deactivate quickly within the first 15 seconds of the anneal, followed by a deactivation on a longer time scale.

using TSUPREM-4 simulations and the resulting values are listed in Table I. After 900°C anneals, B and Sb doped samples showed only slight diffusivity enhancements. Little additional diffusion occurred between five and forty minutes, indicating that all the interstitial injection takes place within the first five minutes of the anneal. Similarly, most of the deactivation happened during the early stages of the anneal. In the P doped samples, boron clustering was observed in the first two boron spikes annealed at 600 and 700°C. The diffused profiles were composed of a narrow, immobile Gaussian curve superimposed on a broadened, mobile Gaussian curve. Similar behavior was reported in ion-implanted samples[7] where excess interstitials from the implant damage formed immobile, inactive clusters with boron atoms. The enhancement factors listed in Table I were determined by fitting only the third peak that did not show any clustering. A two million times enhancement was measured after a five minute anneal at 600°C, indicating a huge number of interstitials are generated by the deactivation of phosphorus. This interstitial flux into the bulk is believed to be distinct from the emitter push effect, since no diffusion of the P profile is seen at this temperature. These B diffusivity enhancements measured for the P doped samples are on the same order of magnitude as those reported by Rousseau[1] for As deactivation. These large diffusivity enhancements for As and P exist despite only a two times lower active fraction compared to the B and Sb cases. This difference suggests that a different mechanism is responsible for the deactivation of Sb and B compared to P and As.

DISCUSSION

Several experimental and theoretical works have shown that As deactivates through the formation of small arsenic – vacancy clusters[1,5,8,9,10,11,12]. In this work, it has been shown that the deactivation behavior of phosphorus is in many ways similar to that of arsenic. Both species deactivate quickly at low temperatures releasing a large number of interstitials. Under these conditions, long-range diffusion of the dopant is not possible. Additionally, X-Ray standing wave analysis of As doped samples[1] and RBS channeling analysis of P doped samples (not shown), reveal that a majority of the inactive dopants remain on substitutional lattice positions. Based on these similarities, we believe that phosphorus initially deactivates through the formation of phosphorus – vacancy clusters,

Table I: Enhancement of Boron diffusivity due to deactivation

Dopant	Temp (°C)	Time (min)	Enhancement	Active fraction
B	900	5	25	62%
	900	40	4.25	60%
Sb	900	5	25	67%
	900	40	5	58 %
P	600	5	2×10^6	33%
	600	40	3×10^5	38%
As[1]	600	2400	1×10^5	41%
	750	4	7513	61%

leading to a kick out of interstitials. Upon annealing at high temperatures or for long times when diffusion of the dopant is possible, these clusters may evolve into precipitates.

Boron and antimony doped samples, however, show a different deactivation behavior. These samples have a much slower deactivation that occurs at much higher temperatures. A small interstitial generation is observed with a similar time transient as the deactivation reaction. In the case of boron, the deactivation also coincides with the observation of diffusion of the dopant profiles into the bulk. The dopant profile diffused 140nm within the first five minutes of the 900°C anneal and only diffused an additional 34nm within the next 35 minutes. Similarly, Table I reveals that 38% of the dopant deactivates within the first five minutes, followed by only an additional 2% within the next 35 minutes. This deactivation occurs despite the fact that the dopant that diffused into the bulk is below the solubility limit and electrically active. It is proposed that the diffusion enables the agglomeration of the boron atoms into silicon boride precipitates.

The electrical data shows that antimony deactivation follows much of the same behavior as the boron deactivation. Based on these results, it would appear that the inactive dopant takes the form of precipitates. This hypothesis is reinforced given the work by Nylandsted Larsen[13,14] where precipitates were observed in Molecular Beam Epitaxy (MBE) – grown, super-saturated samples after annealing for long times at temperatures above 872°C. However, the SIMS profiles of the samples in this work (not shown) demonstrate that there was no appreciable diffusion of the Sb profile at any of the temperatures studied in this work. Without the diffusion of the dopant, precipitates are unlikely to form. The number of interstitials being generated by the deactivation is much smaller in magnitude than that in the As and P doped samples. However, they are similar to the values reported by Fage-Pederson et al.[15] by the formation of Sb_2V clusters in the initial stages of deactivation where precipitation has not occurred. Antimony may then form precipitates at longer times. Further investigation is needed to clarify the exact nature of inactive antimony.

CONCLUSION

We have studied the deactivation behavior of super-saturated concentrations of common dopants in silicon. In order to meet technology needs for shallow junctions, these dopants must remain electrically active through moderate temperature treatments during backend processing. Sb and B prove to be promising candidates for n- and p-type dopants as they maintain electrical activation up to temperatures as high as 700-800ºC. Conversely, P and As doped samples deactivate quickly at temperatures as low as 500ºC. It is proposed that As and P deactivate through a cluster mechanism with a native vacancy, ejecting an interstitial in the process, while B deactivates through the formation of precipitates. The reason for the classifications between stable and unstable species is not understood at a fundamental level. The explanation does not lie in the type, the size, the diffusion mechanism or the diffusivity of the dopant. An understanding of the fundamental deactivation mechanisms of the different dopants will aid in the determination of a means of slowing down or even eliminating deactivation.

ACKNOWLEDGEMENTS

The authors would like to thank S. Talwar, Y. Wang, S. Keene, G. Rimple and C. Gelatos at Verdant Technologies for their discussions and for the laser annealing. Corey Harris performed some of the anneals as a summer intern sponsored by the National Nanofabrication Users Network (NNUN). The Semiconductor Research Corporation (SRC) provided the funding for this work.

[1] P.M. Rousseau, Ph.D. Thesis, Stanford University (1994); P.M. Rousseau, P.B. Griffin, W.T. Fang, J.D. Plummer, *J. Appl. Phys.*, **82** (7) 3593 (1998)

[2] F.A Trumbore, *Bell System Tech. J.*, **39**, 205 (1960)

[3] A. Armigliato, D. Nobili, P. Ostoja, M. Servidori, S. Solmi, *Semiconductor Silicon, 1977*, Electrochemical Society, (1977) p. 638

[4] G. Masetti, D. Nobili, S. Solmi, *Semiconductor Silicon 1977*, Electrochemical Society, (1977), p. 648

[5] S. Solmi, D. Nobili, J. Shao, *J. Appl. Phys.*, **87** (2), 658 (2000)

[6] D. Nobili, A. Carabelas, G. Celotti, S. Solmi, *J. Electrochem Soc.*, **130** (4) 922, (1983)

[7] P.A. Stolk, H.-J. Gossmann, D.J. Eaglesham, D.C. Jacobson, J.M. Poate, H.S. Luftman, *Appl. Phys. Lett.*, **66** (5) 568 (1995)

[8] R.B. Fair, G.R. Weber, *J. Appl. Phys.*, **44**, 273 (1973)

[9] M. Ramamoorthy, S.T. Pantelides, *Phys. Rev. Lett.*, **76** (25) 4753 (1996)

[10] K.C. Pandey, A. Erbil, G.S. Cargill, R.F. Boehme, D. Vanderbilt, *Phys. Rev. Lett.*, **61**, 1282 (1988)

[11] E. Guerroro, H. Potzl, R. Tielert, M. Grasserbauer, G. Stingeder, *J. Electrochem. Soc.*, **129** (8), 1826 (1982)

[12] S. Luning, P.M. Rousseau, P.B. Griffin, P.G. Carey, J.D. Plummer, *Tech. Dig. Int. Electron Devices Meet.*, 457 (1992)

[13] A. Nylandsted Larsen, P. Kringhoj, J. Lundsgaard Hansen, S. Yu. Shiryaev, *J. Appl. Phys.*, **81** (5) 2173 (1997)

[14] P. Gaiduk, J. Fage-Pedersen, J. Lundsgaard Hansen, A. Nylandsted Larsen, *Phys. Rev. B*, **59** (11) 7278 (1999)

[15] J. Fage-Pedersen, P. Gaiduk, J. Lundsgaard Hansen, A. Nylandsted Larsen, *J. Appl. Phys.*, **88** (6) 3254 (2000)

Mat. Res. Soc. Symp. Proc. Vol. 669 © 2001 Materials Research Society

Laser Thermal Induced Crystallization for 20 nm Device Structures

Shenzhi Yang and Michael O. Thompson
Dept. Materials Sciences and Engineering, Cornell University, Ithaca, NY 14853

ABSTRACT:

The melt kinetics of shallow junction formation by laser thermal processes has been studied using transient conductance measurements. The melt and solidification dynamics of 20 nm amorphous layers were measured and shown to follow behaviors predicated by deeper melts, including explosive crystallization and interface bounce back. The effects of surface barrier oxides and metal absorber layers, required for CMOS process integration, were examined and shown to be nearly negligible. Quantitative evaluation of a device process window by these measurements was in good agreement with sheet resistance results. Finally, the effect of the buried oxide in SOI structures was investigated. Solidification velocities in such structures were reduced by a factor of three as compared with bulk silicon.

INTRODUCTION:

The scaling of MOSFET dimensions into sub-100 nm regimes requires increasingly precise control of the critical device dimensions, including source/drain shallow junctions. As the devices shrink, short channel effect and linkup resistances to the channel severely degrade performance [1]. To suppress these effects, the ITRS predicts that junction depths must approach 20 nm coupled with junction abruptness on the nanometer scale [2]. Simultaneously, the sheet resistances cannot increase substantially requiring that activated doping levels in the source and drain increase to beyond solid solubility limits.

While there is continued advancement in fast ramp RTA and spike annealing, these techniques are fundamentally limited by thermodynamic equilibrium and point defect enhanced diffusion and will have difficulties reaching (anticipated) requirements beyond the 100 nm ITRS node [2,3]. Only selective epitaxy and laser thermal processing (LTP) have demonstrated the non-equilibrium doping concentrations and abrupt profiles required [4,5].

In LTP, an amorphous layer formed at the surface determines the junction depth. This layer is formed either by the impurity implant itself (*i.e.* arsenic) or by a pre-amorphization Si or Ge implant to 20-30 nm depth followed by a low energy implant (*i.e.* boron). Pulsed laser irradiation induces a surface melt with solidification trapping impurities on substitutional and electrically active sites [6]. Because of the melting temperature difference between amorphous and crystalline silicon [7], there is a process window that melts only the amorphous layer and not the underlying substrate. As impurity diffusion in the liquid phase is extremely rapid, dopants are uniformly distributed throughout the melt leading to an abrupt junction defined by the amorphous layer. This technique is able to form very shallow abrupt junctions with activated dopant concentrations well above the equilibrium solid solubility [4,5].

However, process integration issues require the addition of a barrier oxide and metal absorber layer over the silicon surface [8]. These films potentially modify the dynamics of the process and their influence is unknown. Fundamental understanding of the thermodynamics and kinetics of LTP, along with quantification of the process integration modifications, is critical to accurate modeling and interpretation of device performance. This work examines the details of the melt kinetics during LTP shallow junction formation.

EXPERIMENTS:

Laser induced melt dynamics were studied using transient conductance techniques (TCM). This technique exploits the 30-fold difference in conductivity of molten and solid silicon (at the melting temperature). As depicted in figure 1, the melt thickness as a function of time can be obtained by monitoring the conductance in real time. Neglecting conduction in the solid (a second order correction), the thickness of the liquid is given by

$$D_{liq} = \rho_{liq} \times \frac{L}{W} \times \frac{1}{R_{coax}} \times \left(\frac{V_{bias}}{V_{scope}} - 2 \right)^{-1}$$

where ρ_{liq} is the liquid conductivity, $\frac{L}{W}$ the conduction path length/width ratio, R_{coax} the cable impedance, V_{bias} power supply, and V_{scope} is the signal after baseline correction.

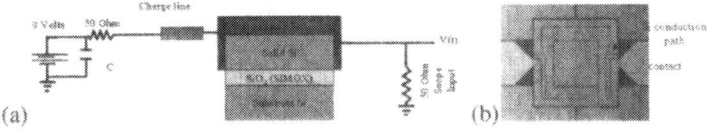

(a)   (b)

Figure 1: Transient conductance measurement (a) Schematics (b) Sample picture. Majority of the conduction contributed from the melt layer.

Transient conductance samples were prepared photolithographically on either 600 nm silicon-on-sapphire (SOS) or 190 nm/370 nm silicon-on-insulator (SOI) wafers [9]. SOS or SOI are required to suppress photocurrents in directly irradiated samples. At 600 nm, SOS samples behave nearly identical to, and are used to model processing in, bulk silicon. The low thermal conductivity of the oxide in SOI leads to reduced solidification velocities and significantly different behaviors. For amorphous Si studies, wafers were subjected to a $10^{15} cm^{-2}$ 10 kV arsenic implant prior to patterning, leading to an $\approx$20 nm amorphous surface layer.

To study the influence of metal absorber structures, more complex structures were also formed. For surface oxide studies, a 20 nm PECVD ($240^{\circ}C$, SiH_4+N_2O plasma) oxide was deposited over the devices. For the metal overlayer studies, a Ti/TiN film was deposited and patterned. To avoid capacitive coupling modifications to the TCM signal, the metal overlayers had to be patterned into disconnected blocks covering approximately 90% of the conductance path.

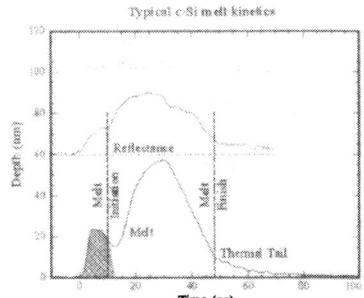

Figure 2: Typical crystal silicon melt kinetics. TCM signal consists of photo peak, melt kinetics and thermal tail. Reflectance and TCM melt signal agree very well.

Samples were irradiated with a homogenized (±5%) 308 nm XeCl excimer laser (pulse duration of ≈35 ns). Laser fluences were calibrated to a bulk Si melt threshold of 650 mJ/cm^2. In addition to TCM measurements, the surface reflectance was monitored *in-situ* using either a 633 or 750 nm CW laser source. Sheet resistances were measured *ex situ* after laser irradiation directly, or inferred from the conductance tails.

RESULTS AND DISCUSSION:

Bulk and amorphous Si melting:

A typical melt trace for bulk-like crystalline silicon (no amorphous layer) is shown in figure 2. Surface reflectance measurements are useful to determine the total melt duration but are difficult to quantify with respect to depth. The TCM trace, in contrast, depicts the entire real-time melt process. Prior to melt initiation, the trace exhibits a photoinduced peak (direct irradiation of crystalline Si) equivalent to ≈20 nm of melt. Once the surface melts, the photoinduced signal decays rapidly. Melt proceeds in monotonically until near the end of the laser pulse; thermal conduction into the solid then extracts the enthalpy of melt and establishes the solidification velocity. After the surface solidifies, a small thermal tail remains from conduction through the hot solid silicon; this signal decays with cooling of the sample over microsecond timescales.

Melt dynamics in amorphized samples exhibit more complex behaviors. At high fluences, the melt extends beyond the amorphous layer into the crystalline silicon substrate as shown in figure 3 at an ≈700 mJ/cm^2 fluence. The melt extends to 50 nm – well beyond the 20 nm amorphous layer. For comparison, the same sample was irradiated a second time – this time from the crystalline state – at a fluence to generate a comparable melt depth; both traces are shown in the figure. This comparison reveals the significant differences during the melt traces. Compared to the crystalline sample, photoinduced currents in the amorphized sample are almost totally suppressed – less than 2 nm melt equivalent compared to 20 nm for c-Si. Due to the thermal conductivity of the amorphous film, melt also occurs approximately 5 ns earlier in the laser pulse. However, once the melt reaches the amorphous/crystalline interface at 20 nm, there is a "bounce-back" in the melt of 2-3 nm. This occurs when the melt, near the amorphous

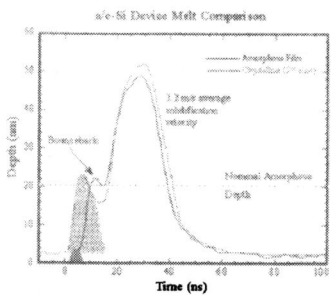

Figure 3: Melt across amorphous layer into crystal region at high laser fluence (~700mJ/cm^2). Dashed curve shows melt of c-Si to a comparable depth.

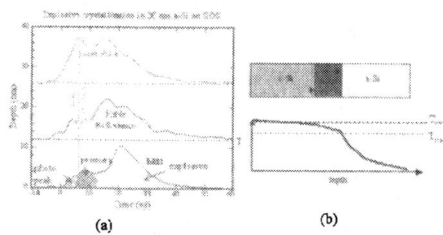

Figure 4: Melt with explosive crystallization for shallow melt (~350 mJ/cm^2). (a) Melt kinetics from TCM. (b) Schematic of explosive crystallization.

melting temperature, reaches the interface and is suddenly undercooled with respect to the crystal seed. The melt retreats until the liquid temperature is raised above the crystal melting point, and the interface can once again progress into the substrate. Melt again terminates near the end of the laser pulse and solidifies with an average velocity near 3.2 m/s. The final baseline level after melt terminates corresponds to both a deep thermal tail, and also conduction from the activated arsenic dopants (from the amorphizing implant). At low fluences, the melt does not proceed beyond the amorphous layer. In this regime, explosive crystallization occurs [10], as shown in figure 4. The melt consists of triple distinct peaks. The first is the small photo-induced current, followed by a primary melt induced directly by the incident laser energy. But when the initial melt begins to solidify, a secondary melt is formed that extends the melt throughout the amorphous zone. This is the explosive crystallization and is illustrated schematically in figure 4(b).

As the interface begins to solidify from the primary melt as a polycrystalline layer, the conditions shown in 4(b) are established. A thin liquid layer is trapped between a crystalline layer with a high melting temperature and the amorphous film with a lower melting temperature. Thermal conditions are thus established permitting self-propagation of a secondary melt throughout the remaining amorphous layer. The maximum thickness of the secondary melt can be much less than the amorphous thickness. In figure 4(a), the secondary melt is never more than 10 nm thick but likely propagates throughout the 20 nm layer.

The experimental melt depth data were correlated with sheet resistances as summarized in figure 5. [It is important to note that the absolute values of the sheet resistivity cannot be compared to bulk silicon due to the large number of stacking faults in SOS. Only the trend is significant.]
The maximum melt depth increases monotonically with the increase of laser fluence and the sheet resistance decreases monotonically. For fluences below 500 mJ/cm^2, the primary melt does not reach the amorphous-crystalline interface, although the total melt approaches it above 400 mJ/cm^2 due to explosive crystallization. In this regime, the sheet resistance improves by activation within the melt. Because of the fine-grain poly left by explosive crystallization, the improvement in sheet resistivity tracks the primary melt.

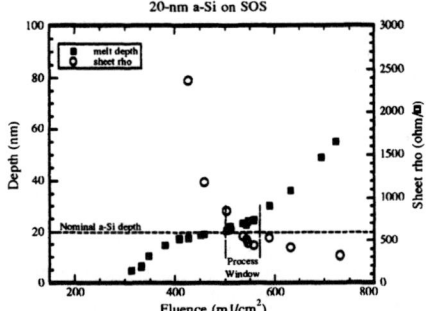

Figure 5: Comparison of melt and sheet resistance data for 20 nm a-Si on c-Si..

Above 600 mJ/cm^2, the melt penetrates into the crystal substrate and the sheet resistance decreases due to junction extension and improved mobility. The intermediate fluences correspond to the interface reaching, but not exceeding, the amorphous/crystalline interface.

Melt under oxide and metal/oxide layers:

LTP processing on patterned wafers requires the use of an oxide barrier and metal absorber layer. The metal layer provides thermal balancing and physical integrity while the oxide isolates active device structures from these metals. Melting under such structures has never been quantitatively measured. Of particular concern is whether the oxide may delay the onset of melting since no free surface is available for nucleation; heterogeneous nucleation must instead occur at the oxide interface.

The effect of the oxide layer was investigated for both crystalline and amorphous silicon. Since oxide is transparent to 308 nm, energy is still absorbed directly by the silicon. Experimentally, the melt kinetics are very similar to those without the oxide cap, except for a rigid shift in fluence, as in figure 6. The detailed melt kinetics for different fluence regions are both shown, including interface bounce-back and

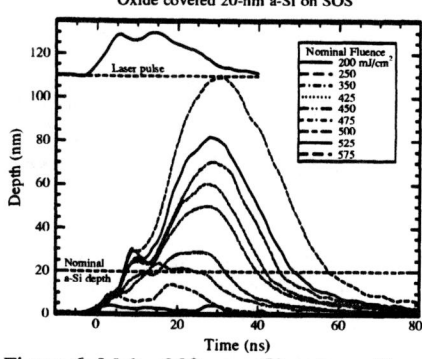

Figure 6: Melt of 20 nm a-Si under oxide

explosive crystallization. With oxide, the melt threshold is reduced by the anti-reflective properties of the oxide; for a 20 nm oxide, the coupled fluence increases by almost 50%.

With a fully implemented metal stack structure, laser energy is absorbed in the metal and conducted through the oxide to melt silicon. Since silicon is never directly irradiated, the photoinduced current in the TCM trace is totally suppressed. Figure 7 shows the melt behavior over a wide range of energies in the full stack. The amorphous melting ledge is clearly visible for deep melts.

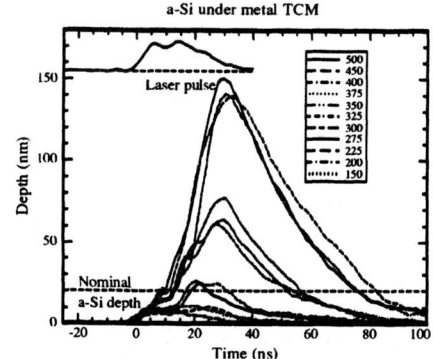

Figure 7: Melt kinetics for samples with fully implemented metal stack.

Melting of silicon-on-insulator:

Figure 8 compares melting dynamics of SOI and bulk silicon. The only significant difference lies in the solidification velocity, which is reduced by nearly a factor of three to 1 m/s. This is a direct result of the thermal barrier oxide that restricts heat flow to the substrate. The melt threshold (<400mJ/cm^2) is reduced for the same reason. The reduced velocity is likely to improve crystal quality both in the full melt and explosive regimes.

CONCLUSION:

In summary, we demonstrated the real time melt monitoring by transient conductance technique for shallow junction formation by laser process. Melt kinetics for thin amorphous layer were studied and are similar to those observed for deeper melts. The process window measured by direct melt measurements are in good agreement with sheet

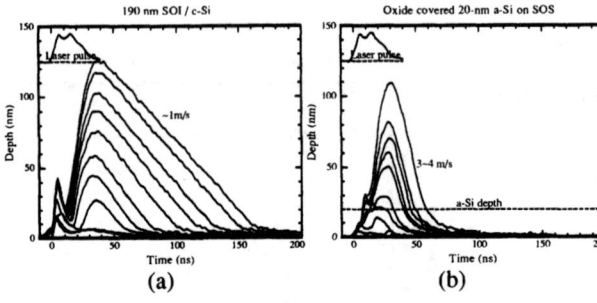

Figure 8: Comparison of melt on SOI and bulk. Solidification in SOI is much slower due to poor heat dissipation through the buried oxide

resistance results. Neither oxide nor metal overlayers show any substantial effect on solidification dynamics. Solidification velocities in SOI were also measured and are reduced, compared to bulk, by a factor of three, potentially leading to better epitaxy.

ACKNOWLEDGEMENTS:

Support for this work was provided through the Front End Processing Center of the Semiconductor Research Corporation. Work was performed in the Cornell Nanofabrication Facility (a member of the National Nanofabrication Users Network), supported by NSF under Grant ECS-9731293, Cornell University and industrial affiliates. We would also like to Verdant Technologies (Ultratech) for their financial support and use of facilities in this work.

REFERENCES:

[1] E. I. Erin C. Jones, Mater. Sci. Eng B Solid State Adv. Technol., **24**, (1) (1998)
[2] *International Technology Roadmap for Semiconductors*, San Jose, CA: Semiconductor Industry Association, 1999.
[3] A. Agarwal, H.-J. Gossmann, A. T. Fiory, Mater. Res. Soc. Symp. Proc., **568**, 19 (1999)
[4] B. Yu, Y. Wang, H. Wang, Q. Xiang, C. Riccobene, S. Talwar, and M.-R. Lin, Int. Electron Device Meet. Tech Dig., **99**, 509 (1999)
[5] S. Talwar, G. Verma, K.H. Weiner, Proc. of Ion Impla. Tech., 74 (1998)
[6] J.M. Poate, J. W. Mayer, *Laser Annealing of Semiconductors*, Academic Press, 1982
[7] M. O. Thompson, G. J. Galvin, J. W. Mayer, P. S. Peercy, J. M. Poate, D. C. Jacobson, A. G. Gullis, and N. G. Chew, Phy Rev Lett, **52**, 2360 (1984).
[8] S. Talwar, Y. Wang, M.O. Thompson, K. Jones, Mater. Res. Soc. Symp. Proc.,(2001)
[9] J.-P. Colinge, *Silicon-on-insulator Technology: Materials to VLSI*, 2nd ed., Kluwer Academic Publishers, 1997
[10] M. O. Thompson, P.S. Peercy, Mater. Soc. Symp. Proc., **51**, 99 (1985)

Advances in RTA

Mat. Res. Soc. Symp. Proc. Vol. 669 © 2001 Materials Research Society

Implant Dose and Spike Anneal Temperature Relationships

K. K. Bourdelle, [1] A. T. Fiory, [2,*] H.-J. L. Gossmann, [2] and S. P. McCoy [3]

[1] Agere Systems, Orlando FL 32819
[2] Agere Systems, Murray Hill NJ 07974
[3] Vortek Industries, Vancouver, B.C., Canada V6P6T7

ABSTRACT

The method of ion implantation and spike annealing for preparing shallow junctions suitable for the extension regions bridging the channel and source/drain contacts of CMOS transistors are studied by annealing blanket implants. Junction depths at a given sheet resistance for low energy B implants are minimized for the combination of a fast ramp with a sharp-spike anneal. This is shown to be physically based on activation energy phenomenology. The fraction of electrically activated B is insensitive to implant dose, unlike the case of transient enhanced diffusion. Arsenic implants show higher activation fraction than comparably annealed P implants, without the large transient enhanced diffusion which is attributed to P and Si-interstitial coupled diffusion. For targeted sheet resistance and junction depth, spiking temperature trends lower with implant dose, concomitant with decreasing fraction of activated dopant.

INTRODUCTION

Given the substantial investment of research, development and capital in ion implantation and rapid thermal annealing, the evolution of these techniques for future technology generations is of keen interest throughout the silicon integrated circuit community. At the previous symposium in this series, Gossmann et al. applied device modeling analysis to determine the components of the series resistance in the on state of MOSFETs, based on specifications in the International Technology Roadmap for Semiconductors [1]. Among potential problem areas that were identified are the contacts to the deep junctions, which may require unrealistically high dopant activation; sufficiently low sheet resistance of junction extensions, which may be already attained with current techniques; and lateral junction abruptness in the link-up region, which also correlates with off-state leakage owing to carrier spill over. It is important to recognize that specifications for future technologies continually evolve, vary with device architecture, and depend on whether designs are for high performance and speed or low power applications.

Generally, however, spike annealing methods have become associated with advancements in implant and anneal strategies for junction formation [2] and also poly-Si gate electrode activation [3]. This paper presents data pointing to a physical basis for spike annealing with fast ramping rates for reducing the junction depth of B implants [4]. Additional data is presented for spike annealing of P and As implants. As annealing methods are necessarily tied to equipment capability, this issue is also investigated in the context of ramp rate and sharpness of the thermal spike. All implants used drift mode operation of production implanters. Rapid thermal annealing was done with incandescent lamp and arc lamp systems.

---

* present address: Physics Department, New Jersey Institute of Technology, Newark NJ 07102

ACTIVATION AND DIFFUSIVITY OF B IMPLANTS

The shallow implant profile of high-dose low-energy B implants diffuses substantially when annealed to electrically activate a signification fraction of the dopant, owing to boron-enhanced diffusion effects [5]. Spike annealing with fast ramp rates was shown to help reduce the diffusion depth [6]. A reduced diffusion depth, however, can also be a simple consequence of having reduced the thermal budget. To examine this point, we take the sheet resistance, R_S, as a measure of a given desired level of electrical activation, and examine the dependence of diffusion depth on the annealing time and temperature relationship. Increasing the soak temperature such as to yield the same targeted sheet resistance can compensate a reduced soak time at temperature [4].

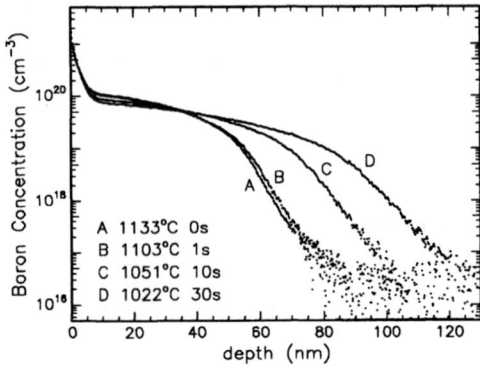

FIG. 1. Boron SIMS profiles of a 500 eV 10^{15} cm^{-2} B implant annealed in 0.1% O_2 at temperatures and times given in the legend inset.

The data in Fig. 1 show B concentrations from secondary ion mass spectroscopy (SIMS) for various anneals of a blanket 500 eV, 10^{15} cm^{-2} B implant in n-type Si (100) wafer material. The results were obtained with an AG Heatpulse system with a ramp rate of 150 °C/s. Anneal soak temperatures were decreased as respective times at temperature were increased from nominally 0 s for a spike to 30 s. Analysis of SIMS and electrical measurements are presented in Table I. The sheet resistance remains nearly, though not precisely constant, while the diffusion depth increases with increasing time and decreasing temperature of anneal. The table shows junction depth parameters, X_J, defined as the depth where the B concentration from SIMS or carrier concentration falls to 10^{18} cm^{-3}. Electrical measurements were made by the Hall van der Pauw method on etched mesa structures. We find that four-point probe techniques that are commonly used for sheet resistance measurements often yield erroneously low R_S for shallow junctions because of junction leakage at the probe sites. A complementary error function model of the carrier profile was used to fit Hall coefficient and sheet resistance data to determine a X_j for the carrier concentration [4]. The results of this experiment confirm that zero dwell-time, or spiking, provides the minimum diffusion [2].

To study the temperature-time relationship at constant R_S quantitatively, anneals were carried out for three B implant energies, 500, 750, and 1000 eV, each at 10^{15} cm^{-2} dose. The annealing times were varied from a spike to 600 s. A fixed R_S value was selected for each implant energy based on the result for a 1050 °C, 10 s soak

Table I. Sheet resistance and junction depth parameters, from SIMS and electrical measurement, for anneals of a 500 eV 10^{15} cm^{-2} B implant at given temperatures and times.

	T (°C)	time (s)	R_S (Ω/Sq)	SIMS X_J (nm)	Elect. X_J (nm)
A	1133	0	268	64	63
B	1103	1	256	66	73
C	1051	10	251	84	76
D	1022	30	245	101	92

anneal. Anneals for the other times were carried out at several temperatures, which were then interpolated to find the temperature corresponding to the selected fixed R_S. For short time anneals, heating and cooling transients comprise most of the anneal cycle, so we define the anneal time self-consistently, based on the notion that the process has a dominant activation energy. To take the thermal transients into account consistently, the anneal time is defined in terms of the temperature vs. time data recorded during the anneal by means of the following expression,

$$\text{time} = t_S \sum_i \exp[\, k_B^{-1} E_A (T^{-1} - T_i^{-1})\,], \quad (1)$$

where t_S is the sampling time interval, E_A is the activation energy, T the anneal temperature (peak temperature for a spike anneal, otherwise mean soak temperature) and T_i are the recorded temperatures. The sum, which extends over the annealing cycle, is dominated by terms where T_i are near T. The value of E_A is determined self-consistently, based on the activation energy that fits the T vs. time relationship. For the three implants the variation of T with time under the constraints of constant R_S falls on a universal curve, as shown in Fig. 2. The fixed R_S values are shown in the inset. The dashed curve is an Arrhenius fit with a thermal activation energy of $E_A = 5.04 \pm 0.03$ eV. This same activation energy is used in the computation of the time (E_A is derived by iterative convergence). This value of E_A for electrical activation of the B dopant turns out to be about 1 eV larger than the activation energy for B diffusion, as

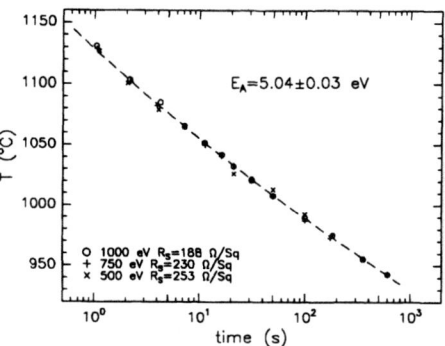

FIG. 2. Relationship between anneal temperature and time, Eq. (1), to activate B implants of three energies at 10^{15} cm^{-2} dose to yield sheet resistances given in the legend.

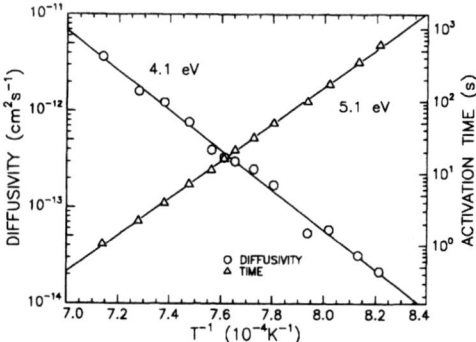

FIG. 3. Arrhenius plot of mean B diffusivity and time to activate a fixed sheet carrier density, $N_S = 5.3 \times 10^{14}$ cm^{-2}, for a 750 eV B implant at 10^{15} cm^{-2} dose. Dashed curves are fits with the activation energies shown.

discussed below. It is also similar to the value reported for activating higher energy implants [7].

Nearly the same result is obtained when the Hall effect data are used to impose a similar constraint of constant activated sheet carrier density, N_S. For the 750 eV implant, the annealing conditions were determined that correspond to fixing $N_S = 5.3 \times 10^{14}$ cm^{-2}, which is the result obtained for 1050 °C and 10 s. Figure 3 shows both the activation time and B diffusivity on an Arrhenius plot. The diffusivity of B was determined by fitting the Hall van der Pauw data to an erfc functional form. One finds for such anneals a relatively small variation in diffusion depths compared to the variation in annealing times. The diffusivity thus determined is interpreted to be

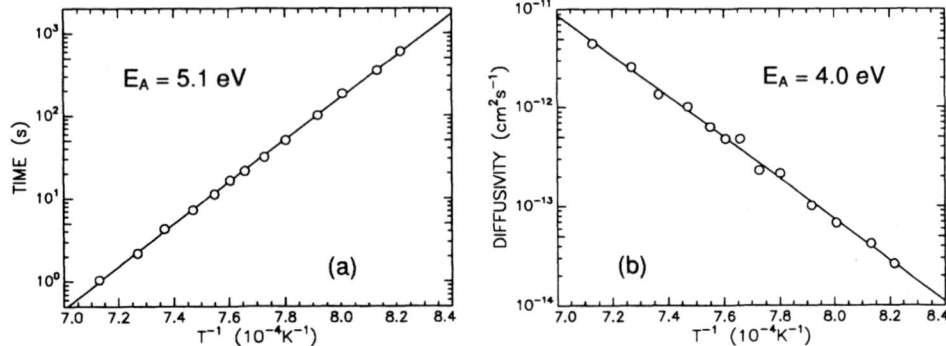

FIG. 4. Arrhenius plots of (a) activation time and (b) mean diffusivity for annealing 1000 eV 10^{15} cm^{-2} implants at temperatures and times to yield constant sheet carrier density $N_S = 6.5$ $\times 10^{14}$ cm^{-2}. Solid curves are fits with the thermal activation energies shown.

an average over the anneal cycle. While the activation energy for N_S activation, at 5.1 eV, is close to that found for R_S activation, the activation energy for B diffusion is significantly lower, at 4.1 eV. A similar analysis of data for the 1 keV implant for constant N_S yields nearly the same 1.1 eV difference in activation energies, as shown in Figs. 4 (a) and (b).

SPIKE ANNEALING METHODS

Figure 5 illustrates some of the commercial methods available for spike annealing. The furnace bell jar method of Fig. 5(a) is largely single sided, although there is underside heating as the wafer is raised. Other techniques with single side wafer heating may use an array of incandescent lamps (Applied Materials), an arc lamp as in Fig. 5 (b), or hot plates (Mattson, ASM Levitor). Incandescent lamp systems with dual sided heating, shown generically in Fig 5 (c), surround the wafer with maximum lamp exposure to achieve high heating rates.   There are tradeoffs in heating and cooling rates, time at maximum temperature, and process uniformity among the methods. Generally, multiple lamp methods require more complex multiple-zone power control and wafer rotation to achieve reasonable heating uniformity, compared to systems with a high power lamp or a stationary heat source.   As with advances in ion implantation, annealing methods are continually evolving with improvements, also. In what follows, we shall ignore uniformity and repeatability issues, although these may be critical for device processing.

Figures 6 and 7 show temperature vs time curves for spike anneals recorded by several commercial annealing systems, representing a recent state of the art. The origin of the time scale in Fig. 6 is the point at which heating commences (region below 800 °C not shown). The data in Fig. 7 are offset and expanded in time to facilitate viewing the differences in heating and cooling rates and the sharpness of the temperature peak.

Attributing the rate limiting effect in a spike annealing process to the mechanism with the highest activation energy, the effective thermal budget will be determined largely by the amount of

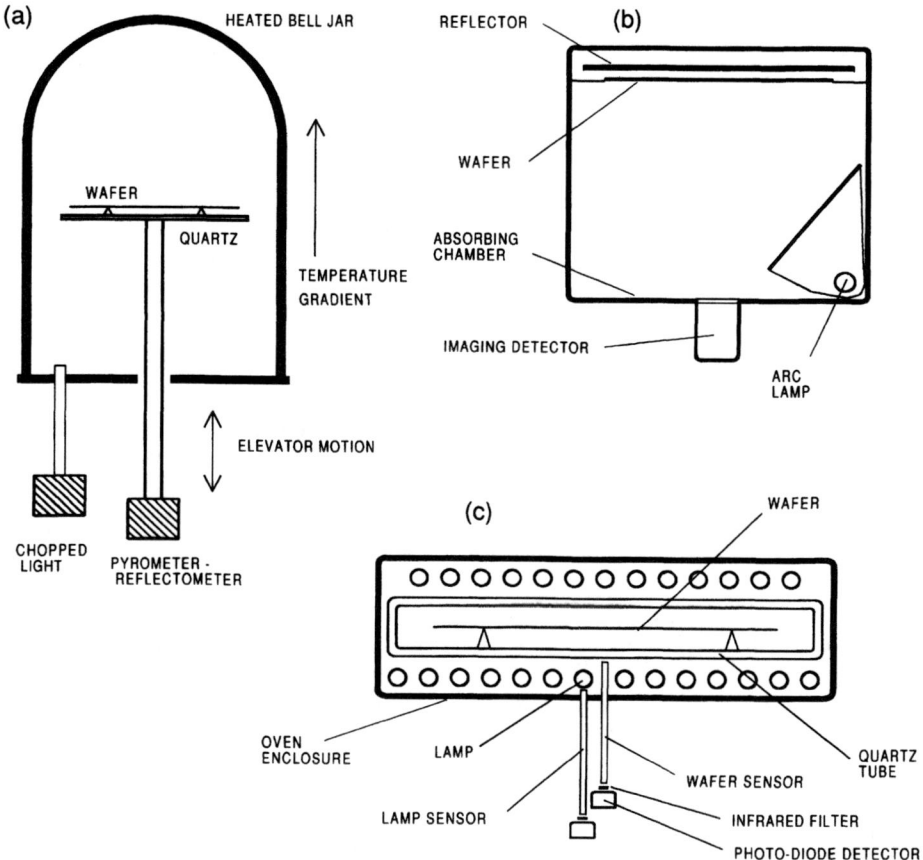

FIG. 5. Schematic illustrations of three methods for spike annealing. (a) Bell jar furnace with vertical wafer motion to control heating and cooling (after Axcelis); (b) arc lamp for heating wafer in absorbing chamber (after Vortek); (c) chamber with tungsten-halogen lamp arrays (after Mattson /AST and Bell Labs).

time spent at the maximum temperature. Thus we use the same approach as in the previous section, and define the effective annealing time in terms of the temperature vs time curve weighted by an effective activation energy. Figure 8 illustrates for the four curves of Fig. 7 the variation of temperature with an effective process cumulative time, defined as time calculated according to Eq. (1) with $E_A = 5$ eV and with origin at the start of the anneal. The total cumulative time of these anneals are 0.27, 0.86, 1.64 and 1.75 s for curves A through D, respectively, and are denoted as the *cycle time*. The minimum cycle time is obtained with a combination of a fast ramp and maximum sharpness of the temperature peak.

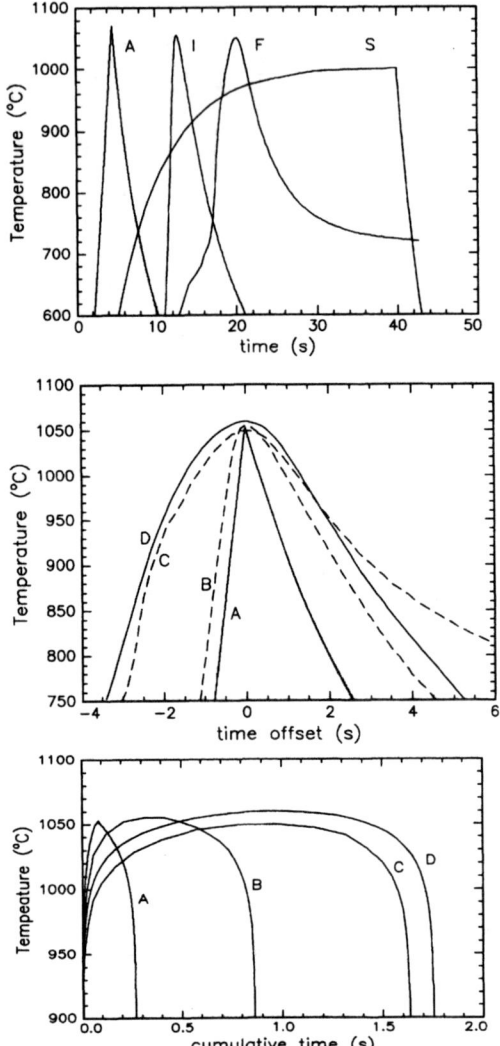

FIG. 6. Temperature vs time curves for 4 methods of spike annealing. Curve A is obtained by arc-lamp heating (courtesy Vortek Industries), curve I is with dual-sided incandescent lamp heating (courtesy Mattson, AST), curve F is with furnace bell jar (courtesy Axcelis, Summit), and curve S is for a hot plate method (courtesy Mattson Technology).

FIG. 7. Temperature vs time relative to time of peak temperature. Solid curve A is obtained by arc-lamp heating (Vortek Industries), dashed curve B is with dual-sided incandescent lamp heating (Mattson, AST), dashed curve C is with furnace bell jar (Axcelis, Summit), and solid curve D is for single side incandescent lamp heating (courtesy Applied Materials, Centura).

FIG. 8. Temperature vs effective process cumulative time computed via Eq. (1) with $E_A = 5$ eV, for the four temperature-time curves of Fig. 7. The origin of cumulative time is the start of the heating cycle.

SPIKE ANNEALS OF BORON IMPLANTS

The sensitivities of sheet resistance and junction depth to both ramp rate and sharpness of spike was tested by annealing a 500 eV B implant. Test samples were cut from the same wafer and annealed in an arc lamp system at Vortek Industries. Since the arc lamp is capable of rapidly switching high optical power output and rapidly modulating temperature, it is suitable for emulating a wide range of ramping rates and sharpness of spike. A series of recipes for temperature vs. time were created to emulate the typical variations among available annealing techniques, some of which were illustrated in the previous section. Figure 9 shows a set of temperature vs. time curves used to test spike variability within a given thermal budget. The rounding at maximum temperature for some of the spike methods was emulated by a finite dwell time at maximum temperature. The maximum temperature in each curve was adjusted to yield the same effective anneal time as for a "soft" spike anneal, as exemplified by curve 1 in Fig. 9. Curve 1 emulates the 1050 °C spike shown as curve D in Fig. 7. A total of 4 recipe sets were prepared with 8 respectively similar recipes in each set. The temperatures of the "soft" spike profiles like curve 1 in Fig. 9 are 1000, 1025, 1050, and 1075 °C.

Some of the results are shown in Figs. 10 through 12. Figure 10 shows the general trend where diffusion and junction depth, X_J, defined at 10^{18} cm^{-3} concentration, increase with annealing temperature. The temperature dependence curves shift towards higher temperature with increasing ramp rate. The data for 80 °C/s heating rate emulate a furnace bell jar method, an example of which is curve 1 in Fig. 9; 110 °C/s data emulate a dual side lamp system, e.g., curve 3 in Fig. 9; and 400 °C/s data are for a sharp arc-lamp spike, e.g, curve 5 in Fig. 9.

To elucidate the dependence on ramp rate, junction depth data were analyzed for constant sheet resistance, R_S, obtained by interpolation of the data matrix. Figure 11 shows the variation of junction depth, derived from both electrical and SIMS measurements, as a function of ramp rate for a fixed $R_S = 550$ Ω/Sq. An implicit variable in these data are the peak temperatures, which vary from ~ 1025 to ~ 1075 °C. The 4 points connected by the dotted line correspond to sharp spike anneals (similar to curves 5 through 8 in Fig. 9) with X_J

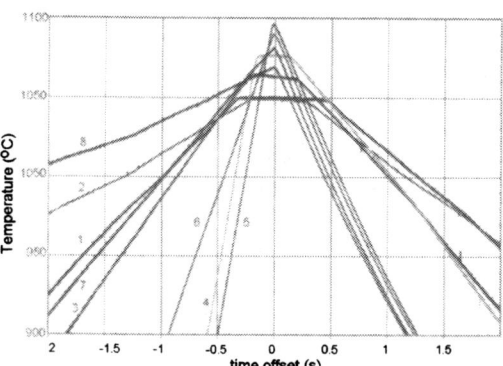

FIG. 9. Temperature vs time, relative to time of temperature peak, emulating 8 spike annealing capabilities. Heating rates are varied from 50 to 400 °C/s, cooling rates from about 50 °C/s to 160 °C/s, and dwells at peak temperature from 0 to 0.8 s. Maximum temperature is varied to maintain a constant thermal budget for effective time, given by in Eq. (1) for $E_A = 5$ eV. Anneals use front side arc-lamp heating in N$_2$ + 0.1% O$_2$ ambient.

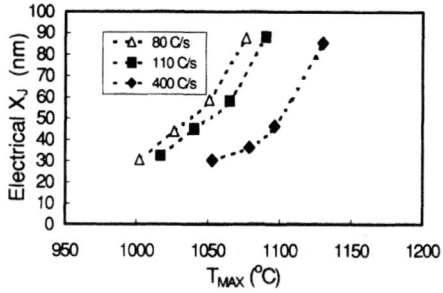

FIG. 10. Junction depths from analysis of electrical measurements (depths at 10^{18} cm^{-3} carrier concentration) as functions of peak temperature in spike anneals for heating ramp rates of 80, 110, and 400 °C/s, and cooling rates of 60, 80, and 150 °C/s, respectively. 500 eV, 2×10^{15} cm^{-2} B implant.

determined by SIMS. The results for the sharp spikes indicate that most of the benefit of fast ramping is reached near 100 °C/s with little advantage accruing at higher ramp rates. The other 4 data points determined by SIMS, corresponding to "soft" spikes (similar to curves 1 through 4 in Fig. 9), yield systematically larger X_J than for the sharp spikes at comparable ramp rates. Similar behavior is shown in X_J for the carrier concentration determined from electrical measurements. Another view of the trends in the data is shown in Fig. 12, where X_J from electrical measurements is plotted against cycle time, as defined via Eq. (1) for the full anneal cycle. The general trend confirms that sharp spikes and short cycle times yield lower X_J at a given constant R_S.

The trends with implant dose for 500 eV B implants are shown in Figs. 13-16. Here, all anneals used sharp spikes at 400 °C/s, arc-lamp heating from the back side, and Ar ambient with 0.1% O_2. Sheet resistance generally decreases and carrier concentration generally increases with implant dose, as shown in Figs. 13 and 14, implying that increased B dose allows one to reduce the spike temperature. The variation in the activated fraction shown in Fig. 15 tends to be monotonic with temperature and not particularly sensitive to implant dose. There does not seem to be the characteristic difference between low and high dose that has been observed in the onset of boron-enhanced diffusion [5]. The reason may be that the rate limiting process for carrier generation is B diffusion close to the implant peak (E_A= 5 eV). Figure 16 shows the sheet resistance as a function of the X_J determined from SIMS. The temperatures corresponding to

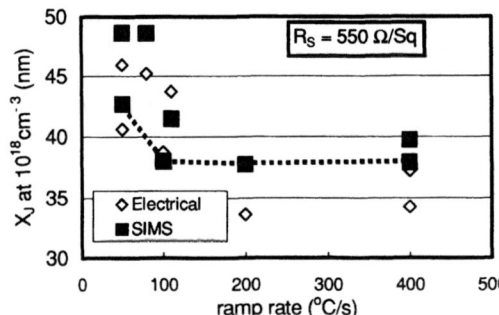

FIG. 11. Junction depths from electrical and SIMS measurements against heating ramp rate. For SIMS: sharp spike anneals are connected by dotted curve; remainder are for "soft" spike anneals. 500 eV, 2×10^{15} cm^{-2} B implant.

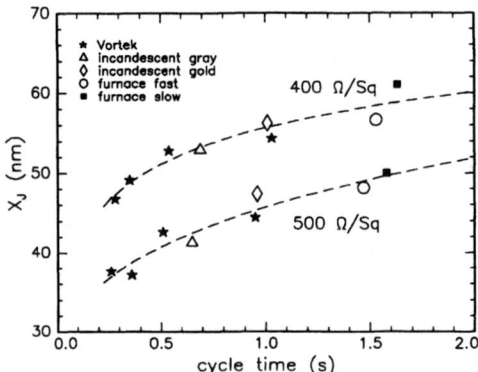

FIG. 12. Junction depths from electrical measurements are shown for two sheet resistances against cycle time of spike anneals. Starred points correspond to sharp spikes similar to curves 5 - 8 in Fig. 9; the remainder are "soft" spikes: circle points, curve 1; square points, curve 2; diamond point curve 3; and triangular points, curve 4 in Fig. 9, respectively. 500 eV, 2×10^{15} cm^{-2} B implant.

these points can be found in Fig. 13. These results show that increasing the implant dose from 4×10^{14} to 10^{15} cm^{-2}, and decreasing the spike anneal temperature from 1105 to 1045 °C, yields significantly smaller X_J at similar R_S values.

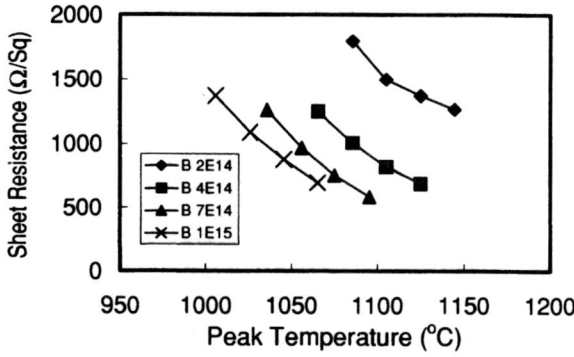

FIG. 13. Sheet resistance vs. spike anneal temperature for 500 eV B implants at four doses indicated (cm^{-2} units).

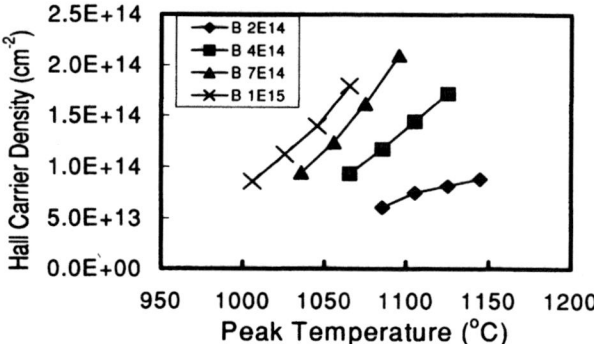

FIG. 14. Hall carrier density vs. spike anneal temperature for 500 eV B implants at four doses indicated (cm^{-2} units).

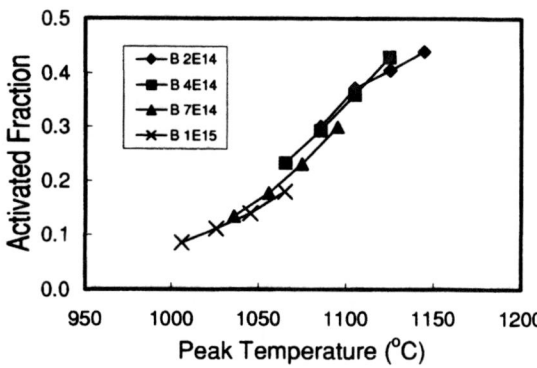

FIG. 15. Activated fraction (Hall carrier density / implant dose) vs. spike anneal temperature for 500 eV B implants at four doses indicated (cm^{-2} units).

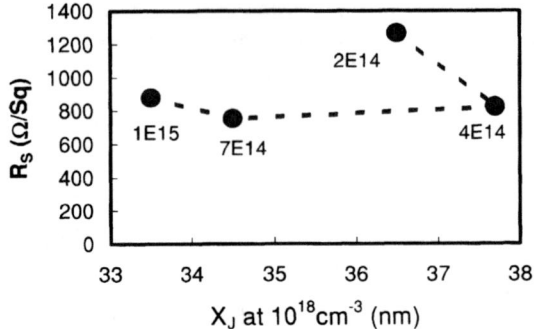

FIG. 16. Sheet resistance vs. junction depth determined by SIMS for selected spike anneals of 500 eV B implants at the four doses indicated (cm^{-2} units).

SPIKE ANNEALS OF ARSENIC AND PHOSPHORUS IMPLANTS

For the formation n-type junctions we examined a 5 keV As implant at a single high dose, 10^{15} cm^{-2}, and four 1.5 keV P implants in the dose range 2×10^{14} to 10^{15} cm^{-2}. Wafers were p-type (100) Si. Anneals used sharp spikes at 400 °C/s, arc-lamp heating from the back side, and Ar ambient with 0.1% O_2. Electrical activation of the As implant is less sensitive to spike temperature than for either P implants, as shown in Fig. 17, or B implants (compare Fig. 13). Notably, the carrier density activated in the As implant exceeds that for the P implants, as shown in Fig. 18. Lower carrier mobilities for the As implant results in comparable R_S for the two species, however.

The activated fraction shown in Fig. 19 is both dose and temperature dependent. The results are quite different from the weak dose dependence found for B activation and shown in Fig. 15. Figure 20 shows that for a given activation of about 25% of the P dopant, the corresponding spike anneal temperature decreases monotonically with dose from 1130 to 1025 °C. SIMS measurements of such anneals with 25% activation were found to exhibit nearly the same diffusion depths for the P dopant, irrespective of dose. The junction depth, shown in Fig. 21, is therefore weakly dependent on the P implant dose. The P diffusion is enhanced, presumably by

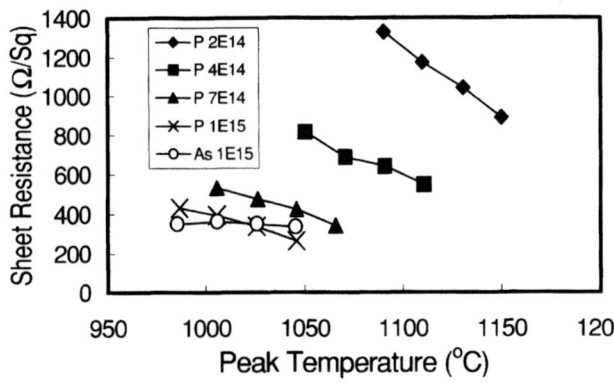

FIG. 17. Sheet resistance vs. spike anneal temperature for 1.5 keV P implants at four doses and a 5 keV As implant.

the presence of Si interstitials (Si$_I$) associated with implant damage. To estimate an effective Si$_I$ concentration, the Prophet process simulator was used to simulate Si$_I$-P coupled diffusion [8]. The ratio of the Si$_I$ and P concentrations, denoted as the plus factor, that allows the process model to match the diffusion depth observed by SIMS is also plotted in Fig. 21. This analysis suggests that a residual concentration of Si$_I$ of 6 to 10 % of the P dose can account for the enhanced diffusion. The near absence of enhanced diffusion and the higher dopant activation makes As the species of choice for shallow n-type extensions.

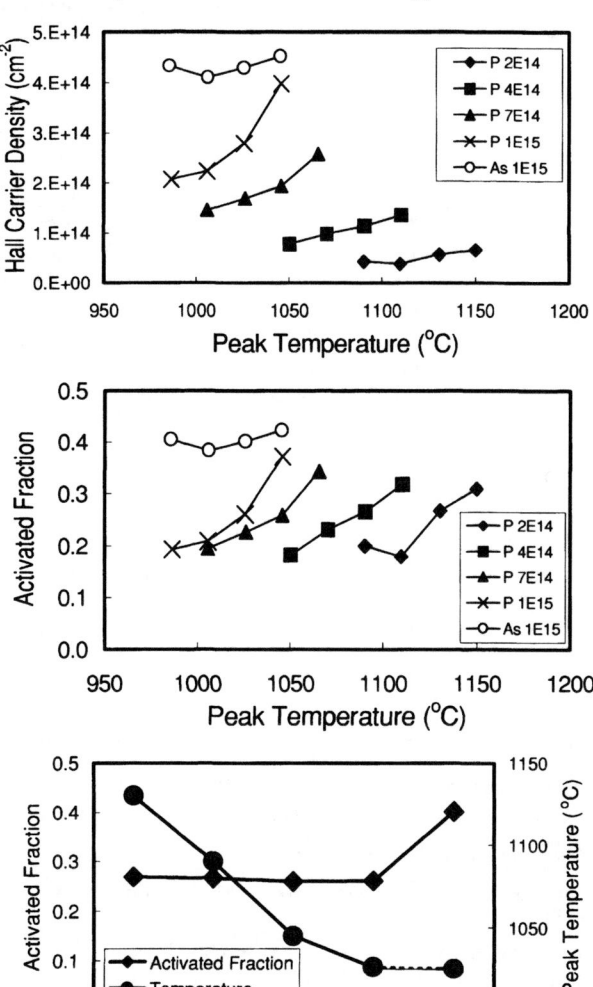

FIG. 18. Hall carrier density vs. spike anneal temperature for 1.5 keV P implants at four doses and a 5 keV As implant.

FIG. 19. Activated fraction (Hall carrier density / implant dose) vs. spike anneal temperature for 1.5 keV P implants at four doses and a 5 keV As implant.

FIG. 20. Activated fraction (Hall carrier density / implant dose) and spike anneal temperature for selected 1.5 keV P implants at four doses and a 5 keV As implant.

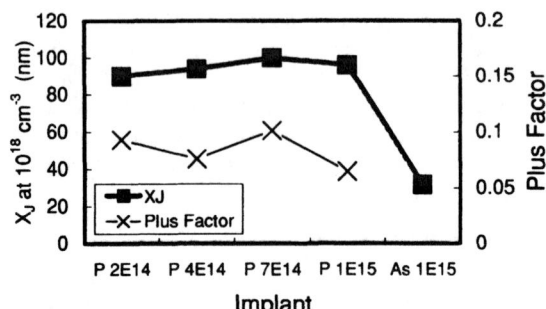

FIG. 21. Junction depths from SIMS for selected 1.5 keV P implants at four doses and a 5 keV As implant, for same anneals as in Fig. 20. Model fits for Si_I plus factor ($[Si_I]$ / $[P]$, coupled diffusion) are shown for the P implants.

CONCLUSIONS

Sharp spike annealing with ramp rates at 100 °C/s or above are shown to aid in the formation of shallow p-type junction extensions by means of high-dose and low-energy B implantation, owing to a 1 eV higher thermal activation energy for carrier creation relative to dopant diffusion. For n-type dopants spike annealing produces higher electrical activation and shallower junctions for As implants, when compared to P implants annealed with comparable thermal budget. The enhanced thermal diffusion of P creates junction depths at a given fraction of electrical activation that are weakly dependent on implant dose. Coupled P and Si-interstitial diffusion modeled with plus factors of 0.06 to 0.10 can account for the observed diffusion depths.

ACKNOWLEDGEMENTS

The SIMS analyses were produced by J. M. McKinley and F. A. Stevie, and by Evans East. Wafer processing and annealing employed research and development facilities of Agere Systems in Orlando, Fla. and Murray Hill, N.J. and Vortek Industries, Vancouver, B.C., Canada. The authors acknowledge stimulating discussions with A. Agarwal, G. S. Higashi, and C. S. Rafferty.

REFERENCES

1. H.-J. Gossmann, C.S. Rafferty, and P. Keys, Mat. Res. Soc. Symp. Proc. **610**, B1.2.1 (2000).

2. S. Saito, S. Shishiguchi, A. Mineji, and T. Matsuda, Mat. Res. Soc. Symp. Proc. **532**, 3 (1998).

3. A.T. Fiory, K.K. Bourdelle, P.K. Roy, Appl. Phys. Lett. **78**, 1071 (2001).

4. A.T. Fiory and K.K. Bourdelle, Appl. Phys. Lett. **74**, 2658 (1999).

5. A. Agarwal, H.-J. Gossmann, D.J. Eaglesham, S.B. Herner, A.T. Fiory, T.E. Haynes, Appl. Phys. Lett **74**, 2435 (1999); *ibid.*, p. 2331.

6. A. Agarwal, H.-J. Gossmann, D. J. Eaglesham, L. Pelaz, S. B. Herner, D. C. Jacobson, T. E. Haynes, R. Simonton, Mater. Sci. in Semicond. Proc. **1**, 17 (1998).

7. T.E. Seidel and U. MacRae, Radiation Effects **7**, 1 (1971).

8. C. S. Rafferty, G. H. Gilmer, M. Jaraiz, D. J. Eaglesham, and H.-J. Gossmann, Appl. Phys. Lett. **68**, 2395 (1996).

Mat. Res. Soc. Symp. Proc. Vol. 669 © 2001 Materials Research Society

Antimony and Boron Diffusion in Silicon and Silicon Germanium under the Influence of Point Defects Injection by Rapid Thermal Anneal

Aihua Dan, Arthur F. W. Willoughby, Janet M. Bonar*, Barry M. McGregor, Mark G. Dowsett*** and Terry J. Ormsby***

Materials Research Group, and *Department of Electronics and Computer Science, School of Engineering, University of Southampton, Southampton, UK
**Department of Engineering, University of Cambridge, UK
***Department of Physics, University of Warwick, Coventry, UK

ABSTRACT

The effect of point defect injection on the diffusion of antimony and boron in silicon and silicon-germanium alloys has been studied by comparison of inert with injection diffusions. In this work, Sb and B in Si were used as control wafers to investigate Sb and B diffusion behavior in $Si_{0.9}Ge_{0.1}$. The point defect injection technique was carried out by rapid thermal annealing (RTA) Sb and B in Si and $Si_{0.9}Ge_{0.1}$ samples with the various surface coatings in either oxygen or ammonia atmospheres to inject either interstitial or vacancy defects. The diffusion profiles for as-grown and RTA annealed samples were measured by Secondary Ion Mass Spectrometry (SIMS). Diffusivities for B in Si and $Si_{0.9}Ge_{0.1}$ were obtained using computer simulations of the measured boron profiles for their annealed samples. Sb diffusion in Si and $Si_{0.9}Ge_{0.1}$ was found enhanced by vacancy injection and retarded by interstitial injection. The enhanced B diffusion in Si and $Si_{0.9}Ge_{0.1}$ was found by interstitial injection. These results confirm that Sb diffusion in $Si_{0.9}Ge_{0.1}$ is primarily dominated by vacancy-mediated mechanism, while B diffuses in $Si_{0.9}Ge_{0.1}$ by an interstitially mediated mechanism. The effect of the RTA diffusion time on the B diffusion in Si and $Si_{0.9}Ge_{0.1}$ has also been investigated. The diffusivity versus diffusion time of B in Si and $Si_{0.9}Ge_{0.1}$ for inert and injection samples is presented. It was found that the shorter annealing time had the faster diffusion. This suggested that it caused by transient diffusion effect arising from point defects.

INTRODUCTION

Antimony is the only dopant generally accepted in which diffusion both in silicon and germanium is dominated by vacancy-mediated mechanism, but its diffusion behavior in silicon germanium alloys needs more investigation[1]. Boron diffusion in Si is generally accepted to diffuse by an interstitial mediated mechanism. It is suggested that the mechanism for B diffusion in SiGe (Ge content up to 20%) primarily involves Si interstitials[2]. Some work has given valuable evidence on the diffusion mechanism of boron in Si and SiGe alloys. Kuo et al[3] used thermal oxidation to inject Si interstitials into the bulk and observed enhanced B diffusion in Si and strained $Si_{1-x}Ge_x$ (x <18%). Fang et al[2] carried out experiments to inject vacancies by nitridation with ammonia and showed a significant retardation by vacancy injection, confirming that B diffuses in $Si_{0.8}Ge_{0.2}$ via an interstitial mechanism as is accepted in silicon. Thus, we consider Sb in SiGe alloys for an excellent test species to study vacancy-mediated diffusion and B to study interstitial-mediated diffusion. Rapid Thermal Annealing (RTA) is a new thermal process to utilize in the Si device fabrication to form shallow junctions. The effects of defect injection by furnace anneal and RTA processes have found no difference[7]. Based on this work, the point defect injection technique has been developed by rapid thermal annealing samples with various coated surface layers in oxygen or ammonia in our study. During RTA annealing in oxygen ambient, bare surfaces should experience interstitial injection[4], and Si_3N_4 coated surfaces should experience vacancy injection due to stress in the film[5]. During RTA annealing in ammonia, bare silicon surfaces should experience vacancy injection, while SiO_2 covered surfaces should experience interstitial injection[6]. In either oxygen or ammonia, SiO_2/Si_3N_4 bilayer covered surfaces should experience non-injection, inert diffusion[4]. Under this assumption, we used Sb and B diffusion in Si as control cases and $Si_{0.9}Ge_{0.1}$ as the unknown, to investigate the effects of defect injection by RTA on Sb and B diffusion in Si and $Si_{0.9}Ge_{0.1}$, and to identify the Sb and B diffusion mechanisms in $Si_{0.9}Ge_{0.1}$ alloy, which is reported in this paper, and to study the transient effects caused by excess point defects on Sb and B diffusion in Si and SiGe, which is our on going work.

EXPERIMENTAL

P-type CZ(100) silicon wafers were used as the substrates of all the B diffusion in Si and SiGe samples and n-type CZ(001) silicon wafers for Sb samples. The sample structures of Sb in Si and SiGe are similar to B in Si and SiGe. The layer structures are shown in Fig.1. These structures are consisting of well-defined narrow distributions of the Sb or B dopants in expitaxial layers of strained $Si_{0.9}Ge_{0.1}$ or Si layers. For preparing B in SiGe samples, substrates were RCA cleaned before loading into the 800°C growth chamber. Silicon buffer layer with thickness of 200nm were grown first on substrate. After deposition of buffer layer, 90nm of $Si_{0.9}Ge_{0.1}$ was grown with boron layer of concentration 1×10^{19} atoms/cm^3 at the centre. Then a 120nm silicon cap was grown. All the B doped layers were deposited by low-pressure chemical vapour deposition (LPCVD) using SiH_4, GeH_4, B_2H_6 and H_2. For B in Si samples, the growth processes were very similar, with the $Si_{0.9}Ge_{0.1}$ replaced by Si. For Sb in SiGe and Si samples, molecular beam epitaxy (MBE) was used for layer growth. Sb δ-layer with concentration of 3×10^{18}cm^{-3} was grown within a 90nm 10% Ge SiGe layer, followed by a 80nm Si capping layer for Sb in SiGe samples, and a 3×10^{18}cm^{-3} Sb δ-layer for Sb in Si control samples. Following growth, 100nm of SiO_2 was deposited at 400°C using SiH_4 and O_2 on samples, which was photolithographically patterned and wet etched in buffered hydrofluoric (BHF) acid solution at 25°C to get SiO_2 coated and bare Si surface. Following this, 120nm of Si_3N_4 was deposited at 740°C in patterned regions to get the samples with the bare silicon surface, Si_3N_4 surface and bilayer SiO_2/Si_3N_4 surface.

Rapid thermal annealing (RTA) was performed. B in Si and SiGe samples with the bare silicon and bilayer SiO_2/Si_3N_4 coated surfaces were annealed in O_2 at temperature of 1000°C for 2,5,10,25 seconds. Sb in Si samples with bare Si, LTO, Si_3N_4 and SiO_2/Si_3N_4 surfaces were annealed at 1100°C for 30,60,90 seconds in O_2 or NH_3. Sb in SiGe samples with bare Si, LTO, Si_3N_4 and SiO_2/Si_3N_4 surfaces were annealed at 1000°C for 45,90,180 seconds in O_2 or NH_3.

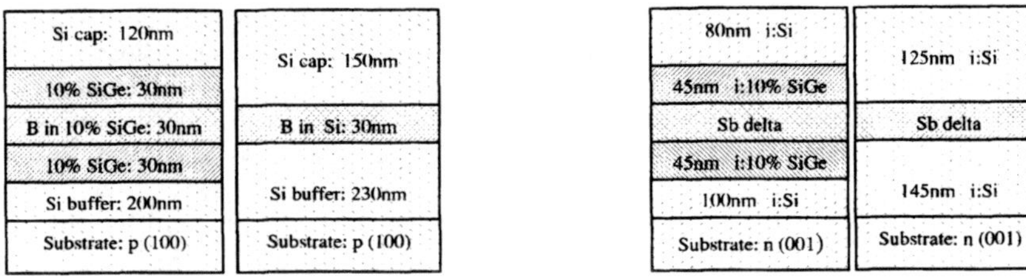

Si cap: 120nm			80nm i:Si	
10% SiGe: 30nm	Si cap: 150nm		45nm i:10% SiGe	125nm i:Si
B in 10% SiGe: 30nm	B in Si: 30nm		Sb delta	Sb delta
10% SiGe: 30nm			45nm i:10% SiGe	
Si buffer: 200nm	Si buffer: 230nm		100nm i:Si	145nm i:Si
Substrate: p (100)	Substrate: p (100)		Substrate: n (001)	Substrate: n (001)

(a) B samples   (b) Sb samples

Figure 1. Schematic of structure of SiGe and Si samples

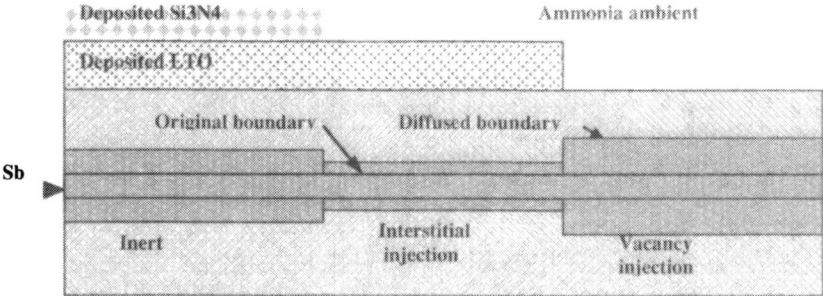

Figure 2. Schematic of the injection experiment, showing the expected effect on Sb diffusion of point defect injection, as well as the reference (no injection) case

During annealing in O_2, bare surfaces should experience interstitial injection, and Si_3N_4 coated surfaces should experience vacancy injection due to stress in the film. During annealing in NH_3, bare silicon surfaces should experience vacancy injection, while SiO_2 covered surfaces should experience interstitial injection. In either oxygen or ammonia, SiO_2/Si_3N_4 bilayer covered surfaces should experience non-injection, inert diffusion. The expected effects of excess point injection by RTA in NH_3 on Sb diffusion are shown in Fig.2. After annealing, SiO_2 and Si_3N_4 layers were removed for diffusion profiling. Secondary Ion Mass Spectrometry (SIMS) was performed on as grown and annealed samples. B in Si and SiGe samples were sputtered by O_2^+ primary beam with energy of 500eV at normal incidence, and Sb in Si and SiGe samples were sputtered by O_2^+ primary beam with energy of 500eV at 20° wrt surface normal incidence.

After SIMS profiling, analysis of as-grown and all annealed SIMS profiles were carried out. The as grown profiles in this study were obtained from SIMS measurements of nonannealed samples, which were grown under conditions identical to those of annealed samples. Although the SiGe samples are from one wafer and all the Si samples from another, since a certain amount of variability existed in the epitaxial growth and SIMS profiling process, dopant peak positions and doses presented by the as grown and annealed profiles could not be exactly the same for a same wafer. To compensate this variability and make the profiles comparable, all Sb and B profiles were shifted so that their peaks would have a common depth and scaled to ensure that all samples would have same doses, since the mass conservation law demands that the does be constant during the diffusion process. After shifting and scaling the profiles, a simulation program was performed by convoluting the initial profiles (SIMS profiles of the as-grown samples) to produce annealed profiles, then compared to the measured annealed curves so that the diffusivities were determined when the obtained simulated profiles matched with the measured curves. Figure 3 shows a typical case and the resulting fits to the annealed boron profile of a B in Si inert sample, which was RTA annealed in oxygen at 1000°C for 25s.

RESULTS AND DISCUSSION

Sb SIMS diffusion profiles for Sb in Si as grown and annealed samples, which RTA in NH_3 at 1100°C for 30s under inert, interstitial injection and vacancy injection, are presented in Figure 4(a). The Sb profile for the interstitial injection case is much narrower than the inert diffusion. It indicates that Sb diffusion in Si is retarded under interstitial injection. The Sb profile for the vacancy injection case is obviously wider than the inert diffusion, which indicates Sb diffusion in

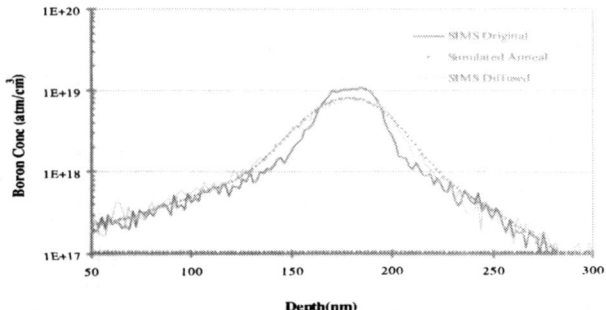

Figure 3. A simulation resulting fits to the boron diffusion profile of B in Si. The line (SIMS Original) is the boron SIMS profile of B in Si as-grown sample. The line (Simulated Anneal) is the simulated SIMS annealed profile obtained by our simulation program. The line (SIMS Diffused) is the SIMS measured boron profile of a B in Si inert sample, which was RTA in oxygen at 1000°C for 25s.

Si is enhanced under vacancy injection. These results confirm that Sb diffuses in Si via a vacancy-mediated mechanism. For Sb in Si as a control case, this result also suggests that our point defect injection technique, which injects point defects by RTA annealing various surfaces in NH₃, was successful, as the expected enhanced and retarded diffusion under vacancy and interstitial injection respectively observed.

Figure 4(b) shows the Sb diffusion profiles of Sb in $Si_{0.9}Ge_{0.1}$ as grown and annealed samples, which RTA in NH₃ at 1000°C for 45s under inert, interstitial injection and vacancy injection, the SiGe region is also shown. Comparing the Sb annealed profiles with the as grown, it was found that Sb has diffused out the SiGe region under vacancy injection and inert anneal, Sb just stayed within the SiGe region under the interstitial injection diffusion for this particular annealing condition. Comparing Sb profiles under the injections with the inert diffusion, Sb profiles under vacancy injection and interstitial injection have similar relative positions as those of Sb in Si case, although the differences between the profiles are much smaller between Sb profiles of Sb in SiGe than those of Sb in Si. This still demonstrates that enhanced diffusion is shown by the

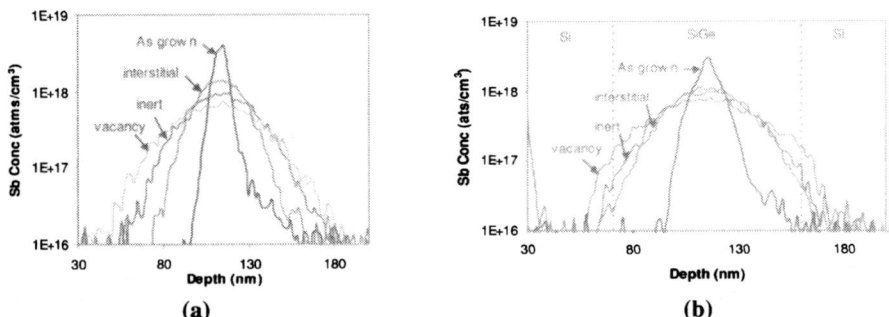

(a)   **(b)**

Figure 4. Effects of excess point defects on (a) Sb diffusion in Si: RTA 30s at 1100°C in NH₃, (b) Sb diffusion in SiGe: RTA 45s at 1000°C in NH₃

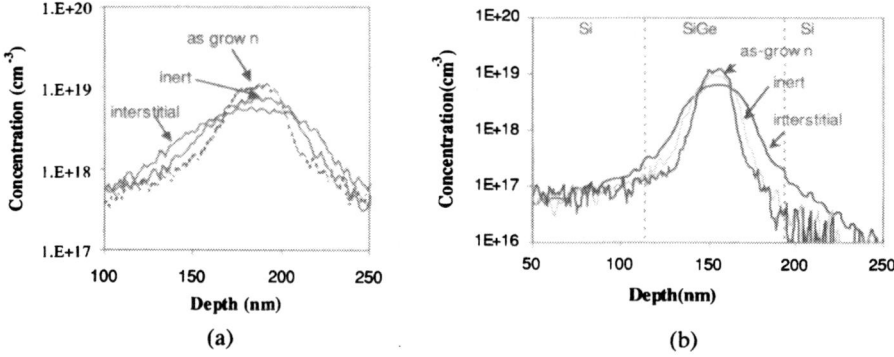

Figure 5. Effect of defect injection on (a) B diffusion in Si: RTA 25s at 1000°C in O_2 (b) B diffusion in SiGe: RTA 25s at 1000°C in O_2

vacancy injection profile, while the retarded diffusion is indicated by the interstitial injection profile. These results suggest that Sb diffusion in SiGe has similar diffusion mechanism as in Si, i.e. Sb diffusion in SiGe primarily involvs a vacancy-mediated mechanism.

The B SIMS concentration depth profiles of B in Si for as grown and annealed samples, which were RTA in O_2 at 1000°C for 25s under inert and interstitial injection, are shown in Figure 5(a). The interstitial injection profile lies outside the inert diffusion profile, indicating that interstitial injection enhanced B diffusion in Si. Figure 5(b) shows the SIMS profiles B in SiGe for as grown and annealed samples, which were RTA in O_2 at 1000°C for 25s under inert and interstitial injection, the SiGe region is also shown. Comparing the annealed profiles with the as grown, B stayed well within the SiGe region for inert case, while B diffused out of the region for interstitial injection. Similar results were found as for B in Si, enhanced B diffusion in SiGe indicated by the interstitial injection profile. The relation of B diffusivities in Si and $Si_{0.9}Ge_{0.1}$ versus the diffusion time is plotted in Figure 6. Comparing the diffusivities for all the B in Si and $Si_{0.9}Ge_{0.1}$ inert or injection cases, higher diffusivities were found in interstitial injection diffusions. It confirms that the mechanism of B diffusion in Si and $Si_{0.9}Ge_{0.1}$ is primarily involved an interstitial-mediated. It was also found higher diffusivity at shorter diffusion time. It suggested that there might be transient diffusion arising from the defects. Investigation of the transient effects is underway.

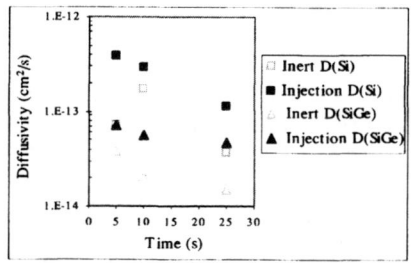

Figure 6. The relation of the diffusivities of B in Si and SiGe versus diffusion time

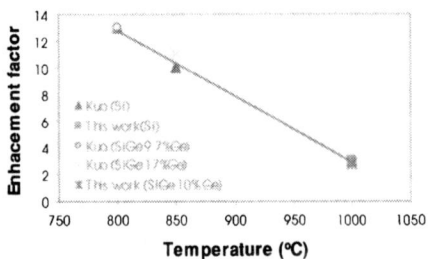

Figure7. The enhancement factor of B diffusion in Si and SiGe by interstitial injection from this work compared with the results of Kuo[3].

The enhancement factor of B diffusion in Si by interstitial injection from this work compared with the results of Kuo[3] is presented in Figure 7. It shows that the interstitial enhancement factors of B in Si and SiGe are very similar and the enhancement factor from this work is consistent with the trends reported by Kuo et al[3]. It also suggests that the enhancement factor for B diffusion in Si and SiGe under interstitial injection decreases as increasing the diffusion temperature.

CONCLUSIONS

In this study, enhanced diffusion for Sb diffusion in Si and $Si_{0.9}Ge_{0.1}$ was observed by vacancy injection, while diffusion was retarded by interstitial injection. B diffusion in Si and $Si_{0.9}Ge_{0.1}$ was found enhanced by interstitial injection. It confirms that Sb diffusion in $Si_{0.9}Ge_{0.1}$ has similar diffusion mechanism as in Si, which diffuses by a vacancy-mediated mechanism; while B diffusion in Si and $Si_{0.9}Ge_{0.1}$ is primarily involved an interstitially mediated mechanism. Investigation of the effect of diffusion time on B diffusion in Si and $Si_{0.9}Ge_{0.1}$ has been begun. It was found that the shorter annealing time had the faster diffusion. This suggests transient diffusion arising from grown in defects. More work is needed to clarify the effect of the transient diffusion.

ACKNOWLEDMENTS

This work was part of the project RAPID funded in part as ESPRIT Long Term Research project No 23481.

REFERENCES

1. A.F.W.Willoughby,J.M.Bonar, and A.D.N.Paine, *Mat. Res. Soc. Symp. Proc.*, **568**, 253 (1999).
2. T. T. Fang, W. T. C. Fang, P. B. Griffin and J. D. Plummer, *Appl. Phys. Lett.*, **68**, 791 (1996).
3. P. Kuo, J. L. Hoyt, J. F. Gibbons, J. E. Turner, D. Lefforge, *Appl. Phys. Lett.*, **67**, 706 (1995).
4. K. Osada, Y. Zaitsu, S. Matsumoto, M. Yoshida, E. Arai, and T. Abe, *J. Electrochem. Soc.*, **142**, 202 (1995).
5. Y.Zaitsu, T.Shimizu, J.Takeuchi, S.Matsumoto, M.Yoshida, T.Abe, E.Arai, *J. Electrochem. Soc.*,**145**, 258(1998).
6. P. M. Fahey, P. B. Griffin and J. D. Plummer, *Rev. Mod. Phys.*, **61**, 289 (1989).
7. J.M.Bonar,B.M.McGregor,N.E.Cowern,A.Dan,G.A.Cooke,A.F.W.Willoughby,*Mat. Res. Soc. Symp. Proc.*(2000).

Mat. Res. Soc. Symp. Proc. Vol. 669 © 2001 Materials Research Society

Influence of low thermal budget pre-anneals on the high temperature redistribution of low energy boron implants in silicon

F. Boucard[1,2,3], M. Schott[2], D. Mathiot[2], P. Rivallin[3], P. Holliger[3], E. Guichard[1]

[1] SILVACO France, 8, Avenue de vignate, F-38610 GIERES, France
[2] PHASE/CNRS,23 rue du Loess, BP 20, F-67037 STRASBOURG Cedex 2, France
[3] LETI, CEA GRENOBLE, 17 avenue des martyrs, F-38054 GRENOBLE Cedex 9, France

ABSTRACT

It is now well established that the transient enhanced diffusion (TED) of ion implanted boron in silicon limits the formation of the ultra-shallow junctions required for the extreme deep sub-micron devices. It is also known that this TED is linked to the fate (elimination and agglomeration) of ion implantation related excess self-interstitials. Thus it can be expected that the final high temperature redistribution is at least partly governed by the effective initial point defect distribution at the onset of the high temperature plateau.

In this contribution we present the experimental evidence that low thermal pre-anneals, by affecting the initial self-interstitials distribution, affects boron redistribution during a subsequent high temperature anneal. Samples implanted with high dose boron at 3 keV were first annealed at 700°C for various durations. These samples, as well as reference samples without the pre-anneal, were then RTA annealed at various high temperatures around 1000°C. The resulting B profiles were measured by SIMS. It is found that the pre-annealed samples exhibits a clear reduction of the TED as compared with the reference ones.

INTRODUCTION

Ion implantation is the common technique to introduce impurities into silicon for doping advanced silicon devices in microelectronics. This technique induces a huge over-saturation of point defects in the Si crystal which leads to an anomalous broadening of the dopant profile during the high temperature activation anneal. This phenomenon, known as transient enhanced diffusion (TED) of dopant, is very noticeable for boron, the diffusion of which in Si is mainly mediated by self-interstitials silicon atoms. In fact, the implantation related initial over-saturation of self-interstitials leads to the nucleation and growth, upon annealing, of a collection of various self-interstitial clusters with the eventual formation of extended defects such as {113} defects or dislocation loops [1-3]. These various defects maintain in their vicinity a self-interstitial supersaturation depending on their exact shape and size [4]. Thus the overall boron TED is governed by the evolution and the annihilation / growth kinetics of the point and extended defects present in the material.

On the other hand, it is well known for a long time that silicon self-interstitials can easily diffuse in silicon, even at room temperature. The purpose of this work is to test if it is possible to take advantage of this fact to modify the initial configuration and distribution of self-interstitial silicon atoms using low thermal budget pre-anneals, with the hope of reducing the subsequent TED. Let us point out that a suspicion of such an influence of a low temperature treatment can be suspected from the observation that the temperature ramping-up rate affects TED [5,6] (a pre-anneal treatment can be seen as the extreme case of an infinitely low ramp rate).

In this article, we present experimental evidence that such low thermal budget pre-anneals reduce the high temperature redistribution of low energy boron implants in silicon.

EXPERIMENTS

Samples were prepared from a <100> n-type (P doped), 7-10Ω.cm Czochralski Si wafer. This wafer was implanted with $^{11}B^+$ at an energy of 3 keV and a dose of 3×10^{15} cm^{-2} under a tilt angle of 7° through a thin oxide film of 20Å. Individual samples were then cut from this wafer and submitted to a two-steps anneal, in an argon inert ambient. First a low thermal budget pre-anneal at 700°C was performed for various times, followed by a rapid thermal anneal (RTA) using xenon lamp at high temperature (from 850°C to 1000°C) during 20s. The ramp rate for the RTA was approximately 150°C/s for all the anneals. Some reference samples were RTA treated without the low temperature pre-anneal.

In a first set of experiments, pairs of samples (one with a low temperature pre-anneal and a reference sample) were annealed together in the RTA set-up to minimize experimental scattering. In this set of experiments, we investigated various RTA conditions (temperature and time) for a given pre-anneal (700°C, 20 s).

In a second set of experiments, we investigated the influence of the duration of the 700°C pre-anneal on the final boron redistribution during a given RTA at 950°C for 20s.

The experimental conditions undergone by all the samples used in this study are summarized in Table I.

Table I : Sample description.

Set	Sample	Pre-anneal	RTA
1	1a	None	850°C – 20s
	1b	700°C – 15 min	850°C – 20s
	2a	None	900°C – 20s
	2b	700°C – 15 min	900°C – 20S
	3a	None	935°C – 20s
	3b	700°C – 15 min	935°C – 20s
	4a	None	1000°C – 20s
	4b	700°C – 15 min	1000°C – 20s
2	5a	700°C – 2 min	950°C – 20s
	5b	700°C – 5 min	950°C – 20s
	5c	700°C – 7 min	950°C – 20s
	5d	700°C – 10 min	950°C – 20s
	5f	700°C – 15 min	950°C – 20s
	5g	700°C – 20 min	950°C – 20s

After the final RTA treatments, the chemical boron profiles of all the samples were measured by secondary ion mass spectrometry (SIMS).

RESULTS

Figure 1 shows a typical results evidencing the influence of the pre-anneal. This specific example corresponds to a final RTA at 935°C for 20s (samples 3a and 3b). The sample 3a, (dashed line) underwent only the high temperature RTA, whereas the sample 3b (solid line) was pre-annealed at 700°C for 15 minutes before the same RTA treatment.

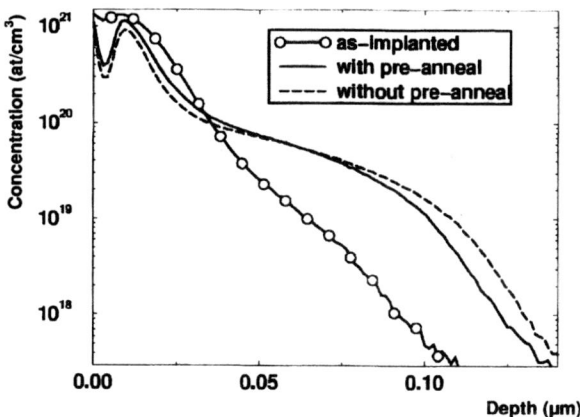

Figure 1 : Comparison of B profiles after a high temperature RTA (935°C, 20 s), with or without a pre-anneal at 700°C for 15 min.

The reduction of the TED in case of sample 3b is evident from the figure. This indicates an apparently paradoxical result : in spite of a larger total thermal budget, the sample having undergone a low temperature pre-anneal exhibits a lower total diffusion length.

In order to get more insights in the underlying mechanism, we first carried out isochronal RTA treatments at various temperatures following the same low temperature pre-anneal. The corresponding results are given in Figure 2. In this figure we plotted, as a function of the final temperature, the reduction of the junction depth (measured at a concentration level of 10^{19} cm^{-3}) extracted from the comparison of the SIMS profiles of the samples with a pre-anneal, compared with the reference ones. From this figure, we can see that the benefit of the low temperature pre-anneal increases with the final RTA temperature : the higher is the RTA temperature, the greater is the reduction in the junction depth, as compared with a reference sample undergoing the same RTA treatment without pre-anneal.

In Figure 3 we report on the influence of the duration of the pre-anneal. This figure shows the various boron profiles obtained after a 950°C, 20 s RTA following pre-anneals at 700°C for various times. As we can observe, for a greater pre-anneal duration, the reduction of TED is more pronounced. The longest pre-anneal time used in this study, i.e. 20 minutes, leads to the greatest reduction of the TED. It is also emphasized that there is no indication of a saturation of the reduction for the pre-anneal times used in this study. Thus it is quite possible that further reduction could be achieved by increasing further the pre-anneal duration.

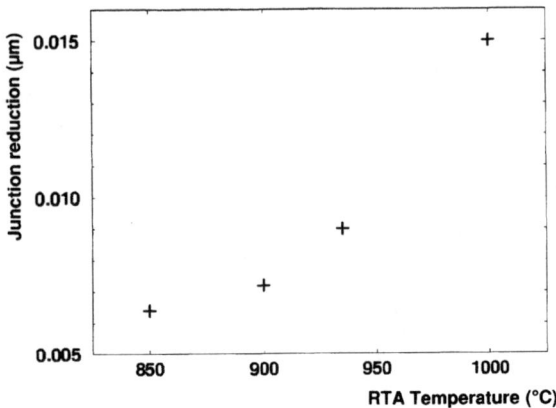

Figure 2 : Reduction of the junction depth for pre-annealed samples, compared to reference samples. The depths are measured at a concentration of 10^{19} at.cm^{-3}.

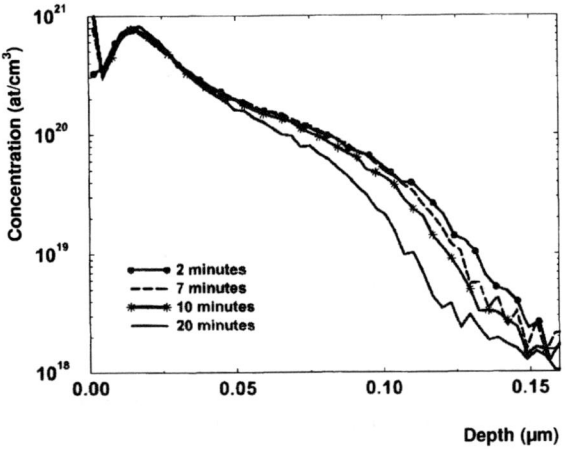

Figure 3 : Comparison of B profiles obtained after RTA (950°C, 20 s) as a function of the duration of the pre-anneal at 700°C.

DISCUSSION

We want first to discuss our results in connection with recent reports concerning the influence of the RTA ramp rate on TED [5,6]. In these studies, it is reported that the higher is the ramp rate, the lower is the TED. These results have lead to the general opinion that in order to minimize TED it is necessary to achieve the high temperature plateau within a time as short as possible. Thus, if the low temperature pre-anneal is viewed as a test for an infinitely low ramp rate, direct extrapolation of these results would lead to the expectation that such a pre-anneal would induce a maximum increase of the TED during the subsequent RTA treatment, in clear contradiction with our experimental results.

However, as already noted in [5], the origin of the TED reduction during high ramp-rate RTA's is mainly linked to the reduction of the overall thermal budget, i.e. the time during which the temperature is high enough to allow dopant diffusion. Thus, our apparently conflicting results can be easily reconciled by considering that in our particular experimental condition the initial pre-anneal happens at a temperature low enough to avoid boron diffusion, even in presence of a high self-interstitial supersaturation. In other words it is only necessary to accept that boron diffusion is significant only above a threshold temperature. During a one step RTA, this temperature is rapidly achieved and B atoms can diffuse during nearly the whole thermal cycle, with the necessity to reduce the total duration (i.e. increasing the temperature ramp rate) in order to reduce the global diffusion. On the other hand, if the pre-anneal takes place at a temperature below this threshold, it is then possible to eliminate part of the mobile free self-interstitials induced by the ion implantation process without starting to redistribute the dopant. Then the final RTA starts with a reduced initial self-interstitial supersaturation and it will thus induce less TED.

Before concluding we simply comment on some possible mechanisms by which the free self-interstitial supersaturation can be decreased during the low temperature pre-anneal. All of these suggestions rely on the fact that the self-interstitial are extremely mobile even at low temperature, and that the self-interstitial supersaturation at the end of the pre-anneal (i.e. at the onset of the final RTA treatment) is the result of the competition of various events occurring during the pre-anneal.

i) The first possible contribution is simply self-interstitial annihilation at the surface of the sample which can act as a sink for the defects. Of course this mechanism is expected to be of more and more importance with the necessary decrease of the implantation energy leading to as-implanted B profiles closer and closer to the surface.

ii) Self-interstitial diffusion towards the bulk of the sample with the eventual annihilation on pre-existing sites (impurities, grown-in defects,...) can also partially contribute to the reduction of the supersaturation.

iii) It is also quite possible that self-interstitial clustering starts during the low thermal treatment, inducing the nucleation and growth of the extended defects. This defects act thus as sinks for the free self-interstitials, decreasing their supersaturation. Moreover this mechanism would lead to the presence at the onset of the final RTA of a distribution of clusters of larger mean size, as compared to the distribution in absence of pre-anneal. Since the free self-interstitial concentration in equilibrium with a cluster distribution is reduced for larger clusters [4], this will further reduce the resulting TED. However it must be emphasized that if this latter scenario is the dominant one, the diffusion enhancement, although less intense, could last during longer time, until the larger defects are totally annealed.

Of course the various suggestions made above are only tentative explanations. Moreover it is more than likely that all the suggested mechanisms occur at the same time with relative influences depending on the detailed experimental conditions. Further experiments, including TEM observations to follow the extended defects evolution, will be necessary to conclude on the exact mechanism involved.

CONCLUSION

In this study we have clearly established that a low temperature pre-anneal, by affecting the self-interstitial distribution, can effectively reduce the transient enhanced diffusion of low energy implanted boron during RTA. Although it is still to early to conclude on the exact physical mechanism involved, this observation indicate that new strategies can probably be considered to minimize TED and thus achieve the ultra low junctions required for the next generation of deep submicron devices.

ACKNOWLEDGEMENTS

The authors are indebted to Dr. J.J. Grob at PHASE, A.M. Papon and C. Lavirron at LETI laboratory for many enlightening discussions.

REFERENCES

1. A. Claverie, L.F. Giles, M. Omri, B. de Mauduit, G. Ben Assayag, and D. Mathiot, Nucl. Instrum. Methods Phys. Res. B **147**, 1 (1999)

2. K. S. Jones, S. Prussin, and E.R. Weber, Appl. Phys. A **45**, 1 (1988)

3. F. Cristiano, J. Grisiola, B. Colombeau, M. Omri, B. de Mauduit, and A. Claverie, J. Appl. Phys. **87**, 842 (2000)

4. A. Claverie, B. Colombeau, G. Ben Assayag, C. Bonafos, F. Cristiano, M. Omri, and B. de Mauduit, Mater. Sci. In Semicond. Process. **3**, 269 (2000)

5. A. Agarwal, H.J. Gossmann, and A.T. Fiory, J. Electron. Mater., **28**, 1333 (1999)

6 G. Mannimo, P.A. Stolk, N.E.B. Cowwern, W.B. de Boer, AZ.G. Dirks, F. Roozeboom, J.G.M. van Berkum, and P.H. Woerlee, Appl. Phys. Lett. **78**, 889 (2001)

Mat. Res. Soc. Symp. Proc. Vol. 669 © 2001 Materials Research Society

REVERSE DIODE LEAKAGE IN SPIKE-ANNEALED ULTRA-SHALLOW JUNCTIONS

Hans-Joachim L. Gossmann
Agere Systems, Murray Hill, NJ

Tao Feng
University of Florida, Gainesville, FL

Aditya Agarwal, Peter Frisella and Leonard M. Rubin
Axcelis Technologies, Beverly, MA

ABSTRACT

We have investigated diode leakage in junctions produced by ion-implantation of B with energies of 0.5 - 2 keV and doses of 2×10^{14} — 2×10^{15} cm^{-2} into n-type wells of $\sim 1 \times 10^{18}$ cm^{-3}, after rapid-thermal anneals (RTA) in lamp-based and hot-wall furnaces. Junctions are as shallow as 30 nm and were directly probed to avoid complications arising from metalization. The leakage current, I_{lkg}, was found to be independent of the implant dose at any reverse voltage (-1 and -5 V). This implies that the electrically active defects are sufficiently far removed and on the surface-side of the junction. In both systems, a spike anneal (no intentional dwell time at peak-temperature) resulted in higher I_{lkg} than a soak anneal (dwell time of several seconds at peak-temperature). However, for the same spike annealing recipe, the hot-wall RTA produces tighter distributions than the lamp-based RTA. The width of the distribution is a measure of the temperature uniformity across the wafer. Best leakage currents are of the order 1×10^{-6} A/cm^2, in good agreement with device simulations The shallowest junctions exhibit $I_{lkg} \sim 5 \times 10^{-4}$ A/cm^2, still well below the specification of even the low power transistor of a 100 nm technology.

INTRODUCTION

The 1999 International Technology Roadmap for Semiconductors (ITRS-99)[1] specifies for a 100 nm Si technology junction depths of less than 30 nm at the channel and less than 70 nm at the source/drain contact. For many reasons, such as uniformity, reproducibility, and cleanliness, ion implantation is presently the method of choice for the formation of these junctions. On the other hand, ion implantation creates defects, giving rise to a host of undesirable effects, such as dopant clustering and transient enhanced diffusion.[2] Hence, during processing, every attempt is made to minimize diffusion, with the consequence that the dopants are not driven very far beyond the region of ion-implantation-induced damage.

The Roadmap also demands for the 100 nm node an extension sheet resistance as low as 200 Ω/□. While is has recently been argued[3][4] that a sheet resistance of ~1kΩ/□ suffices, this still requires very high dopant concentrations. The demand for small diffusive distances translates into a low thermal budget, whereas high electrically active dopant concentrations imply a high annealing temperature, where the solid solubility is large. Therefore, "spike"-anneals, i.e. rapid thermal anneals (RTA) with nominally zero dwell time at the peak temperature, have become the method of choice for activating the junctions.[5][6]

The temperature uniformity across the wafer is an issue with spike-anneals and together with the low thermal budget and the very high dopant concentrations may lead to excessive junction leakage current, I_{lkg}. In this paper we report on an investigation of junction leakage in p^+n junctions as they would be found in the extension and the deep source drain of a 100 nm technology PMOS transistor.

EXPERIMENTAL

Diodes were produced in a 1-lithography-level process-flow on 200 mm p/p^+ epi-substrates. The process flow is summarized in Fig. 1.

```
     300 nm FOX growth
        lithography
oxide RIE, 20 nm remaining
       resist strip
    15:1 HF timed etch
       well implants:
    P 700 keV 1.5×10^13 cm^-3
    P 370 keV 6×10^12 cm^-3
    P  80 keV 3×10^12 cm^-3
    P  20 keV 3×10^12 cm^-3
 4 nm thermal oxide growth
    15:1 HF timed etch
various low energy B implants
  various thermal treatments
```

Figure 1. Process flow.

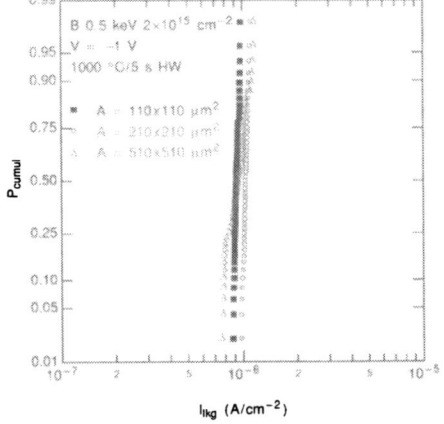

Figure 2. Cumulative probability distribution of reverse current density of a junction implanted with 0.5 keV B 2×10^{15} cm^{-2}, after annealing in a hot-wall (HW) reactor at 1000 °C, 5 s. Shown are three different diode sizes at -1 V.

Diodes are square and have sizes of 110×110, 210×210, and 510×510 μm^2. Low energy implants of B were done on an Axcelis ULE in drift-mode. Anneals, RTA with zero (spike-anneals) and finite dwell-times, were performed in a lamp-based RTA system (L) and a hot-wall furnace (HW). Ramp-rates for the spike RTA were 140 °C/s in the lamp-based system and 80 °C in the hot-wall reactor. Junction-depths for the various implants and anneals are shown in Table 1.

To prevent complications arising from metalization, junctions were directly probed with the wafer-backside as one of the contacts. Immediately before probing, wafers were dipped in 50:1 H_2O : HF to remove the native oxide. The load on the probe tip was carefully increased until a good contact was made without punching through the very shallow junction. For good junctions, such as shown in Fig. 2, the leakage current at -1 V scales very well with the area of the diode, indicating that perimeter effects are negligible. The reverse current of ~10^{-6} A/cm^2 is in excellent agreement with a process and device simulation.

Table 1. Junction-depths at 1×10^{18} cm^{-3} for the implants and anneals [in a lamp-based system (L) or a hot-wall reactor, (HW)] used in this work. The label 'sim' denotes a calculated value.

IMPLANT	ANNEAL	x_j (nm)
B 0.5 keV 2×10^{14} cm^{-2}	1000 °C 5 s	35 (sim)
B 0.5 keV 2×10^{15} cm^{-2}	1000 °C 5 s	75 (sim)
B 0.5 keV 2×10^{14} cm^{-2}	1050 °C SPIKE (L)	25
B 0.5 keV 2×10^{15} cm^{-2}	1050 °C SPIKE (L)	45
B 0.5 keV 2×10^{15} cm^{-2}	1050 °C SPIKE (HW)	50
B 2 keV 1×10^{15} cm^{-2}	1000 °C 5 s	75 (sim)
B 2 keV 1×10^{15} cm^{-2}	1050 °C SPIKE	75 (sim)

IMPLANT DOSE DEPENDENCE

For the ultra-shallow implants discussed here, peak volume concentrations can easily exceed 10^{21} cm^{-3}. Furthermore, for 0.5 keV B, a dose above about 6×10^{14} cm^{-2} leads to the formation of a silicon boride phase[7] and boron-enhanced diffusion of boron (BED).[8] Figures 3 and 4 compare the leakage current for implants below and above this critical dose, after an anneal at 1000 °C for 5 s in a lamp-based RTA system.

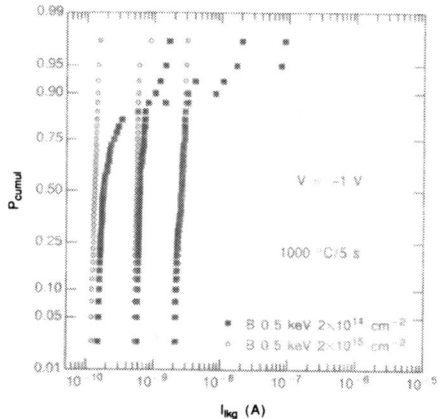

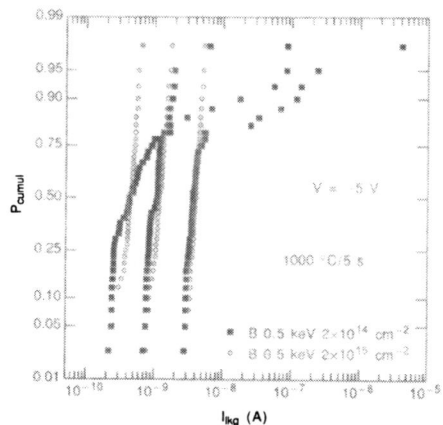

Figure 3. Cumulative probability distribution of reverse current of a junction implanted with 0.5 keV B to doses of 2×10^{14} cm^{-2} (squares) and 2×10^{15} cm^{-2} (circles) after annealing in a lamp-based RTA system at 1000 °C, 5 s. Shown are three different diode sizes at -1 V.

Figure 4. As Fig. 3, but at -5 V.

The leakage current is apparently independent of dose, for -1 V and -5 V. As can be seen from Fig. 5, diffusion during the anneal drives the metallurgical junction about 20 nm beyond its position right after implant for the 2×10^{14} cm^{-2} dose, but about 60 nm for the 2×10^{15} cm^{-2} dose. The fact that this does not impact the leakage current implies that either there are no electrically active defects left after the anneal or that they are all on the surface side of the junction outside of the depletion region. Nevertheless, the shallowness of the junction produced by the

2×10^{14} cm^{-2} implant does have an effect: The leakage current distribution has some number of data-points in the high current tail.

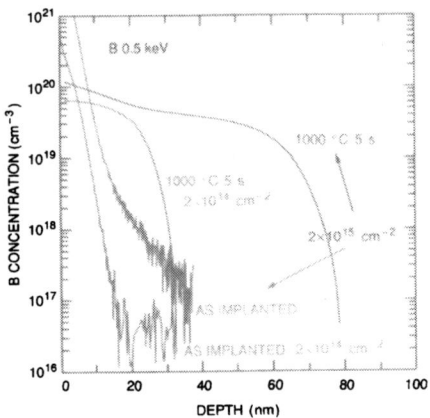

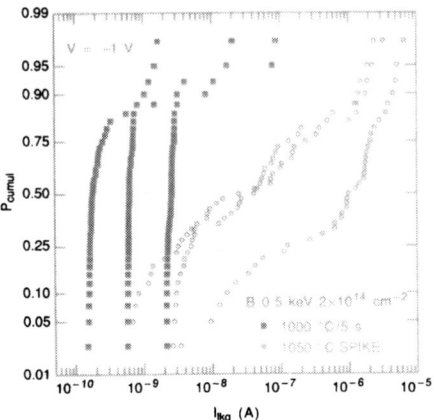

Figure 5. Concentration vs depth of a 0.5 keV B implant, as-implanted and after a 1000 °C, 5 s anneal, for a dose of 2×10^{14} cm^{-2} and 2×10^{15} cm^{-2}.

Figure 6. Cumulative probability distribution of reverse current of a junction implanted with 0.5 keV B 2×10^{14} cm^{-2} after annealing in a *lamp-based* RTA system at 1000 °C, 5 s (squares) and with a 1050 °C spike (circles). Shown are three different diode sizes at -1 V.

SPIKE ANNEALS

For anneals in the lamp-based system, junctions degrade significantly for a RTA anneal to 1050 °C with nominally zero dwell time at the peak temperature (Fig. 6 and 7). For the 2×10^{14} cm^{-2} implant not only does the width of the distribution increase substantially, but even the best diodes exhibit larger I_{lkg} than the diodes that were annealed at 1000 °C, 5 s, even though the latter's junctions are only 10 nm deeper. Leakage improves for the the 2×10^{15} cm^{-2} implant, in particular there are now quite a few diodes with leakage equal to diodes annealed at 1000 °C, 5 s, despite the fact that the latter have junctions 30 nm deeper. Yet, the I_{lkg} distribution is much *broader* than diodes implanted with 2×10^{14} cm^{-2} and annealed at 1000 °C, 5 s that have junctions 10 nm *shallower*.

We conclude that the leakage in spike-annealed diodes is not just due to the shallowness of the created junction but also due to the uniformity of the temperature across the wafer during the spike anneal. This is not entirely surprising, since it is very difficult to achieve good temperature uniformity during a spike anneal in a lamp-based system without special design changes, and no attempt was made to optimize uniformity.

That is indeed the temperature uniformity that plays a major role for the broad leakage distribution in Figs. 6 and 7, is born out by Fig. 8. There we show the results of spike anneals carried out on diodes implanted with 0.5 keV B 2×10^{15} cm^{-2} in a *hot-wall* RTA system. The distribution is now as tight as that of the corresponding 1000 °C, 5 s anneal with a much deeper

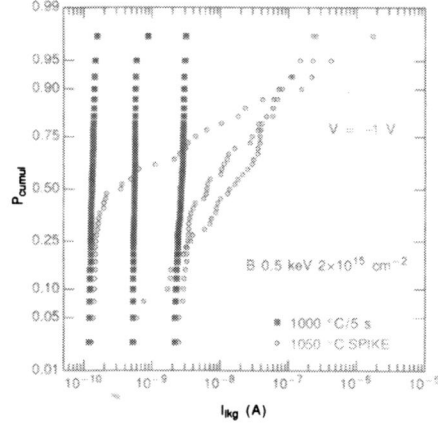

Figure 7. Cumulative probability distribution of reverse current of a junction implanted with 0.5 keV B 2×10^{15} cm^{-2} after annealing in a *lamp-based* RTA system at 1000 °C, 5 s (squares) and with a 1050 °C spike (circles). Shown are three different diode sizes at -1 V.

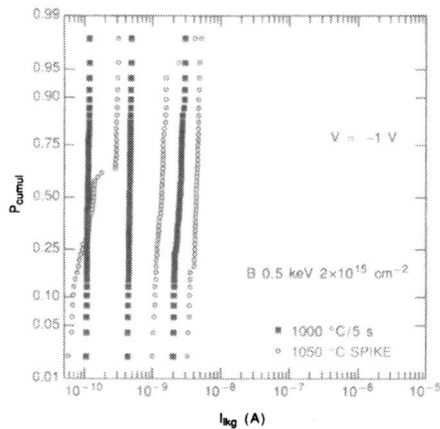

Figure 8. Cumulative probability distribution of reverse current of a junction implanted with 0.5 keV B 2×10^{15} cm^{-2} after annealing in a *hot-wall* RTA system at 1000 °C, 5 s (squares) and with a 1050 °C spike (circles). Shown are three different diode sizes at -1 V.

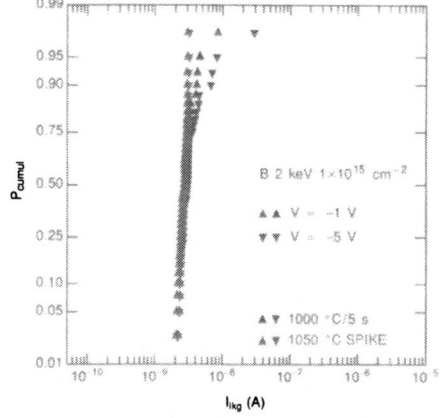

Figure 9. Cumulative probability distribution of reverse current of a junction implanted with 2 keV B 1×10^{15} cm^{-2} after annealing in a *lamp-based* RTA system at 1000 °C, 5 s (squares) and with a 1050 °C spike (circles). Shown are results for $(510 \ \mu m)^2$ diodes at -1 V and -5 V.

junction.

Deeper implants, such as 2 keV, as they would be used for the SD-contact, yield good diode behavior after a 1050 °C spike anneal, even in the lamp-based RTA (Fig. 9).

According to the ITRS-99 a 100 nm technology would have a window-1 size of 115 nm.[1] The length of the source or drain area will at most be twice that. The worst junction measured in this study, with a leakage current of about 0.05 A/cm^2, would then contribute $\approx$ 50 pA/μm, still well below the off-current of the high performance transistor (20 nA/μm). An improvement by just one order of magnitude, easily attainable for example by optimization of the temperature uniformity, would bring that down to well below the off-current of even the low power transistor (20 pA/μm). The comparison tends to become more favorable for later technologies since the area decreases while off-currents are projected to increase.

SUMMARY

We have investigated junction leakage in SD junctions typical of a 100 nm technology.

- Leakage current is independent of the dose of the 0.5 keV B implant, 2×10^{14} cm^{-2} and 2×10^{15} cm^{-2}, resp..

- RTA with zero-dwell time (spike RTA) leads to significant degradation of the leakage current distribution, due to a combination of temperature non-uniformity in the lamp-based RTA and very shallow junction depth.

- Spike anneals in the hot-wall RTA yield excellent junctions.

- Junction leakage currents are in general well below the off-currents of the associated transistor, as projected by the ITRS99.

REFERENCES

1. Semiconductor Industry Association, *The International Technology Roadmap for Semiconductors*, SIA, San Jose, CA (1999). http://public.itrs.net/files/1999_SIA_Roadmap/Home.htm

2. P. A. Stolk, H.-J. Gossmann, D. J. Eaglesham, D. C. Jacobson, C. S. Rafferty, G. H. Gilmer, M. Jaraiz, J. M. Poate, H. S. Luftman, and T. E. Haynes, "Physical mechanisms of transient-enhanced dopant diffusion in ion-implanted silicon," *J. Appl. Phys.* **81**, 6031-50 (1997).

3. H.-J. Gossmann, C. S. Rafferty, and P. H. Keys, "Junctions for deep sub-100 nm mos: How far will ion implantation take us?," *MRS Proc.* **in press**(2000).

4. P. H. Keys, H.-J. Gossmann, K. K. Ng, and C. S. Rafferty, "Series resistance limits for 0.05 μm MOSFETs," *Superlattices and Microstructures* **27**, 125-36 (1999).

5. A. Agarwal, A. T. Fiory, H.-J. Gossmann, C. S. Rafferty, and P. Frisella, "Ultra-shallow junction formation by spike annealing in a lamp-based or hot-walled rapid thermal annealing system: effect of ramp-up rate," *Mater. Sci. Semicond. Process.* **1**, 237-41 (1998).

6. A. Agarwal, A. T. Fiory, and H.-J. Gossmann, "Effect of ramp rates during rapid thermal annealing of ion implanted boron for formation of ultra-shallow junctions," *J. Electron. Mater.* **28**, 1333-9 (1999).

7. A. Agarwal, H.-J. Gossmann, D. J. Eaglesham, S. B. Herner, A. T. Fiory, and T. E. Haynes, "Boron-enhanced diffusion of boron from ultralow-energy ion implantation," *Appl. Phys. Lett.* **74**, 2435-7 (1999).

8. A. Agarwal, H.-J. Gossmann, and D. J. Eaglesham, "Boron-enhanced diffusion of boron: Physical mechanisms," *Appl. Phys. Lett.* **74**, 2331-3 (1999).

Mat. Res. Soc. Symp. Proc. Vol. 669 © 2001 Materials Research Society

Boron Solubility Limits Following Low Temperature Solid Phase Epitaxial Regrowth

C. D. Lindfors and K. S. Jones
Department of Materials Science and Engineering, University of Florida,
Gainesville, FL 32611-6130 U. S. A.

M. J. Rendon
Semiconductor Products Sector, Motorola Inc.
Austin, TX 44548 U. S. A.

ABSTRACT

The work described herein focuses on examining the effect of solid phase epitaxial regrowth (SPER) on boron implanted silicon. It is shown that boron levels within the silicon can greatly enhance or reduce the regrowth rate of the silicon. Electrical measurements show optimum sheet resistances for 5 keV, 2×10^{15} cm^{-2} implant conditions yielding sheet resistance values of ~140 Ω/sq at 500 °C annealing to ~120 Ω/sq at 650 °C. Results using Hall effect and four-point probe show lower doses of boron will become fully active but levels will drop significantly as dose is increased. Lastly, maximum active concentrations of boron appear to reach values of ~3-4$\times10^{20}$ cm^{-3} for a boron dose of 1×10^{15} cm^{-2} after SPER. Lower SPER anneal temperatures or higher doses tend to activate less boron.

INTRODUCTION

As the semiconductor industry continues to scale down device dimensions there is an increased need to activate higher levels of dopants to maintain low sheet resistance layers. The ITRS roadmap[1] put forth by the Semiconductor Industry Association indicates the 35 nm technology node will need junctions as shallow as 13-17 nm with sheet resistances in the 100-400 Ω/sq range. However, conventional processing techniques, such as ion implantation and rapid thermal anneal, seem to be hitting a limit for the ability to generate shallow active layers. The major drawbacks arise from the phenomenon known as transient enhanced diffusion (TED) which has been shown to significantly increase junction depths[2,3]. To circumvent the effect of TED one needs to either anneal with a very high ramp rate for short periods or anneal at low temperatures which allows the underlying crystalline silicon to act as a seed to reorder the amorphous layer. The dopant atoms will rest on substitutional sites and become electrically active upon regrowth. There has been recent work[4-6] and work done in the late 1970s to early 1980s[7-10] that verifies high levels of active dopant can be achieved using SPER. The goal of this experiment was to study shallow junctions formed by SPER.

EXPERIMENTAL

N-type silicon wafers with <100> orientation were amorphized using a 30 keV, $1x10^{15}$ cm^{-2} Si^+ implant and then implanted with B^+ at 5 keV in the dose range of $5x10^{14}-8x10^{15}$ cm^{-2}. These wafers were then cut into 15x15 mm^2 pieces so samples could be annealed at various times at temperatures. Anneals at 500 °C were carried out in a quartz tube furnace with a N_2 purge at times of 20-45 minutes while the 550-650 °C anneals where performed using a rapid thermal anneal (RTA). Samples were annealed at 550 °C for 3-5.5 minutes, 600 °C for 25-50 seconds, and 650 °C for 4-15 seconds.

To measure the success of the SPER process, several characterization techniques were employed. Variable angle spectroscopic ellipsometry (VASE) measured the amorphous layer thickness on the surface of the silicon pieces to determine when regrowth is complete. Electrical characterization utilized four-point probe to measure sheet resistance and limited Hall effect measurements where taken to determine the active boron doses. Finally, secondary ion mass spectrometry (SIMS) was used from the as-implanted samples to compare active dose to implanted dose.

RESULTS

Since the boron dose is varied it is expected that the time to complete regrowth will be affected by the boron concentration. Previous work[11] using various dopants indicates there is a maximum regrowth rate achieved at a particular concentration, and above or below this concentration the regrowth rate decreases. This is seen in our work and is illustrated in figure 1 for 500 °C anneals. Figure 1 shows that as the boron dose increases up to $4x10^{15}$ cm^{-2} the regrowth is being completed at shorter times, while above this dose it is taking longer for complete regrowth. It is also interesting to note that at the highest dose the regrowth is halted around ~30 Å. This is also seen for anneals at 550 °C but the higher temperature anneals show complete regrowth at all doses.

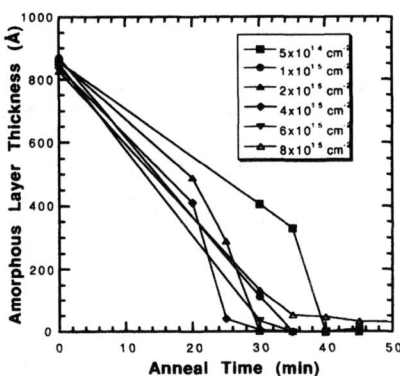

Figure 1: Plot of amorphous layer thickness remaining versus anneal time at 500 °C.

Sheet resistance measurements show optimized dose conditions for all the anneal temperatures. Figure 2 plots the sheet resistance versus the boron implant dose for samples annealed at 500 °C. For this and all the other anneal temperatures a minimum sheet resistance is found for the implant dose of 2×10^{15} cm^{-2}. Figure 2 also illustrates how regrowth is still occurring for the 8×10^{15} cm^{-2} dose and as it continues the sheet resistance is lower due to further activation. At 500 °C the minimum sheet resistance reaches ~140 Ω/sq while a minimum sheet resistance of ~120 Ω/sq occurs for samples annealed at 650 °C. Therefore there seems to be a slight temperature dependence on the optimum sheet resistance. The junction depth, defined at 10^{17} cm^{-3}, is 108 nm for these samples.

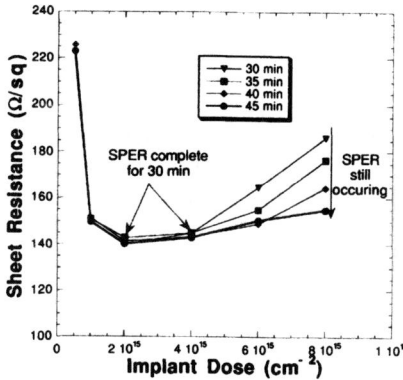

Figure 2: Sheet resistance versus implant dose for 500 °C anneals

Reliable Hall effect results were more difficult to obtain, therefore only a limited number of samples are reported. The Hall system was measuring electron carriers at room temperature instead of holes but if samples were cooled below ~250 K the transition to measuring holes would occur. Thus samples were measured at 200 K to determine active carrier doses. The contacts made to the sample were ohmic and an oxide etch was performed on a sample to rule out the possibility of surface contamination. A mass spectrum using SIMS was also run on a sample to determine if any residual elements may have entered the samples but only the boron was detected. The final test was to take a sample with a 2×10^{15} cm^{-2} B$^+$ implant and measure after SPER and compare the results after a 900 °C, 30 minute post anneal that ensures the boron is activated. After the SPER at 500 °C for 45 minutes the sample measured a sheet resistance of 141 Ω/sq on the four-point probe while the Hall system measured electrons at 80 Ω/sq and room temperature and holes at 141 Ω/sq and 200 K. The results following the post anneal measured a sheet resistance of 135 Ω/sq on the four-point probe while the Hall system measure 125 Ω/sq at room temperature and 128 Ω/sq at 200 K but both measurements were holes. This suggests that the results from the four-point probe are believable but there is obviously some difficulty obtaining accurate Hall measurements.

Since results from the four-point probe were much more easily accomplished and trustworthy there were two approaches to determining percent boron and maximum active boron concentration. For Hall effect the active dose measured is simply compared to the implant dose, however for the four-point probe it is harder to extract the active dose. To determine the active dose a program was set up to calculate the sheet resistance using the SIMS profile which has been modified such that above a certain boron concentration it is assumed the boron is inactive. Then estimating the mobility at each concentration using an empirical formula from Schroder[12], the SIMS profile can be modeled as a set of resistors in parallel. The SIMS profile is modified until the theoretically predicted sheet resistance matches the experimental values and thus the active dose and maximum active boron concentration are extracted. The drawback to using four-point probe data occurs for the low doses where the predicted sheet resistance is higher than the actual sheet resistance; therefore 100% activation has to be assumed.

Combined results from the Hall effect system and four-point probe are shown in figure 3. This figure plots the percent active boron versus the implanted boron dose and clearly shows that a much smaller fraction of boron is being activated as higher implant doses are reached. The figure also shows that the Hall system is measuring slightly higher activation than that predicted by the four-point probe but the general trend is still

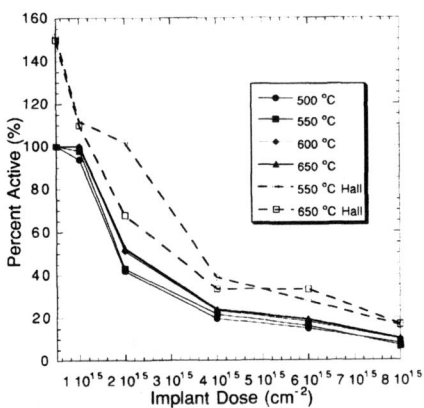

Figure 3: Percent activation versus implant dose for samples tested using 4-point probe and Hall effect. Samples were annealed at 500 °C for 45 min, 550 °C for 3 min, 600 °C for 35 sec, and 650 °C for 4 sec.

the same. If we now take the maximum boron concentration determined from the four-point probe and plot it versus boron implant dose we see there is a maximum active concentration achieved for the 1×10^{15} cm^{-2} dose which has a maximum concentration of $\sim 4 \times 10^{20}$ cm^{-3}. This plot is given in figure 4 and also shows that the highest active concentrations are achieved in the samples implanted with 1-2×10^{15} cm^{-2} doses of boron.

As mentioned before, the lower boron dose samples, such as the 1×10^{15} cm^{-2} set, may have measured a sheet resistance lower than the predicted sheet resistance so 100 % activation is assumed and may explain the sudden increase in maximum activation, but lowest overall sheet resistance values were seen at the 2×10^{15} cm^{-2} implant dose. Finally, if we take the data in figure 4 and now plot it versus anneal temperature it is clear there is a slight advantage to annealing at higher regrowth temperatures as is seen in figure 5. All doses show highest active concentrations at 650 °C annealing except for the lowest dose, which activates all the boron at all temperatures.

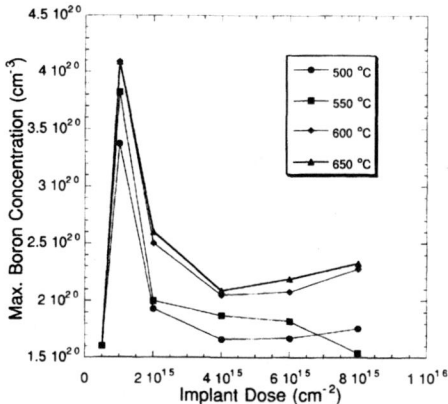

Figure 4: Maximum active boron concentration as a function of implant dose. Samples were annealed at 500 °C for 45 min, 550 °C for 3 min, 600 °C for 35 sec, and 650 °C for 4 sec.

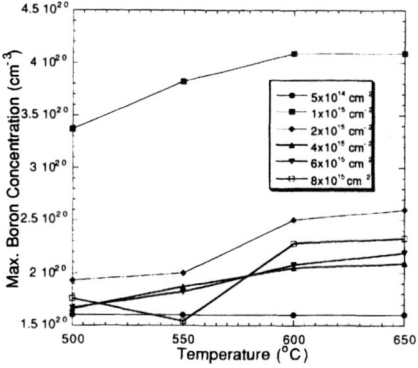

Figure 5: Maximum active boron concentration as a function of anneal temperature. Samples were annealed at 500 °C for 45 min, 550 °C for 3 min, 600 °C for 35 sec, and 650 °C for 4 sec.

CONCLUSIONS

Solid phase epitaxial regrowth appears to be a sufficient method to reach future technology nodes. The results of this work show that indeed boron is activated at these lower temperature anneal conditions where TED is suppressed. We have shown that optimum implant conditions exist for reaching the lowest sheet resistance and producing the highest maximum activated concentration levels. It appears that concentrations of $\sim 3\text{-}4 \times 10^{20}$ cm^{-3} boron can be activated using the SPER technique under the right conditions. Finally, there is evidence that indicates slightly better activation occurs at the highest regrowth temperature.

ACKNOWLEDGEMENTS

The research in this paper was supported by the Semiconductor Research Corportation under contract 632.001. The authors would like to thank Martin Giles and Paul Packan for their support and John Borland for insite into this research. We would also like to thank Mark Clark for the valuable SIMS work performed.

REFERENCES

1. International Technology Roadmap for Semiconductors (ITRS), 1999 Edition, Dec., 1999, Semiconductor Industry Association, 181 Metro Drive, Suite 450, San Jose, CA 95110 (http://www.itrs.net/1999 SIA Roadmap/Home.htm)
2. A. E. Michel, Nuclr. Inst. and Meth. In Phys. Res. B, **37/38**, 379 (1989)
3. D. J. Eaglesham, P. A. Stolk, H. J. Gossman, and J. M. Poate, Appl. Phys. Lett., **65**(18), 2305 (1994)
4. C.D. Lindfors, K.S. Jones, M.E. Law, D.F. Downey, and R.W. Murto Si Front End Processing – Physics and Technology of Dopant-Defect Interactions II, Mat. Res. Soc. Proc., **610**, B.10.2.1 (2000)
5. W. L. Harrington, C. W. Magee, M. Pawlik, D. F. Downey, C. M. Osburn, and S. B. Felch, J. Vac. Sci. Technol. B, **16**(1), 286 (1998)
6. C. M. Osburn, D. F. Downey, S. B. Felch, and B. S. Lee, 11th Intl. Conf. on Ion Imp. Tech., 607 (1996)
7. L. Csepregi, E. F. Kennedy, J. W. Mayer, and T. W. Sigmon, J. Appl. Phys. **49**(7), 3906 (1978)
8. R. Drosd and J. Washburn, J. Appl. Phys. **53**(1), 397 (1982)
9. J. S. Williams (J. M. Poate, G. Foti, and D. C. Jacobson, eds.), *Surface Modification and Alloying*, (Plenum Press, New York, 1982) p. 133
10. J. S. Williams and K. T. Short (S. T. Picraux and W. J. Choyke, eds.), *Metastable Materials Formation by Ion Implantation*, (North Holland, New York, 1982) p. 109
11. J. S. Williams and J. M. Poate, eds., *Ion Implantation and Beam Processing*, (Academic Press, New York, 1984) p. 27
12. D. K. Schroder, *Semiconductor Material and Device Characterization*, (John Wiley & Sons Inc., New York, 1990) p. 232

Simulation and Modeling

Mat. Res. Soc. Symp. Proc. Vol. 669 © 2001 Materials Research Society

Modeling of Dopant Defect Interactions

C. Camarce[2], L. Radic[2], P.Keys[1], R. Brindos[1], K.S. Jones[1], and M. E. Law[2]
[1]Department of Materials Science and Engineering
[2]Department of Electrical and Computer Engineering
University of Florida, Gainesville, FL 32611

ABSTRACT

This paper presents a model for {311} defects based on in-situ experiments. The model fits the 311 dependence on silicon implant energy and doses. The surface dependence of the model is described in detail, and compared to previous literature data. New data is presented on the surface effect on {311} dissolution and the model is compared to that data. In addition, the model is also used to explain the effects of doping on {311} defect behavior. Doping does not influence the dissolution of {311} defects, but only influences their nucleation behavior.

INTRODUCTION

Transient Enhanced Diffusion (TED) has been the primary modeling challenge of the process simulation community. Qualitatively, the damage from the implant creates an enhanced level of point defects which substantially increases the diffusion of dopants for a short time until the implant damage is fully annealed. This qualitative explanation of the phenomena has been accepted for some time. Quantitative explanations, however, have been difficult to develop.

The standard model of TED requires that the Frenkel pair distribution generated by the implant recombines very quickly. This leaves behind the excess atoms that were implanted - the "+1" model. This dose of interstitials quickly forms extended defects, which in turn dissolve relatively more slowly and create the enhanced diffusion of dopants that is observed. Rafferty *et al.*[1] have suggested that the dissolution is controlled by a strong surface sink. The surface annihilation is responsible for eliminating the excess defects. The rate of surface recombination, therefore, becomes a critical component of the model.

Quantitative Transmission Electron Microscopy (TEM) has been used to help investigate and quantify this model. Eaglesham, *et.al*, [2]have suggested that for low dose silicon implants, interstitials are stored in {311} defects. This defect slowly dissolves and maintains an interstitial excess. Initial investigations have shown that this defect dissolves with a time constant approximately the same as for TED, and that this defect is "the source of the interstitials"[2]. This confirms the first part of the standard model, that roughly a "+1" of defects is contained in the {311}'s and that this defect dissolves slowly with a time constant similar to that of TED. Because of this work, we now have confirmation of part of the standard model.

In this paper, a model is presented and compared to a variety of data showing the dependence of dissolution of the {311} defects on the surface. We also compare the model to dissolution studies performed in heavily doped backgrounds, and show that the model can explain these phenomena as well.

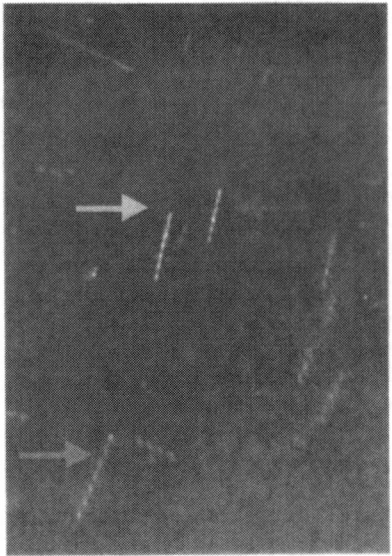

Figure 1 – Left picture is the start of the in-situ annealing. The right picture is after 15minutes at 770°C. The large defect in the lower left has completely dissolved in 15 minutes. The left of the bright pair in the center of the picture has shrunk in length by about a factor of 2, while the neighboring defect has only changed about 10% in size.

311 DEFECT MODEL

In-Situ annealing in the TEM allows individual defect behavior to be observed and monitored[3]. A 100keV Si 10^{14}cm^{-2} implant was used to damage a silicon wafer. These samples were then preannealed at 750°C for two hours in a conventional furnace. The samples were then annealed in-situ in the TEM at a variety of temperatures. Figure 1 shows the evolution at 0 and 15 minutes. In this time period, note that the defect in the lower left has almost completely dissolved. The longer defects are not more stable energetically than smaller defects. This work clearly shows that longer defects can dissolve much faster than shorter defects.

Analysis of the dissolution of different defects in the in-situ study demonstrate that the defect decay is best fit by a linear decay rather than an exponential. The defects tend to lose a constant number of interstitials per time, and the rate does not vary with the length of the defect. Figure 2 shows the linear decay rate as a function of the defect size. There is not a strong trend of decay rate as a function of length, further supporting that longer defects are not more stable than short defects.

A {311} model has been developed based on the results of the in-situ study[3]. This model solves for the total number of interstitials in the defects (C_{311}) and the total number of defects (D_{311}). Capture and release of intersititals on the {311} defects occurs only at the end of the defects, and therefore is proportional to the number of defects, D_{311}. This provides two distinct results. First, individual defects dissolve at a nearly constant rate, since the dissolution is proportional only to the end size. The length of the defect does not determine the dissolution

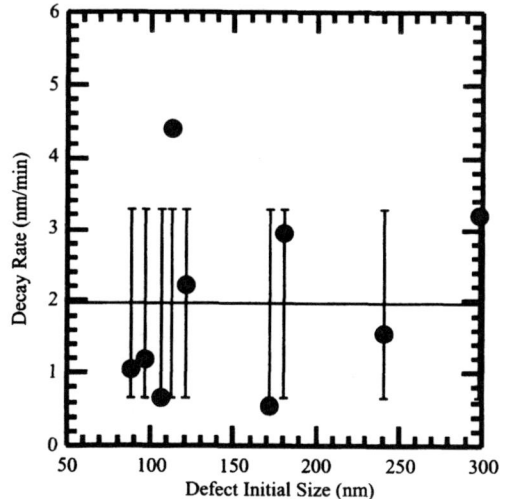

Figure 2 - Decay rates of defects. Error bars are plus and minus one standard deviation about the mean. The model value at 770°C is 2.3nm/min, only about 20% higher than the mean and well within the one standard deviation error bars.

rate. This is similar to the in-situ result observed in Figure 2. The model predicts a decay rate of 2.3nm/min at 770°C, within the one standard deviation error bars in Figure 2. The rate-limiting step for the dissolution is release of interstitials from the defect, rather than diffusion to the surface as in previous models.

The second interesting result is that the defect ensemble decays at a rate that is dependent on size. The number of defects determines the decay rate. Consider two ensembles with the same number of interstitials, but with different sizes. The ensemble with the smaller size has more defects than the one with the larger size, and therefore will decay faster. The {311} defect population decays at a rate that is proportional to the interstitial loss rate and inversely proportional to the size of the defect.

NUCLEATION BEHAVIOR

Nucleation of the defects occurs during the implant process. Initial distributions of defects are from UT-Marlowe using the kinetic accumulation damage model[4]. Vacancies and interstitials are allowed to recombine and complex during a room temperature anneal of the full damage profile. Since the vacancies are mobile at room temperature, they either recombine with interstitials or complex into di-vacancies. The remaining interstitials form small interstitial clusters (SMIC's) or sub-microscopic {311} defects. This agrees with MD simulations that show most of the damage cascade in small clusters after implantation[5]. The small cluster energies are similar to those used by Cowern[6] for his most stable defect. The proto-311's provide nucleation site for the growth of the {311} defects. The SMIC's dissolve by releasing interstitials to the surface.

Figure 3 shows the measured plus factor as a function of silicon implant energy for a constant 10^{14}cm^{-2} dose as well as the result from the model. These points were at the end of 15minute, 750°C anneal. The initial defect and interstitial concentrations in {311}'s are predicted well. This is difficult to obtain with a homogenous nucleation model, since the homogeneous

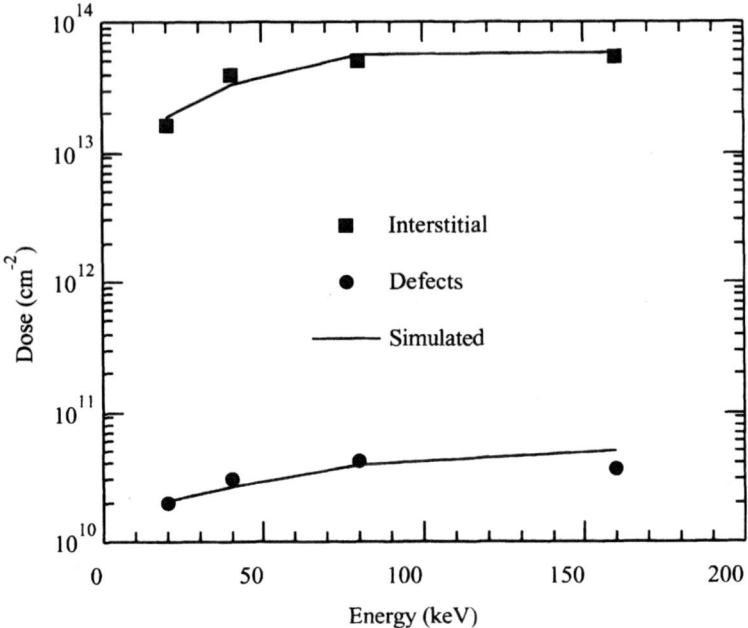

Figure 3 – Dependence on the initial {311} concentration of defects and interstitials on Si implant energy. The implant dose was 10^{14}cm^{-2}.

nucleation is proportional to a power of the peak damage concentration. Since the peak damage concentration drops with energy, it is difficult to get an increase in the number of defects with energy. This model nucleates the {311} defect heterogeneously on SMIC's, so it does not have the same difficulty. Since the SMIC's dissolve to the surface, an increase in the implant energy makes them longer lasting defects. This allows a greater opportunity to form {311} defects, resulting in an increased defect density with increasing energy.

SURFACE EFFECTS ON 311 DEFECTS

The surface is handled differently than most prior models. Interstitials and vacancies have relatively weak recombination rates at the surface. Vacancies can recombine at the surface and leave behind interstitial traps. The interstitials continue to recombine at the surface until all vacancy damage is annihilated. This is similar to what was observed experimentally[7] by Bedrossian *et al.* They showed that vacancies appeared at the surface first after an implant, and then this damage was repaired as interstitials arrived. This is incorporated into the model with a vacancy traps. Vacancies come to the surface during the room temperature anneal and accumulate. At annealing temperatures, interstitials flow to the surface and can recombine on the vacancy site. In addition, di-interstitials can also recombine at the surface[8]. Previous work has

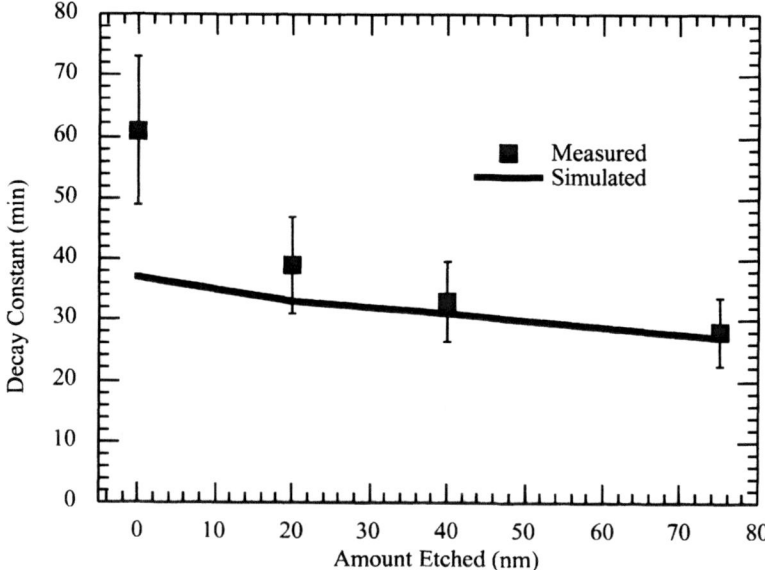

Figure 4 – Dependence on the {311} interstitial decay constant as a function of amount of material etched. Decay rates are extracted from exponential fits to the data or model.

demonstrated that this model predicts the observed energy dependence of the {311} decay well[3].

Energy dependence can be a dangerous way to investigate the surface dependence since an a change in the energy also changes the total damage. A new experiment was performed with a silicon implant with a dose of 10^{14} cm^{-2} and energy of 50 keV. A post-implant anneal at 750 °C for 15 minutes creates a band of {311} defects. The distance from the silicon surface to the top of the {311} defect band is approximately 60nm, and the width of the defect band is 90nm. Chemical Mechanical Polishing (CMP) was used to create samples with 20nm, 40nm, 75nm of surface material removed. This effectively brought the surface closer to the defect band. Cross-section transmission electron microscopy (XTEM) analysis confirms these distances. Furnace anneals are performed on the samples at 750 °C for 15, 45, 90, and 135 minutes.

The decay rate of the {311} defects were extracted assuming an exponential decay of the data. The decay constant is compared to the simulated value of the decay constant in Figure 4. Good agreement is observed for the samples that have been etched, but there is considerable mismatch in the unetched sample. The decay rate found for this implant condition is larger than that observed for similar energies in the past. The surface dependence that is observed in the model arises from a change in the rate interstitials are converted to di-interstitials. Since the di-interstitial surface recombination velocity is fairly large they are effected by surface proximity. The decrease in di-interstitals is then compensated by an increased conversion of interstitials. This results in faster removal of interstitials from the system and a slight increase in the defect dissolution decay.

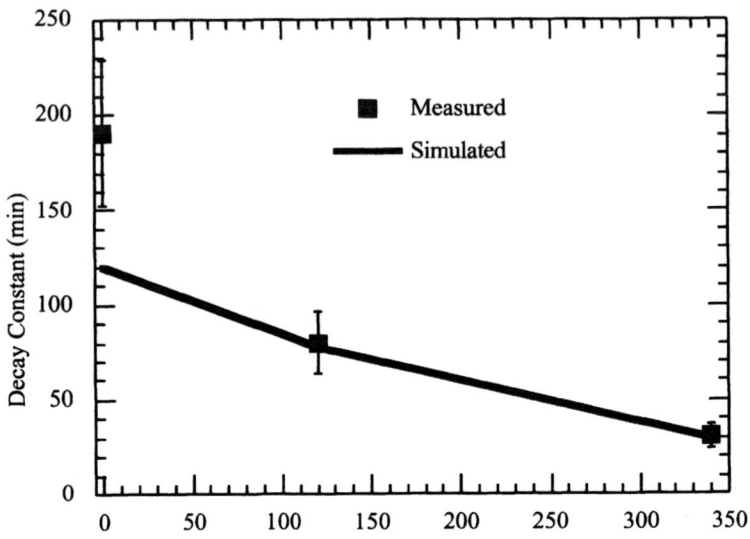

Figure 5 – Measured and simulated decay constants as a function of the amount of silicon polished from surface. Decay constants were extracted using exponential fits to the measured time dependence.

To test the effect on deeper layers, an amorphization study was also performed. Silicon was implanted with silicon at 4 x 10^{13} cm^{-2} at 40 keV, followed by another Si$^+$ implant of 1 x 10^{15} cm^{-2} at 225 keV. The second implant was done at liquid nitrogen temperature (77 K) to prevent self-dynamic annealing and to minimize the formation of type II dislocation loops. The double implant of Si$^+$ forms a 4300 Å continuous amorphous layer. Using CMP, the thickness of the amorphous layer is polished back to depths of 1100 and 3100 Å. Next, furnace anneals are performed at 750 °C on all three amorphous depths in a flowing N$_2$ ambient. The anneal times for the polished samples are 15, 45, 90, and 135 minutes for the 1100 Å sample, 15, 45, 90, 135, and 180 minutes for the 3100 Å sample, and 15, 45, 90, 135, 180, 225, and 270 minutes for the 4300 Å (unpolished) sample. Cross-section transmission electron microscopy (XTEM) and spectroscopic ellipsometry (SE) verify the amorphous layer thickness. Quantitative analysis of {311} defect dissolution at the end-of-range (EOR) region is acquired through plan-view transmission electron microscopy (PTEM). Only {311}'s are observed in the final samples.

Figure 5 plots the measured and simulated decay rates for these samples. A stronger dependence on distance to the surface is observed here, because at these large depths the dissolution does become surface diffusion limited. The surface is so far away that diffusion to the surface becomes the rate limiting step for the dissolution. Consequently, these samples show a much stronger dependence on distance to the surface than observed in Figure 4. The model does a good job of fitting the data for the etched samples, but over predicts the dissolution rate for the unetched sample. The problem may arise due to the width of the {311} layer predicted by the model. UT-Marlowe predicts a very wide band of damage, and the {311} defect band

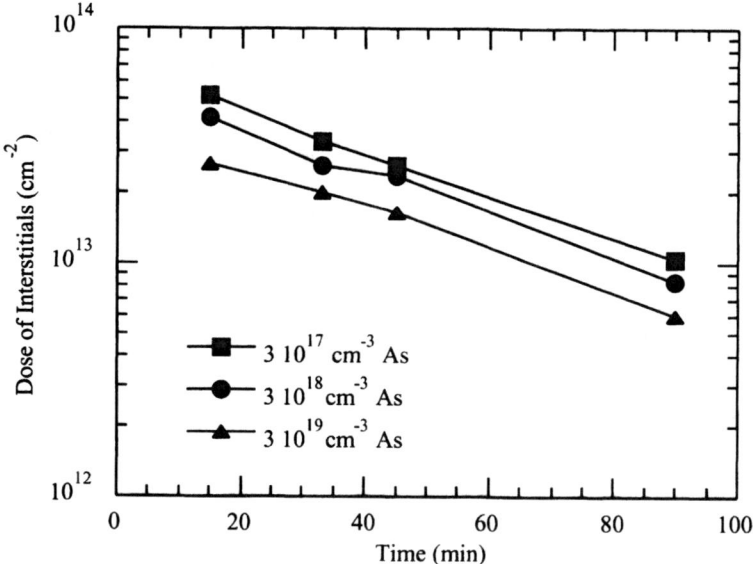

Figure 6 – The number of interstitials in {311} defects as a function of time and arsenic concentration. The decay rate is unchanged with arsenic, but the initial number of interstitials is altered.

that forms is nearly 600nm wide. This is considerably larger (by a factor of 5) than is observed in the experimental structure. This results in more defects being predicted by the model than is observed in the experiment, which has an effect on the decay rate.

A similar problem is observed when simulating recent results from Venezia, *et al*.[9]. They used a silicon implant at 350keV with a dose of 10^{14}cm^{-2}. These samples received a preanneal and then were polished to remove the surface material, similar to the experimental result shown in Figure 4, but at a much higher energy. The defect band predicted by the model extends from 420nm to 1.2um below the surface. The observed defect band was from 520nm to about 800nm below the surface. Agreement between the model and data is found with the sample receiving the most polishing. However, Venezia reported that the thickness dependence is much stronger than that observed with the model.

DOPANT AND DEFECT INTERACTIONS

A series of Arsenic wells were fabricated with peak concentrations ranging from $3 \cdot 10^{17}$ to $3 \cdot 10^{19}$ cm^{-3}. Into each well a 40keV, 10^{14}cm^{-2} silicon implant was performed to make {311} defects. Figure 6 shows the measured defect interstitial doses contained in the {311} defects as measured by TEM. It can clearly be seen that the decay constant of the three different well concentrations are the same. The decay constant is 49min +/- 1minute. The only effect observed is the initial number of interstitials in the {311} defects.

The model agrees with this behavior. The vacancies are annihilated at low temperature leaving behind something roughly equivalent the a "plus 1" dose of interstitials. During the ramp up, there is a competition between the arsenic and the defects for interstitials. A binding energy of the arsenic-interstitial of 0.95eV is sufficient to match the data[10]. During dissolution, the arsenic alters the rate of interstitial removal. However, this doesn't change the dissolution of the {311} defects measurably. The model isn't very sensitive to the removal rate of interstitials from the system.

A similar experiment was performed with phosphorus[11], with similar results. Figure 7 shows the interstitial dose in {311} defects as a function of time and phosphorus concentration. The decay constants for the various phosphorus doses are nearly identical. As with the arsenic wells, the primary influence is on the initial number of interstitials contained in the defects. The model again does a good job predicting this behavior for the same reasons. The model is more complex than that for arsenic, since the phosphorus diffusion is not captured adequately by a simple pair model.

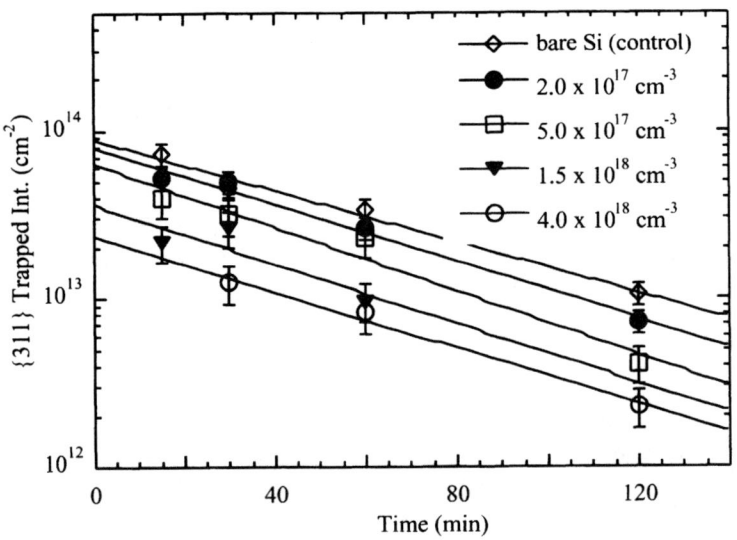

Figure 7 - The number of interstitials in {311} defects as a function of time and phosphorus concentration. The decay rate is unchanged with phosphorus, but the initial number of interstitials is altered.

CONCLUSIONS

A model for {311} defects has been described and compared to data on surface proximity and doping level. Surface proximity has an effect on defect dissolution that is accounted for in the model. Doping influences only the nucleation of {311} defects and not their dissolution. The model accounts for this behavior, because the dissolution of the defects is not strongly dependent on the environment.

ACKOWLEDGMENTS

The authors would like to thank Dale Jacobson for performing the liquid nitrogen implants. We also wish to acknowledge the support of the Semiconductor Research Corporation of this work.

REFERENCES

[1]   C. S. Rafferty, G. H. Gilmer, M. Jaraiz, D. Eaglesham, and H.-J. Gossmann, "Simulation of Cluster Evaporation and Transient Enhanced Diffusion in Silicon," *Appl. Phys. Lett.*, vol. 68, pp. 2395-2397, 1996.

[2]   D. J. Eaglesham, P. A. Stolk, H.-J. Gossmann, and J. M. Poate, "Implantation and Transient B Diffusion in Si: The Source of the Interstitials," *Appl. Phys. Lett.*, vol. 65, pp. 2305-2307, 1994.

[3]   M. E. Law and K. S. Jones, "A New Model for {311} Defects Based on In-Situ Measurements," presented at International Electron Devices Meeting, San Francisco, 2000.

[4]   S. Tian, M. F. Morris, S. J. Morris, B. Obradovic, G. Wang, A. F. Tasch, and C. M. Snell, "A Detailed Physical Model for Ion Implant Induced Damage in Silicon," *IEEE Transactions of Electron Devices*, vol. 45, pp. 1226-1238, 1998.

[5]   M. J. Cuturla, T. D. delaRubia, L. A. Marques, and G. H. Gilmer, "Ion-Beam Processing of Silicon at keV Energies: A Molecular Dynamics Study," *Phys Rev. B*, vol. 54, pp. 16683-95, 1996.

[6]   N. E. B. Cowern, M. Jaraiz, F. Cristiano, A. Claverie, and G. Mannino, "Fundamental Diffusion Issues for Deep-Submicron Device Processing," presented at International Electron Devices Meeting, Washington, DC, 1999.

[7]   P. J. Bedrossian, M.-J. Caturla, and T. D. d. l. Rubia, "Damage Evolution and Surface Defect Segregation in Low-Ebergy Ion-Implanted Silicon," *Appl. Phys. Lett.*, vol. 70, pp. 176-178, 1997.

[8]   M. E. Law, Y. M. Haddara, and K. S. Jones, "Effect of the Silicon/Oxide Interface on Interstitials: Di-Intersitial Recombination," *Journal of Applied Physics*, vol. 84, pp. 3555-60, 1998.

[9]   V. C. Venezia, R. Kalyanaraman, H.-J. L. Gossmann, C. S. Rafferty, and P. Werner, "Depth dependence of {311} defect dissolution," *Applied Physics Letters*, Submitted.

[10]   R. Brindos, P. Keys, K. S. Jones, and M. E. Law, "Effect of Arsenic on Extended Defect Evolution in Silicon," presented at Materials Research Society Symposium, San Francisco, 2001.

[11]   P. Keys, K. S. Jones, M. E. Law, M. Puga-Lambers, and S. M. Cea, "Dopant-Defect Clustering in Phosphorus Implant Silicon: Experimentation and Modeling," presented at Materials Research Symposium, San Farncisco, 2001.

Mat. Res. Soc. Symp. Proc. Vol. 669 © 2001 Materials Research Society

Atomistic Modeling of Amorphization in Silicon

Lourdes Pelaz, Luis A. Marqués, George H. Gilmer[1], Juan Barbolla
Dept. Electricidad y Electrónica, Universidad de Valladolid,
47011 Valladolid, Spain.
[1] Agere Systems (Formerly Bell Laboratories, Lucent Technologies)
600 Mountain Avenue, Murray Hill, NJ 07974, U.S.A.

ABSTRACT

We discuss atomistic simulations of ion implantation and annealing of Si over a wide range of ion dose and substrate temperatures. The DADOS Monte Carlo model has been extended to include the formation of amorphous regions, and this allows simulations of dopant diffusion at high doses. As the dose of ions increases, a continuous amorphous layer may be formed. In that case, most of the excess interstitials generated by the implantation may be swept to the surface as the amorphous layer regrows, instead of diffusing through the crystalline region. This process reduces the amount of transient enhanced diffusion during annealing. This model also reproduces the dynamic annealing during high temperature implants.

INTRODUCTION

When energetic ions strike a silicon substrate, they create zones of disorder, populated by Si self-interstitials and vacancies. The lattice in these disordered regions exhibits different damage configurations going from isolated point defects or point defect clusters in crystalline silicon, to continuous amorphous layers, as the dose of the implanted ions increases and the damage from the ions accumulates. The heating of the wafer during ion implantation can impact the implantation damage and damage healing may occur as the implantation proceeds. This can prevent amorphization even at high doses.

The diffusion of impurities in implanted Si is complicated as a result of the presence of implantation damage. As an example, the diffusivity of B in implanted single crystal Si is anomalously high compared to equilibrium values. This transient enhanced diffusion (TED) of B and other interstitial diffusers results as a consequence of the supersaturation of Si self-interstitials generated during ion implantation [1]. This phenomenon has important consequences for Si processing, since it causes the dopant profiles spread significantly compared to the as-implanted profile during the subsequent annealing steps required for dopant activation.

Although the damage generated during ion implantation is quite complex and several Frenkel pairs are generated by each implanted ion, the "+1" model [2] has been very successful in explaining TED for most technologically relevant process parameters. The "+1" model implies that the Frenkel pairs generated during ion implantation rapidly recombine, and TED is caused by the Si self-interstitial generated as the implanted ion becomes substitutional. This excess interstitial has no vacancy partner to recombine with, and is likely to diffuse through the Si lattice interacting with dopants such as B until finally it is annihilated at the surface.

In a previous paper we have discussed atomistic Monte Carlo simulations of implantation and annealing, that include the influence of Frenkel pairs and defect clustering on TED [3]. In this work we describe an extension of the model which accounts for the effects of amorphization.

SIMULATION MODEL

For this study we carry out atomistic simulations according to the following scheme. The implantation cascades are simulated with MARLOWE [4], which uses the binary collision approximation to generate coordinates of the displaced atoms in the lattice along with those of the implanted atom. The coordinates of the Frenkel pairs and those of the implanted ion, are transferred to the three dimensional kinetic Monte Carlo diffusion code DADOS [5]. They are given random hops according to their respective diffusion hop rates at the implant temperature. New implantation cascades are added at intervals of time determined by the dose rate until the specified dose is reached. Then, annealing can be carried out for a specified temperature and time. The surface is considered to be a perfect sink for point defects. In DADOS only the defect atoms are included in the simulation and their diffusion and interactions have to be explicitly specified by parameters such as diffusion coefficients for each defect or the binding energy between them. The detailed information about the atomic interactions between defects and other parameters is provided by ab initio, tight binding [6] or molecular dynamics [7-9] studies of implantation damage and defects. Also, because DADOS can reproduce experiments at a macroscopic scale, experiments are a valuable source of a number of parameters and mechanisms included in the model [10].

Tight binding and Molecular Dynamics calculations have been used to study the kinetics of the interstitial-vacancy recombination [6,7]. As the vacancy approaches the interstitial, the interstitial and vacancy form a metastable defect structure defined as interstitial-vacancy (I-V) complex, which is a local distortion of the lattice with no excess or deficit of atoms in the lattice. This defect is very unlikely to dissociate into a self-interstitial and a vacancy and eventually it annihilates. However, there may be an energy barrier for the isolated I-V complex to annihilate, and simulations indicate that its average lifetime at room temperature is only of a few microseconds [8].

Molecular dynamics simulations of the damage evolution of cascades show that implantation cascades generate not only point defects but also amorphous pockets; i.e., small amorphous or disordered regions surrounded by crystalline Si [7-9]. The amorphous pockets recrystallize on annealing, leaving only the net number of defects in the form of small interstitial or vacancy clusters when the disordered region disappears. This indicates that there is some rearrangement of the atoms as they recrystallize and that the excess of interstitials and vacancies are confined inside the amorphous pockets. The activation energy for the regrowth of these amorphous pockets depends on their size and shape [9].

In order to implement these ideas in our simulation scheme, when interstitials and vacancies are within the interaction radius of each other, they are assumed to form the metastable I-V complexes, instead of undergoing instantaneous annihilation. An isolated I-V complex has an energy barrier of 0.43 eV for recombination. When it is surrounded by other I-V complexes, i.e, it is in a local disordered or amorphous region, its recombination rate decreases as its coordination number increases. This implies that small amorphous pockets recombine faster than a continuous amorphous layer, where the number of I-V complex neighbors approaches the Si atomic density. This model of amorphous Si correctly exhibits the experimentally observed result that recrystallization occurs at the amorphous-crystal (a-c) interface, and accounts for the cooperative nature of amorphization and recrystallization [11]. Interstitials or vacancies surrounded by I-V complexes are treated as different defects, "amorphous interstitials" or "amorphous vacancies". The rearrangement of defect atoms within the amorphous region during the regrowth is simulated by allowing the diffusion of the amorphous defects within the amorphous pockets or amorphous regions, but confining them mostly to the amorphous region by imposing a large increase in energy if an amorphous defect diffuses into crystalline material.

THE CRYSTALLINE/AMORPHOUS TRANSITION

Ion induced amorphization experiments show that amorphization starts in the regions of high damage concentration, near the ion end-of-range, where lattice damage and a net excess of Si interstitials coexist [12]. Under conditions where the defect production rate (influenced by ion flux and mass) and dynamic defect annealing rate (influenced by temperature) are nearly balanced, damage accumulation is extremely non-linear with ion fluence [13,14]. Despite differences in damage production rates resulting from varying the irradiating ion mass, this critical regimen has been observed for different ions at a characteristic transition temperature which also depends on dose rate. The activation energy of the defect processes controlling amorphization have been found to change with irradiation conditions, varying from 0.7 eV for light ions such as C, to 1.69 eV for Xe [14]. A number of models have been proposed to explain the c-a transition, based on point defect accumulation [15] or the overlapping of local amorphous zones [16]. However, none of these models accounts properly for the dynamic anneal, and therefore they do not reproduce the characteristic critical regime.

In our model, dynamic annealing takes place as the implant proceeds and so, we can capture the characteristic critical regimen, and its dependence on the ion mass. Light ions produce cascades with small amorphous pockets and isolated point defects. Small amorphous pockets are represented in our model by I-V complexes surrounded by few defect neighbors. Its recombination (or recrystallization) rate is high and many of them anneal out. Since most of them do not survive between successive cascades, it is very difficult to accumulate the damage and amorphization is only reached at very high doses. On the contrary, heavy ions generate more concentrated and stable damage. Most of the damage produced is accumulated and the amorphization level is reached at lower doses. The amorphous pockets formed by heavy ions are bigger (I-V complexes completely surrounded by other I-V complexes) and its activation energy for recrystallization is larger. Therefore, the critical regime (where the rate of defect production and annihilation are comparable) occurs at higher temperature, as shown in Fig.1. These results agree very well with experimental data [12-14]. The dramatic damage decrease lies within a narrow temperature window near room temperature for Si self-ion implantation. The transition temperature for lighter ions, such as B or C, lies below room temperature. At room temperature defects annihilate quickly, damage does not survive between cascades and amorphization cannot take place.

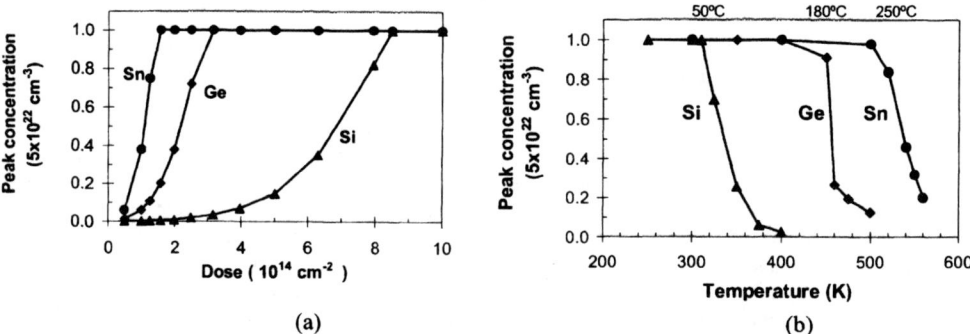

(a)   (b)

Figure 1. Peak concentration of the I-V complexes normalized to the atomic Si density (unity corresponds to complete amorphization). (a) 80 keV, 10^{12} cm^{-2}s^{-1} room temperature implants for Si, Ge, Sn ions versus dose. A superlinear dependence is observed near the transition dose. (b) 80 keV, 10^{15} cm^{-2}, 10^{12} cm^{-2} s^{-1} implants for Si, Ge, Sn ions versus temperature. Amorphization is not possible above the transition temperature.

TED FOR AMORPHIZING AND NON-AMORPHIZING IMPLANTS.

This model treats amorphizing and non-amorphizing implants in a consistent way. Amorphization is a result from the simulation and not an input parameter. Figure 2 represents different annealing steps for a sub-amorphizing implant. The concentration of I-V complexes does not reach the atomic Si density, which implies that a continuous amorphous layer has not been formed. Each implantation cascade generates local amorphous pockets or aggregates of I-V complexes, but they are disconnected. As the annealing proceeds and they regrow, they leave behind the local excess of self-interstitials or vacancies. Although locally an excess of interstitials or vacancies may exist, in each cascade equal numbers of interstitials and vacancies are generated and eventually, they find each other and recombine. Only the excess interstitial generated by the implanted ion survives until it is annihilated at the surface. This excess interstitial is responsible for TED of dopant atoms, giving validity to the "+1" model (or "+n", if the recombination is incomplete [17]).

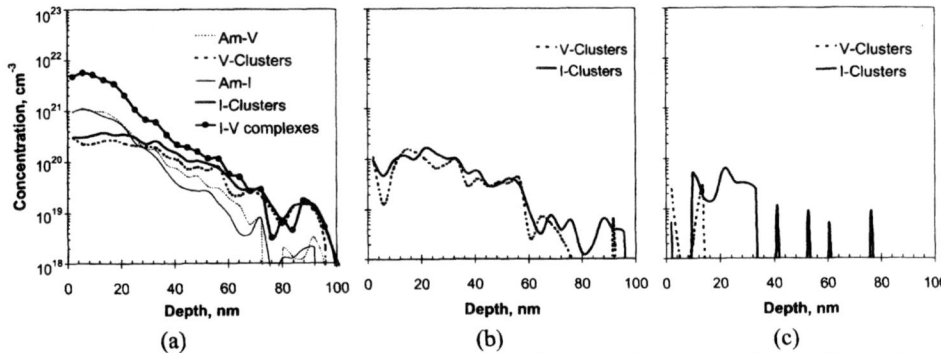

Figure 2. Simulation results from the MC model of ion implantation and annealing. Depth distribution of defects resulting from an implantation of 10^{14} Si ions/cm^2 at 10keV. (a) Profile after room temperature implantation. (b) Profile after annealing at 500°C for 1000sec. (c) Profile after second anneal at 800°C for 1000sec.

The defect evolution for an amorphizing implant is plotted in figure 3. A continuous amorphous layer extends from the surface into the bulk of the crystal (the I-V complex concentration reaches the Si atomic density), and some excess or deficit of atoms (amorphous interstitials or vacacies) exists within this layer. The amorphous defects in this region are mostly annihilated as the layer recrystallizes. This occurs because the amorphous vacancies and interstitials are given large diffusion coefficients, and recombine or annihilate at the surface at a rate faster than that of recrystallization. In the experiments, the regrowth of an amorphous layer that extend to the surface proceeds from the c-a interface and few or no defects are observed in the region of the original amorphous layer [18]. Again, this indicates that the defects were confined to the amorphous layer during recrystallization and are mostly annihilated by recombination or at the surface if they have access to it. The amorphous region starts in the areas of denser damage, which are also where most of the implanted ions (the "+1" interstitials) lie. They diffuse within the amorphous region as the amorphous layer regrows and reach the surface where they are annihilated, not causing any TED in the crystalline region beyond the a-c interface. Only a small fraction of the implanted ions are beyond the a-c interface. They survive after the regrowth of the amorphous layer and are responsible for TED.

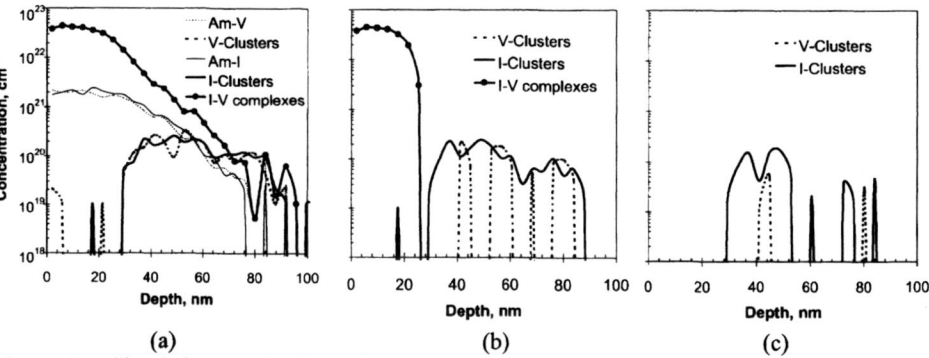

(a)   (b)   (c)

Figure 3. Simulation results from the MC model of ion implantation and annealing. Depth distribution of defects resulting from an implantation of 10^{15} Si ions/cm^2 at 10keV. (a) Profile after room temperature implantation. (b) Profile after annealing at 500°C for 1000sec. (c) Profile after second anneal at 800°C for 1000sec.

TED can be described by the "+1" (or "+n") model for sub-amorphizing doses. However, for amorphizing implants, only a small fraction of the implanted ions contribute to TED. Once the amorphization threshold is reached, an increase in dose causes an increase in the thickness of the amorphous layer, and therefore in the depth of the remaining defects, but not a proportional increase in the number of defects beyond the amorphous interface. TED saturates or increases very slightly with dose once the amorphization regime is reached, as shown in figure 4.

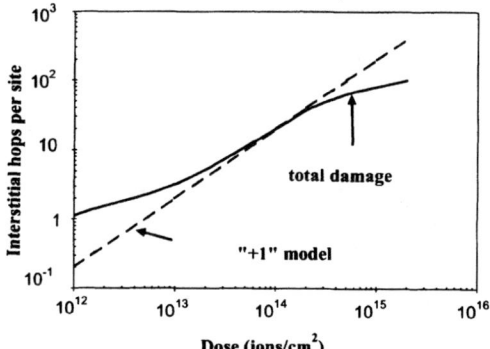

Figure 4. Simulations of transient enhanced diffusion as a function of dose for 10keV Si implants over a range of doses. The diffusion hops are measured by the number of times a site is visited by an interstitial during an anneal at 800°C that eliminates all implant damage and point defect clusters.

CONCLUSIONS

We have developed an atomistic model for amorphization compatible with MC models such as DADOS. The model is based on the formation of the I-V complex as elementary unit, whose recrystallization rate depends on the local density of these defects. Unpaired interstitials and vacancies within the amorphous regions have high mobility. The model encompasses damage ranging from individual defects to continuous amorphous layers. It reproduces, explains and predicts all the essential features of the crystalline-to-amorphous transition, including ion-mass, temperature, dose and dose-rate dependence, and also TED for amorphizing and non-amorphizing implants.

ACKNOWLEDGEMENTS

We acknowledge M. Jaraiz for the provision of the DADOS code.

REFERENCES

1. P.A. Stolk, H.-J. Gossmann, D.J. Eaglesham, D.C. Jacobson, C.S. Rafferty, G.H. Gilmer, M. Jaraiz, J.M. Poate, H.S. Luftman, T.E. Haynes, *J.Appl. Phys.* **81**, 6031 (1997).
2. M. D.Giles, *J. Electrochem. Soc.* **138**, 1160 (1991).
3. L. Pelaz, G.H. Gilmer, V.C. Venezia, H.-J. Gossmann, M. Jaraiz, J. Barbolla, *Appl. Phys. Lett.* **74**, 2017 (1999).
4. M.T. Robinson and I.M. Torrens, *Phys. Rev. B* **9**, 5008 (1974).
5. M. Jaraiz, L. Pelaz, E. Rubio, J. Barbolla, G.H. Gilmer, D.J. Eaglesham, H.J. Gossmann, J.M. Poate, *Mater. Res. Soc. Symp. Proc.* **54**, 532 (1998).
6. M. Tang, L. Colombo, J. Zhu and T. Diaz de la Rubia, *Phys. Rev. B* **55**, 4279 (1997).
7. G.H. Gilmer, T. Diaz de la Rubia, D.M. Stock, M. Jaraiz. *Nucl. Instr. and Meth. In Phys. Res. B* **102**, 247 (1995).
8. L.A. Marqués, L. Pelaz, J. Hernandez, J. Barbolla, G.H. Gilmer, *Phys. Rev. B* **64**, 045214 (2001).
9. M.-J. Caturla, T. Diaz de la Rubia, L.A. Marqués, G.H. Gilmer, *Phys. Rev. B* **54**, 16683 (1996).
10. L. Pelaz, M. Jaraiz, G.H. Gilmer, H-J. Gossmann, C.S. Rafferty, D.J. Eaglesham and J.M. Poate, *Appl. Phys. Lett* **70**, 285 (1997).
11. J.S. Williams, R.G. Elliman, W.L. Brown, and T.E. Seidel, *Phys. Rev. Lett.* **55**, 1482 (1985).
12. O.W. Holland, C.W. White, *Nucl. Instrum. Methods Phys. Res. B* **59/60**, 353 (1991).
13. T. Motooka and O.W. Holland, Appl. Phys. Lett. 61 (25), 3005 (1992).
14. R.D. Goldberg, J.S. Williams, R.G. Elliman, *Nucl. Instrum. Methods Phys. Res. B* **106**, 242 (1995).
15. M.L. Swanson, J.R. Parson, C.W. Hoelke, *Radiat. Eff.* **9**, 249 (1971).
16. F.W. Morehead, B.L. Crowder, *Radiat. Eff.* **6**, 27 (1970).
17. L. Pelaz, G.H. Gilmer, V.C. Venezia, H.-J. Gossmann, M. Jaraiz, J. Barbolla, *Appl. Phys. Lett.* **74**, 2017 (1999).
18. L. Csepregi, J.W. Mayer, T.W. Sigmon, Phys. Lett. 54A (1975), 157.

Mat. Res. Soc. Symp. Proc. Vol. 669 © 2001 Materials Research Society

Atomistic simulations of extrinsic defects evolution and transient enhanced diffusion in silicon

A. Claverie, B. Colombeau, F. Cristiano*, A. Altibelli and C. Bonafos
Ion Implantation Group, CEMES/CNRS, BP 4347, F-31055 Toulouse Cedex 4
* LAAS/CNRS, 7 Avenue du Colonel Roche, F-31077 Toulouse Cedex

ABSTRACT

We have implemented an atomistic simulation of the Ostwald ripening of extrinsic defects (clusters, {113}'s and dislocation loops) which occurs during annealing of ion implanted silicon. Our model describes the concomitant time evolution of the defects and of the supersaturation of Si interstitial atoms in the region. It accounts for the capture and emission of these interstitials to and from extrinsic defects (defined by their formation energy) of sizes up to thousands of atoms and includes a loss term due to the interstitial flux to the surface. This model reproduces well the dissolution of {113} defects in Si implanted wafers. We have subsequently studied the characteristics of TED in the case of B implantation at low and ultra low energy. In such cases, the distance between the defect layer and the surface plays a crucial role in determining the TED decay time. The simulations show that defect dissolution occurs earlier and for smaller sizes in the ultra-low energy regime. Under such conditions, TED is mostly characterized by its "pulse" component which takes place at the very beginning of the anneal, probably during the ramping up. In summary, we have shown that the physical modelling of the formation and of the growth of extrinsic defects leads to a correct prediction of the "source term" of Si interstitials and at the origin of TED.

INTRODUCTION

The predictive simulation of dopant diffusion is essential for the controlled reduction of the dimensions of future IC's. Among other dopants, boron certainly deserves special attention not only because of its technological importance but also because its diffusive behavior has been much more experimentally studied than any other impurity and thus a reliable set of data exists. There are two distinct components in the anomalous diffusion of boron in Si. On one hand, for both high concentrations of B and Si interstitial atoms, boron-Si interstitial clusters (BIC's) are formed and tend to immobilise a fraction of B. On the other hand, the coupling of free-interstitial Si atoms with (probably) substitutional boron atoms enhances B diffusivity by a factor which is proportional to the supersaturation of Si(int)'s in the region. Thus, describing and modelling the transient enhancement of diffusivity that boron encompasses during annealing only requires the knowledge of the time and space evolution of the Si(int)'s supersaturation in the region where B stands and this, at the given temperature. This supersaturation initially results from the injection of Si atoms in the crystal during ion implantation and/or dopant activation. Upon annealing, these Si atoms condense to form defects of various types which, when large enough, can be detected by TEM. Earlier works have ascribed the diffusion anomalies to the release of Si atoms by one such type of extended defects, namely the {113}'s [1,2]. Since then, it has been shown that the formation and dissolution of these defects is not a prerequisited condition for the existence of TED [3].

Recently a better knowledge of the thermal behavior of the extended defects formed after ion implantation has been gained and the driving force for their evolution has been identified [4,5]. The goal of this paper is firstly to show how the modelling of the Si interstitials evolution based on such new concepts allows one to describe well-known experimental results concerning the kinetics of extended defects during annealing. Subsequently, the application of our model to TED experiments will be discussed. In particlular, we will use our simulations to explain the variations of TED amplitude and time scale which are observed after classical high energy and Ultra Low Energy (ULE) implants.

MATERIAL SCIENCE BACKGROUND

Depending on experimental conditions, up to 4 types of extrinsic, i.e., interstitial-type extended defects can be detected by TEM [4]. They range depending on the number of Si atoms they contain from clusters of about 2 nm in size, to {113} defects and to dislocation loops of two types. All these defects are "precipitates" of Si atoms which evolve in size but also in shape and crystallographical characteristics during annealing. The basic mechanism at the origin of their evolution is a competitive growth in which the defects interchange Si atoms through matrix diffusion with rates that depend on their stability i.e., their formation energy, and their ability to capture Si atoms. Because of this interchange, the region where the defects stand is oversaturated with free Si interstitial atoms. A dynamical equilibrium exists between this supersaturation "mean field" and the population of defects. The phenomenon can be conservative or not depending on the eventual presence of a strong sink for Si interstitial in the vicinity of this mean-field. In such cases, for example when the initial excess of Si atoms is close enough to the surface of the wafer and/or when the annealing ambient promotes fast recombination of Si(int)'s at the surface, the loss of interstitials, i.e., the dissolution of the defects before their transformation into more stable forms, can occur. Such a phenomenon is named Ostwald ripening and has been initially developped to describe the growth of liquid droplets in equilibrium with their vapor. The Gibbs-Thomson equation expresses the key concept needed to model this phenomenon and it is being currently used today to describe the nucleation and growth of various particles in solid matrices [6].

ATOMISTIC SIMULATION OF DEFECT GROWTH

The atomistic approach describes the growth of "precipitates" in a matrix by calculating the difference between the capture (F_n) and emission (R_n) rates of every cluster of size n. The growth rate is classically written as the product of the capture area of the particle by the net flux of atoms towards it. In the case of a diffusion limited growth, this growth rate can be written [7],

$$\frac{dn}{dt} = |F_n - R_n| = D_i C_i^* A_n \left.\frac{dS}{dR}\right|_{R=r} = D_i C_i^* \frac{A_n}{R_{eff}} (\overline{S} - S(n)) \qquad (1)$$

where $D_i C_i^*$ is the self-diffusivity of the Si interstitials in the Si matrix. A_n is the capture area of the precipitate. Because R_{eff} represents the radial extension of the diffusion field, A_n/R_{eff} is characteristics of the capture efficiency of the defect. $\overline{S}$ is the mean supersaturation of these Si interstitials within the matrix and $S(n)$ the supersaturation of interstitials in equilibrium with a precipitate containing n atoms. $S(n)$ is given by the Gibbs-Thomson equation and can be written,

$$S(n) = \exp\left[\frac{E_f(n)}{kT}\right], \qquad (2)$$

$E_f(n)$ being the formation energy (the derivative of the total energy) of a precipitate containing n atoms. The emission rate (R_n) is thus proportional to the formation energy of the particle and is given by,

$$R_n = D_i C_i^* \times \frac{A_n}{R_{\mathit{eff}}} \times \exp\left(\frac{E_f(n)}{kT}\right) \qquad (3)$$

The capture rate (F_n) is a function of the environment of the precipitate and is proportional to $\overline{S}$, the mean supersaturation of Si interstitials between the precipitates,

$$F_n = D_i C_i^* \times \frac{A_n}{R_{\mathit{eff}}} \times \overline{S}$$

Both the capture efficiency (A_n/R_{eff}) and the formation energy (E_f) depend on the geometry of the defect.

The atomistic model is based on a set of (n+1) coupled differential equations. The first n equations describe the flux of atoms from particles of size n to particles of size n+1 and n-1,

$$\frac{dN_n}{dt} = F_{n-1} N_{n-1} - F_n N_n + R_{n+1} N_{n+1} - R_n N_n \qquad (4)$$

This equation drives the growth of the defects both in terms of sizes and densities. The last equation describes the "free component" of the Si atoms i.e., the concomitant evolution of the free Si interstitials supersaturation in dynamical equilibrium with the extended defects,

$$\overline{S} = \frac{\sum_{n=2}^{\infty} \beta_n R_n N_n}{D_i C_i^* \left(\left(\sum_{n=2}^{\infty} \frac{A_n}{R_{\mathit{eff}}} N_n\right) + \frac{1}{L_{surf} + r_p}\right)} \qquad (5)$$

N_n is the number of precipitates of size n. L_{surf} is the recombination length at the surface and r_p the depth position of the defects. The quantity β is the number of Si atoms released by the break-up of a cluster (β=2 for n=2, β=1 otherwise). This last equation describes the time evolution of the supersaturation mean-field centered on the defects and at the origin of TED.

The difficulty when trying to simulate the nucleation and growth of interstitial type defects in silicon lies in the various different geometries these defects can adopt. Indeed, from clusters to {113}'s then to dislocations loops of two types, four main different types of "precipitate" may coexist with their own capture efficiencies and size-dependent formation energies. However, for the purpose of this paper, we restrict ourselves to the situation where "only" up to large {113}'s can be formed i.e., the case of non-amorphising implants.

For clusters of size up to 10 atoms, we take the experimentally deduced oscillating formation energies from Cowern [8] with stable configurations for 4 and 8 atoms. They tend, for larger sizes, towards values expected for small {113} defects. In order to access to the formation energies of the {113}'s, we have calculated the total energy of these defects based on their crystallographical characteristics i.e., taking into account the two edge dislocations plus the two mixed dislocations plus the stacking fault energy which altogether define this defect (9). They have been assumed to be planar, rectangular and of constant width (4 nm). They contain 20 atoms per nm. Fig. 1 shows the overall variations of the formation energy of these defects as a function of their size expressed in atoms. This energy curve gently tends towards its asymptotical limit at 0.65 eV [9].

Since little is known of the structure of the clusters, we have reasonably assumed that they are spherical and thus that their capture area is $A_n = 4\pi r^2$ where r is the geometrical radius of the cluster. Approaching the capture area of a {113} defect is far more difficult. Gencer and Dunham [7] have proposed that capture occurs only through the edges of these elongated defects. We prefer to assume that the capture area offered by this type of defect

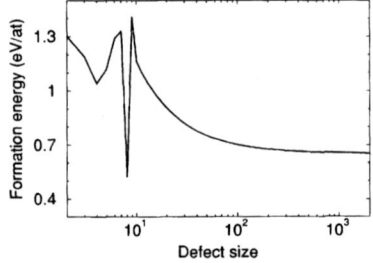

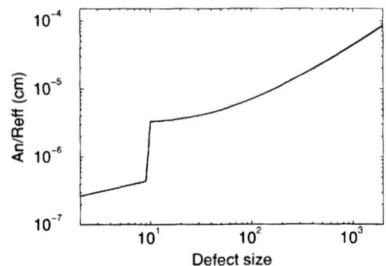

Figure 1: *Variation of the formation energies of clusters and {113}'s as a function of the number of Si atoms they contain.*

Figure 2: *Variation of the capture efficiency of clusters and {113}'s as a function of the number of Si atoms they contain.*

still increases as they become longer. Thus, our capture area is the sum of three terms i) the two cylinders at the width sides, ii) the two cylinders along the length sides and iii) the four hemispheres at the corners. The overall variation of the capture efficiency A_n/R_{eff} is shown on Fig. 2. The exact amplitude and position of the abrupt jump corresponding to the transition from spherical to elongated precipitates have little impact on the results shown in the next section. All simulations were run with the $D_iC_i^* = 2\times10^{24}\exp(-4.52/kT)$ cm^{-1} s^{-1} as deduced from Cowern's experiments [8].

RESULTS

We have tested the model using two notorious experimental studies of defect evolution. The first set concerns the dissolution of the {113}'s initially observed by Eaglesham [1]. Fig. 3 shows the comparison between our simulation and a compilation of their experimental results [1,2]. An excellent fit is obtained by adjusting only $(L_{surf} + r_p)$ at 80 nm. Clearly, these simulations show that dissolution occurs or not depending only on the distance and sink efficiency of the surface. The growth of the defects i.e., the size increase and density decrease they experience during this non-conservative Ostwald ripening, perfectly matches the experimental observations (not shown) and this evolution strongly depends on the size dependence of the formation energy of the defects.

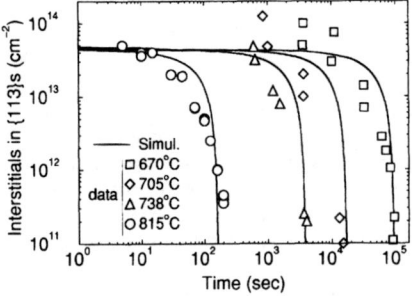

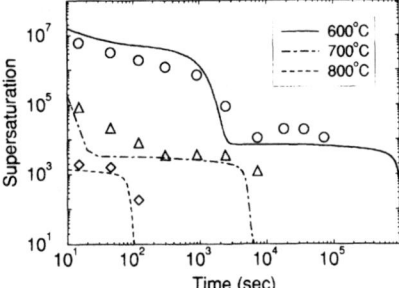

Figure 3: *Dissolution kinetics of {113}'s : comparison between TEM results and simulations.*

Figure 4: *Time variation of the supersaturation of Si(int)'s during annealing : comparison between "marker" experiments and simulations.*

While the model gives satisfaction to describe the "frozen" component of the Si interstitials i.e., the growth of defects, its ability to describe the diffusive behavior of boron has to be tested. For this purpose, we recall a recent experiment by Cowern [8] in which the diffusivity of CVD grown boron delta layers after low dose Si implantation was studied during low temperature annealing. The time variation of the boron diffusivity enhancement i.e., of the Si supersaturation, was extracted from these experiments and are plotted in Fig. 4 along with the results of our simulations. At low temperature, our simulations reproduce well the two observed plateaus involving firstly the stable clusters at 4 and 8 atoms and then the {113} defects. At higher temperature, this evolution is faster and only the {113}'s survive long enough to be detected at this time scale. It is noticeable that such fits cannot be obtained without assuming the presence of at least two efficient traps in the 2-10 atoms range.

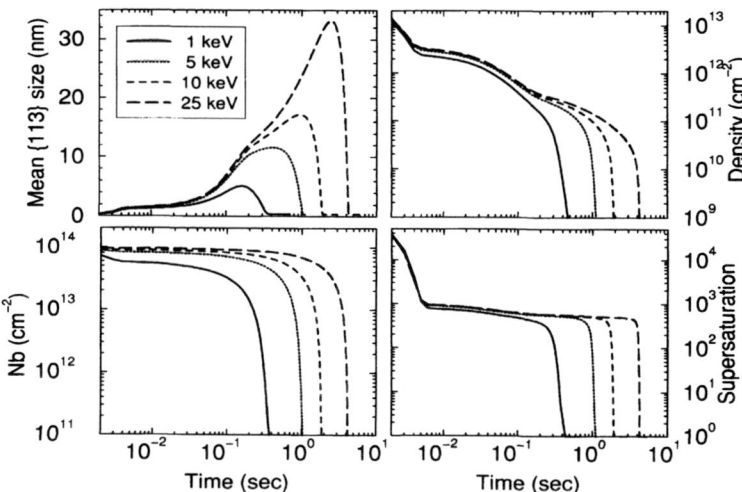

Figure 5: *Simulation of the annealing at 950°C of 10^{14} ions/cm^2 boron implants in the 25-1 keV range. Time evolutions of a) mean size of the defects, b) defect density, c) total number of Si atoms stored in the defects and, d) Si(int)'s supersaturation in the defect region.*

Having gained some confidence on the ability of our model to describe both defect growth and Si supersaturation evolution, we now study through computer simulation the effect of reducing the energy of a 10^{14} ions/cm^2 boron implant in Si from 25 keV to 1 keV. For this, we assume the "+1" model to hold and we "anneal" these implants at 950°C. Fig. 5a and 5b show the time evolution of the mean size and of the density, respectively, of the defects for the four different implants. Up to 35 nm-long {113}'s are produced after 2 seconds annealing of the 25 keV implant, which quickly reduce in size and density for longer annealing times. This agrees well with TEM observations. Moreover, as the implant energy decreases, the maximum size the defects can reach before dissolution decreases. While defects large enough to be possibly detected by TEM can still be formed after lower energy implants, they typically dissolve after 0.1-1s i.e., probably during the ramping up of the anneal. Fig. 5c shows the time evolution of the number of interstitials stored in the defects. Reducing the implant energy from 25 to 5 keV mostly affects the time after which dissolution starts. The overall number of atoms initially stored in the defects is almost independent of the energy in this energy range and close to the implanted dose. In contrast, after 1 keV implant, the surface pumps the free interstitial supersaturation at the very beginning of the cluster growth and thus the total number of Si(int)'s stored in the defects is significantly smaller than observed after higher energy implants. Finally, Fig. 5d shows the evolution of the Si(int)'s supersaturation in the defect region after these implants. At very short times (<0.01 s), the supersaturation shows a "pulse" when only clusters of small sizes exist and rapidly evolve towards {113}'s or dissolve. For longer times, a large plateau is observed and corresponds to the dynamical equilibrium between the supersaturation mean field and the {113} defects. It is worth to be noted that for the 1 keV implant, the "pulse" component weights near about 40% of the overall diffusivity enhancement boron will experience. Since this pulse is driven by small clusters of high formation energies, this nicely explains why the measured activation energy for TED is much smaller after ULE implants (ultrafast TED) than after higher energy implants.

CONCLUSIONS

A physical based modeling of the growth of extended defects resulting from the precipitation of excess Si interstitial atoms has been presented. This model describes the non-conservative Ostwald ripening of precipitates in a matrix in presence of a strong sink. It allows the famous dissolution of {113}'s to be simulated and moreover gives access to the concomitant evolution of the supersaturation of Si(int)'s in the defect region. This supersaturation is the "source term" which is needed in process simulators to account for the diffusivity enhancements observed during annealing of ion implanted silicon.

ACKNOWLEDGEMENTS

This work was finalised shortly after the completion of the RAPID Project (ESPRIT 23481) and just before the beginning of the IST/FRENDTECH Project. However, we are glad to thank the almost continuous support from the European Commission.

REFERENCES

1. D.J. Eaglesham, P.A. Stolk, H.J. Gossmann and J.M. Poate, *Appl. Phys. Lett.*, **65**, 2305 (1994).
2. P.A. Stolk, H.J. Gossmann, D.J. Eaglesham, D.C. Jacobson, C.S Rafferty, G.H. Gilmer, M. Jaraiz, J.M. Poate, H.S. Luftman, T.E. Haynes, *J. Appl. Phys.*, **81**, 6031 (1997).
3. L.H. Zhang, K.S. Jones, P.H. Chi and D.S Simons, *Appl. Phys. Lett*, **67**, 2025 (1995)
4. A. Claverie, B. Colombeau, G. Ben Assayag, C. Bonafos, F. Cristiano, M. Omri and B. de Mauduit, *Mat. Sci. in Semic. Proc.*, **3**, 269 (2000).
5. A. Claverie, F. Cristiano, B. Colombeau and N.E.B Cowern, *Mat Res. Soc. Proc.*, (2000) in print.
6. C. Bonafos, B. Colombeau, M. Carrada, A. Altibelli, G. Ben Assayag, B. Garrido, M. Lopez, A. Perez-Rodriguez and A. Claverie, *Nucl. Instr. Meth. in Phys. Res. B*, (2001) in print.
7. A.H. Gencer and S.T. Dunham, *J. Appl. Phys.*, **81**, 631 (1997).
8. N.E.B. Cowern, G. Mannino, P.A. Stolk, F. Roozeboom, H.G.A. Huizing, J.G.M. van Berkum, W.B. de Boer, F. Cristiano, A. Claverie, and M. Jaraiz, *Phys. Rev. Lett*, **82**, 4460 (1999).
9. B. Colombeau, F. Cristiano, C. Bonafos and A. Claverie, to be published.

AUTHOR INDEX

SUBJECT INDEX

CPSIA information can be obtained at www.ICGtesting.com
Printed in the USA
LVOW06s1015220514

386805LV00011B/424/P